APPLIED CLINICAL ENGINEERING

APPLIED CLINICAL ENGINEERING

BARRY N. FEINBERG, Ph.D., PE, CCE

Senior Scientist
The Kendall Company
Barrington Research Center
Barrington, Illinois

PRENTICE-HALL, INC., Englewood Cliffs, New Jersey 07632

Library of Congress Cataloging in Publication Data

Feinberg, Barry N.
  Applied clinical engineering.

  Includes bibliographical references and index.
  1. Biomedical engineering. 2. Medical instruments
and apparatus. I. Title.
R856.F45 1984   610'.28   85-6547
ISBN 0-13-039488-2

Editorial/production supervision and
  interior design: *Ellen Denning*
Cover design: *20/20 Services, Inc.*
Manufacturing buyer: *Rhett Conklin*

Printed in the United States of America

10 9 8 7 6 5 4 3 2 1

ISBN 0-13-039488-2 025

PRENTICE-HALL INTERNATIONAL (UK) LIMITED, *London*
PRENTICE-HALL OF AUSTRALIA PTY. LIMITED, *Sydney*
PRENTICE-HALL CANADA INC., *Toronto*
PRENTICE-HALL HISPANOAMERICANA, S.A., *Mexico*
PRENTICE-HALL OF INDIA PRIVATE LIMITED, *New Delhi*
PRENTICE-HALL OF JAPAN, INC., *Tokyo*
PRENTICE-HALL OF SOUTHEAST ASIA PTE. LTD., *Singapore*
EDITORA PRENTICE-HALL DO BRASIL, LTDA., *Rio de Janeiro*
WHITEHALL BOOKS LIMITED, *Wellington, New Zealand*

Dedicated to

My Mother, Mitzi
  for her dedication to excellence and the concept of
  "You *Can* Do It"

My Father, Harold
  for his gentle humanity

My Wife, Susan
  professor, scholar, mother, wife, friend
  for her aid, counsel, and encouragement in every phase of
  this arduous task

My Children, Leah and Joel
  who understood why their father didn't have all the time
  he wanted to share with them

CONTENTS

10 ENGINEERING PRINCIPLES OF MEDICAL X-RAY SYSTEMS 237

11 ENGINEERING ASPECTS OF RADIATION 345

16 ELECTRICAL PERFORMANCE AND SAFETY TESTING OF ELECTROMEDICAL EQUIPMENT 489

17 CODES AND STANDARDS THAT PERTAIN TO THE CLINICAL ENGINEER 501

INDEX

PREFACE

There is no doubt that technology has always affected the way man does things. Some technology influences the way in which we live our lives and other technology affects us in more subtle ways. As an example of this, we can see how architecture has been drastically affected by the elevator, which has allowed the construction of multistory buildings and skyscrapers.

The initial focus of medical electronics and instrumentation was to extend the senses in the area of observation and measurement of physiological events. From the turn of the century to the 1940s, a limited amount of medical equipment was available and consisted mainly of the ECG, x-ray, sphygmomanometer, and the thermometer. Thus, for the most part, diagnosis was based more on symptom than on testing and treatment of the ill was done mostly through the administration of drugs. At this time, the physician was the focus and major portion of health care delivery. Since the hospitals had little to offer in the way of general care, much of this century's early health care was done in homes where the patient's family provided nursing care with instructions from the physician.

Engineering contributions to health care through biomedical engineering research and development in the university and in the medical device industries has brought about a change in how health care is delivered. As new equipment for diagnosis and treatment and the development of hospital laboratory facilities grew, the hospital became the focal point for the delivery of health care, with an associated development of allied health specialties. The vast majority of medical technology used in health care delivery is centered around medical devices and equipment which tend to be electronic or electromechanical in nature. It is the introduction of this equipment into the environment of the medical practitioner, who, by education and training is ill-prepared to deal with the highly technical electromedical systems, that have helped spawn a new medical specialty, clinical engineering.

Since it is engineering creativity that has brought this level of science to the bedside of the patient, it is natural for engineers to provide the expertise required for proper utilization of these engineering systems for the delivery of health care. The clinical engineer practices his or her profession within clinics, hospitals, medical centers, or other health care institutions. The specific duties of the clinical engineer depend on the health care setting in which he or she is working.

In the broad technical sphere of professional practice of clinical engineers, they are involved with providing technical consultations to the hospital administrator and medical personnel, as well as education and training relative to the operation and safety of medical equipment. They are also involved with the electronic diagnosis and performance testing of electromedical equipment. Clinical engineering is a profession that applies the principles of the science of engineering directly to the devices and operation of the complex medical devices used in the diagnosis and treatment of the sick and injured.

In today's modern hospital, the delivery of health care has become heavily dependent on complex engineering systems. Thus the person best suited to assume the responsibility for these systems in terms of performance and safety in a health care environment is a professional engineer, one who is educated in the science and profession of engineering. The concept "engineer" as the janitor or the handyman technician has about as much reality in health care today as the concept of "surgeon" being the same as the surgeons of old who were barbers and practiced blood-letting.

Since health care technology management requires knowledge, special education, and experience in the area of clinical engineering beyond traditional educational areas of engineering, a peer-group review system has been established by the International Certification Commission for the certification of engineers in the speciality of clinical engineering. The medical parallel to certification in clinical engineering is board certification of a medical speciality. The goal of the certification system is to provide to the hospital and the public the assurance that an engineer is qualified to undertake the responsibility of medical and hospital technology management.

The clinical engineer accepts the responsibility for the surveillance of the hospital electrical environment and electronic instrumentation in all patient care areas, including nursing floors, intensive care units, operating rooms, outpatient clinics, emergency room, and cardiac catheterization laboratories. To carry out this essential function, a clinical engineer or clinical engineering department must:

1. Prepare performance function and test specifications for each instrument and must define inspection and calibration cycles.
2. Provide procurement advice on all contemplated instrument purchases, particularly with regard to safety, standardization, power requirements, performance, and compatibility with existing equipment.
3. Initiate investigation of unsafe equipment and electrical accidents.
4. Perform control quality of performance testing and preventive maintenance by regular inspection of records, devices, and systems.
5. Accept overall responsibility for all devices and instruments, including planning for purchase and installation, acceptance criteria, certification of performance, and surveillance of warranty and service agreements for all patient-oriented equipment in the hospital.

Another area of concern for the clinical engineer in the management of medical technology is the interpretation of and compliance with the large body of quasi-legal and technical codes and standards set forth by city, state, and federal governments as well as accrediting bodies. A large number of the codes and standards are generated by the National Fire Protection Association (NFPA), which is responsible for the voluntary standards for the safe use of electricity in hospitals and the *National Electrical Code*®. This Code contains sections pertaining to health care facilities, medical x-ray equipment, and therapeutic high-frequency electrical equipment. Many of the standards and codes have the force of law in many parts of the United States. Thus the clinical engineer must be familiar with and understand these codes and standards as well as their implications to the hospital. Therefore, the clinical engineer assumes the role of advisor and counselor in the areas of the technical codes and standards as well as acting as a supervisor to see that they are adhered to within the confines of the hospital.

To meet the educational requirements of the engineering student and the practicing clinical engineer, this book is written to be a university-level engineering book dealing with the main topics of concern to the clinical engineering student and practicing professional alike. It consists of material and problems which were developed from the author's experience as director of the clinical and medical radiation engineering programs and in teaching courses in clinical and medical radiation engineering in the School of Electrical Engineering at Purdue University, West Lafayette, Indiana, as well as serving as chairman of the Board of Examiners in Clinical Engineering Certification for the International Certification Commission and industrial experience in the medical device field.

This book should be accessible to university seniors or first-year graduate students who have completed or are concurrently taking courses in circuits, electronics, linear systems analysis, and electromagnetic fields. A knowledge of anatomy and physiology is desirable but not essential, since the text includes a brief review of the appropriate topics to provide background for the medical instrumentation being discussed.

A survey of the contents of this book will show that whereas some of the more important areas of clinical engineering have been covered in depth, other areas have been summarized only. This arrangement was dictated by the desire to present only enough material for a two-semester classroom course followed by a one-semester laboratory. For a one-semester course, the introductory chapters dealing with physiology and the chapters dealing with the more central of the hospital areas can be used. A detailed study of all the chapters, on the other hand, should provide enough material for a two-semester laboratory course. A student who has mastered this material should be ready for a clinical engineering practicum or hospital internship.

It is hoped that this book will go some way toward establishing the professional identity of this fast-emerging discipline of clinical engineering.

BARRY N. FEINBERG

ACKNOWLEDGMENTS

Many people have helped, directly or indirectly, make this book a reality. It is my pleasure to acknowledge and thank them for their contributions, which are embodied one way or another in the thoughts, concepts, and goals of this book.

In the beginning, there was Don Herbert, TV's wonderful "Mr. Wizard," who opened the door of science and engineering for me and made the experience of learning and thoughtful analysis an art form.

It is my good fortune to have been associated with Drs. Gerald Saidel, Edward Chester, James Schoeffler, David Fleming, and Robert Plonsey at Case Western Reserve University, innovators in the application of engineering to problems in the biomedical field.

I also thank the following people: Dr. Julian Lauchner, past dean of engineering at Cleveland State University, who made possible the first university-based clinical engineering program in the midwest.

Drs. Vernon Newhouse and Paul Stanley, of Purdue University, whose interest and efforts in clinical engineering helped bring about the Purdue Clinical Engineering Program.

Daniel McWhorter of the Kendall Company, director of the Barrington Research Center, for the support and encouragement to accomplish and achieve higher goals.

Dr. William Jarzembski, professor and chairman of the biomedical engineering department, Texas Tech University, for his comments and suggestions on the original manuscript, which added to the overall quality of this book.

1

HOSPITAL ORGANIZATION AND ADMINISTRATION

THE HOSPITAL AS A SYSTEM

It is not the purpose of this section to discuss fully the organizational arrangement of a hospital or other health care facility. However, because the clinical engineer must function within such a system it is important that he or she have some awareness of the complexity of the facility in which they work. Although there are manifold variations of the organizational chart of any hospital, the general arrangement is shown in Figure 1-1. It can be seen from the figure that the hospital administrator serves as the chief officer of the hospital. He or she serves at the discretion of a governing board, which may be called a board of directors, or a board of trustees, or even the owner of the hospital. It is frequently true that the hospital administrator devotes a major portion of his or her time to such matters as the public image of the hospital and its fiscal affairs. He or she is responsible for preparing budgets and otherwise satisfying the numerous accrediting agencies and other organizations, including the third-party payment organizations, such as insurance companies, which are a part of the total health care system.

The functioning of the hospital is supervised by one or more associate administrators, as shown in Figure 1-1. Here, one of these associate administrators is responsible for the medical and professional services of the hospital; a second is concerned with the nursing services, a third is responsible for the many facets of fiscal administration, and a fourth is responsible for

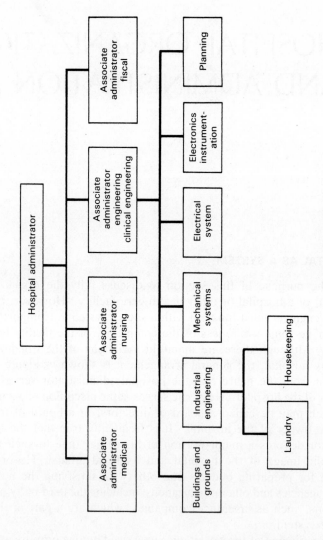

Figure 1-1 Hospital organizational chart.

the functioning of the physical plant in its many ramifications. These four major branches of the hospital are very closely interrelated, yet each must function somewhat independently.

The chief of the medical staff is responsible for the clinical facilities and the personnel, such as other physicians, anesthesiologists, neurologists, radiologists, and so on. He or she will also be responsible for the general medical departments, which will include ear, nose, and throat; general dentistry; urinary; internal medicine; and many other divisions. A third function of the chief of staff will be the supervision of several staff committees which serve to maintain the quality of the services rendered as well as to solve some of the technical and ethical problems that may arise in the hospital. For this reason, surgical policy committees, medical policy, medical records, disaster committees, and so on, will exist for the purpose of coping with the problems that arise in operation of the facility.

The associate director of nursing will, of course, have first responsibility for the handling of the entire nursing staff. This is usually done through a number of managers of individual departments, such as operating rooms, emergency room, delivery rooms, and the various medical and surgical wards of the hospital. The associate director of fiscal matters will do that which is obvious, in that he or she is responsible for the general accounting of the funds handled by the hospital and for credit management and other matters of this kind. The associate director of engineering, a title that may have other names, such as materials handling or general services, will be responsible for the functioning of the physical plant. Each of these divisions could be discussed at quite some length; however, only that of the engineer will be examined in a bit more detail.

CLINICAL ENGINEERING

Although Figure 1-1 implies that the clinical engineer is at the same level as the other associate administrators, he or she is, in fact, a service agent to the primary function of the hospital: the medical activities and services offered to the patient. Professionally, this is the same function as that performed by the other employees at the hospital, that is, to provide medical service to the patient, who is the consumer of these medical services. The clinical engineer in the hospital works in an environment that is very different from the industrial sector in terms of the workplace. For one thing, the burden of competence is thrust on the clinical engineer almost immediately upon obtaining his or her position. This comes about because the persons the clinical engineer is working with are medical technicians, nurses, and physicians instead of other engineers in an engineering group as would be found in industry. Thus the clinical engineer must have the appropriate educational background to allow him or her to be knowledgeable as a clinical engineer because there will usually not be other engineers in the hospital with which to consult, as would be the case in industry. Also, the clinical engineer is constrained by stringent regulations which originate from several agencies; medical, federal, state, and local governments and hospital accrediting organizations. Because of the professional nature of the clinical engineer's position, there is carried with it the specter of liability in the form of malpractice suits. Because of this and the fact that every clinical engineer is providing engineering services to the public that affect the health and well-being of that public, it would seem a requirement of the position that a clinical engineer should be a registered professional engineer in the state in which one is practicing.

Because of the influx of advanced engineering systems into the health care delivery institutions in the forms of electronic monitoring, diagnostic, therapeutic, prosthetic, and orthotic devices, the Facilities Division of the Public Health Service, the Joint Commission for Accreditation of Hospitals, the American Hospital Association, and numerous physicians have strongly recommended that a clinical engineer be part of the staff of every hospital of over 300 beds.

The clinical engineer is under the supervision of the administrative engineer of the hospital and has full responsibility and authority on all matters of equipment effectiveness and electrical safety, both in the electromedical instrumentation systems, including the electrical supply system, and in the area of radiology. The clinical engineer acts as a technical advisor to the medical and paramedical staff on matters of specifications, capabilities, and so on, and in the design or purchase of instruments and electrical and electronic devices. He or she should be authorized to give final approval on equipment purchase with regard to the safety and other technical features of the device.

A partial list of some of the more common duties that are performed by a clinical engineer is as follows:

1. Provide consultation to the medical staff on capabilities and limitations of medical instrumentation and aid in the planning of medical instrumentation systems, radiologic systems, and computer systems for the hospital.
2. Provide design and research capabilities in cooperation with the staff on the development of new and/or specialized devices, instrumentation, or computer systems.
3. Develop, initiate, and operate an educational program in medical equipment utilization for all hospital personnel, including medical, nursing, and maintenance staff.
4. Train and supervise technicians involved in the testing, calibration, preventive maintenance, and repair of electromedical equipment.
5. Develop, initiate, and operate a preventive maintenance and calibration program for the medical instrumentation systems in the hospital.

6. Develop, initiate, and operate a safety test program on medical instrumentation.
7. Establish and monitor purchasing standards and practices for electromedical equipment.
8. Establish, initiate, and operate test practices and procedures for new devices coming into the hospital.

INTENSIVE CARE AREAS

Intensive care areas are defined as those areas where the critically ill are given special attention as a group. This does not mean that they are given group therapy but that they are given attention in a concentrated area. Examples of this would be the post-anesthesia recovery room and the coronary care room.

The post-anesthesia or post-operation recovery room is an area to which the patient is taken from the operating room to be watched carefully until he or she has come out of the effects of the anesthetic. When the patient is removed from the recovery room, he or she may be taken to an intensive care unit if seriously ill or be returned to one of the rooms in a ward.

The Objective of Intensive Care

Monitoring of the patient is the major objective of the intensive care facility. How this is done may be quite varied. For example, monitoring may simply mean continuous watchfulness by the nurses to see how the patient is progressing. In hospitals that make use of electronic monitoring equipment, a patient's vital signs may be measured, recorded, and reported by various forms of electronic medical equipment. Because these new forms of automated patient monitoring systems are convenient to use, it is important that the medical staff not lose sight of the fact that it is still good medical practice, in this age of technology, to go to the patient's bedside occasionally to assess visually the state of the patient. In fact, good hospital practice requires continuous visual contact with the patient by the nursing staff.

Another purpose of patient monitoring is to sound an alarm if an impending serious condition is developing. The vital signs of heart rate, blood pressure, electrocardiogram (ECG), and respiration rate are monitored electronically and an alarm is sounded either at the bedside or more likely at a central station to indicate that corrective action is required.

A third reason for monitoring is to provide retrospective data that can be studied to evaluate the effectiveness of treatment and medication given to the patient. It is not at all unreasonable to assume that within a few years the effect of medication will be measured in terms of the physiological response to the patient's medication, which in turn will be part of a feedback control system that will modify the administration of the medication. As an example, the appearance of atypical or ectopic beats in an ECG may indicate that the more serious condition of ventricular fibrillation may be developing. Lidocaine is one of the medications frequently administered to prevent the development of such atypical beats. One can see that in the future, an automated drug delivery system will be used to administer such appropriate drugs as nitropresside in the proper dosage to deal with the particular condition of the patient.

One of the main physiological parameters monitored in the intensive care unit is heart rate and the morphology or shape of the electrical waveform produced by the heart. This is done to observe the presence of arrhythmias, or to detect changes in heart rate that might be indicative of a serious condition. Thus a cardiac monitor is one of the first electronic devices installed in an intensive or coronary care unit.

Respiration rate may also be monitored and in some hospitals the depth of respiration is measured. Temperature is an important parameter and is particularly important in the care of the neonatal premature infant. It is necessary, however, to be very careful in the measurement of temperature, as the temperature measured may be a function of where it is measured. For example, in the isolette used for premature infants, the temperature sensor that controls the heating unit is usually attached to the abdomen of the infant. That temperature may be 4 or 5 degrees below the core body temperature, especially if the relative humidity is low in the isolette.

Another monitoring device in common use today is a blood pressure monitor that measures central arterial pressure. For this purpose, a catheter is inserted into the arm or into another location, such as the femoral artery, so that a continual measure of blood pressure is obtained. Also, automated arm cuffs are used to monitor blood pressure in the patient when only periodic pressure measurements are required. In the future, it will be desirable to perform blood pressure monitoring noninvasively using either acoustic energy in the form of ultrasound, magnetic techniques such as nuclear magnetic resonance (NMR), or a dynamic force balance method such as that used in the vascular unloading technique.

Other Devices in Intensive Care

With monitoring devices providing information concerning the general state of health of the patient and indicating the possible or actual onset of a serious condition, there should be other devices within the intensive care unit to take care of a particular problem as it develops. For example, if ventricular tachycardia is observed on the monitor, the cardiotachometer alarm is sounded. Tachycardia is very likely to be the forerunner of ventricular fibrillation, a condition of asynchronous beating of different portions of the heart which results in nonpumping of blood to the rest of the body. Unless this condition is corrected within a matter of a minute or two, death will probably result. The countermeasure used to correct this condition is to administer an intensive pulse of electrical current to the heart by applying electrodes to the patient's thorax and delivering a high current pulse across it. Generally, this will cause the heart to revert back to a rhythm that is not so fatal. After the patient has been defibrillated, appropriate medication is administered.

One other important piece of equipment in the intensive care area is a respirator. Such a device, also called a ventilator, provides assistance to a patient who is having difficulty breathing. A variety of such devices have been developed for use in intensive care.

Education of Medical Staff

One of the first things that needs to be done when setting up an intensive care facility is to educate the medical staff about the proper use of the intensive care equipment. It is felt by many hospital administrators and heads of engineering for hospitals that this educational function should be performed primarily by the clinical engineer.

Several areas exist in the education of the staff. One of the first of these is the development of a proper technique for attaching leads to the patient. This is important in ECG equipment because, as will be explained in more detail later, a relatively small imbalance in the resistance of the contacts or electrodes attaching the leads to the patient can result in high noise artifacts in the ECG signal. Elimination of this problem requires teaching the staff two things: proper cleansing of the skin at the point of electrode attachment and the use of electrodes of sufficient quality to make certain that good contact is made with the body.

A second item of importance is teaching the staff to be aware of the proper methods of calibration of the equipment used in intensive care and the importance of performing calibrations. It is important that the clinical engineer see to it that the medical staff is well acquainted with the calibration procedures of the equipment they use. This is true not only for electronic equipment but also for the transducers used in conjunction with the electronics package.

A third important item is that of detection of faulty equipment operation. It may seem obvious, but it is important to see that the user of the equipment understands the proper procedure to be followed in determining whether or not reasonable data are being obtained from the equipment that is being used.

An important item to be taught to members of the intensive care staff, and perhaps the most difficult to give instruction in, is the taking of responsibility in reporting anything doubtful. If a nurse, for example, feels that a piece of equipment has given her or him a slight electrical shock, that fact should be reported to the clinical engineer immediately. In fact, the equipment should be "red tagged" and taken out of service so that it will not be used until it is checked by the engineer. The user of intensive care equipment should report abusive or hard use of a piece of equipment or any falls that the equipment may have undergone. This may seem obvious, yet experience indicates otherwise. It shoud be impressed on the staff that reporting hard use of equipment or other items will not result in action against the person reporting the incident and will be of direct benefit to the patient and/or the medical personnel using the equipment. On the other hand, failure to report an incident might call for corrective action.

A second important problem that appears very frequently in the use of intensive care equipment is that of providing safe electrical power sources in intensive care areas. A number of standards exist which are intended to provide guidelines and in some cases exact procedures for establishing a facility that can be considered safe. However, many of these standards are not studied by the people involved in the installation or maintenance of electrical power. It should be the duty of the clinical engineer in the hospital to be familiar with the standards and to ascertain whether the electrical power system within the facility meets the standards that pertain to the area.

The foremost standard and the one that spells out the majority of the requirements for the hospital is the *National Electrical Code*®, published by the National Fire Protection Association. Underwriters' Laboratories is charged with testing equipment to determine whether it meets the requirements set forth in the standards that pertain to the particular device. Details of these standards and their application are discussed in further detail in later chapters.

CONCLUSION

This chapter is only a brief introduction to certain aspects of the hospital as it affects the clinical engineer. Hospitals are a study in themselves and only a sample of the complex operation of this health care delivery institution has been given here.

In only a little over 10 years, clinical engineering has grown to become a vital part of the health care delivery team and is an essential professional staff position in the hospital. The responsibility of professionalism will become increasingly important as the activities of clinical engineers become even more crucial to the successful and safe operation of a health care institution. Just as with the medical profession, more than a professional school diploma is required to show competence. It is important to the profession of engineering and to the hospital that all persons seeking jobs as clinical engineers be registered professional engineers, PEs. It should be mentioned that hospitals do not come under the ill-conceived "industrial exemption" clause of state engineering registration acts. Thus all 50 states and the District of Columbia require that people who sell engineering services to and practice the profession of engineering in a hospital be registered.

In addition to registration there is an added dimension to the professional qualifications of the clinical engineer: certification in clinical engineering. Certification in this engineering specialty is to clinical engineering what board certification is to the physician. It is a way of demonstrating competence in a specialty area that is above the average skills and knowledge of practitioners of the profession who are not certified.

Health care delivery today is difficult and complex. It is the goal of this book to provide technical understanding of this branch of biomedical engineering to meet the engineering needs of students and practitioners in this new area of the engineering profession.

REFERENCES

1. Feinberg, Barry N., "Professionalism in clinical engineering," *J. Clin. Eng.* Vol. 3, No. 2, 1978.
2. Ray, C. D. (ed.), *Medical Engineering* (Chicago: Year Book Medical Publishers, 1974).
3. Webster, John G., and Albert Cook (eds.), *Clinical Engineering Principles and Practices* (Englewood Cliffs, N.J.: Prentice-Hall, 1979).

2

ELECTRONICS
FOR BIOMEDICAL
INSTRUMENTATION

In an era when solid-state electronics has produced the integrated circuit and computer central processing units (CPUs) as small as two end-to-end postage stamps, one may ask what a topic like electron tubes (or "vacuum tubes," as they are sometimes called) doing in this chapter or even in this book. The rationale is that vacuum-tube characteristics generally are not taught any more as part of the electronics background of the electrical, clinical, or biomedical engineer. Yet the clinical engineer must deal with a substantial amount of older vacuum-tube electronic equipment in today's hospitals. And, believe it or not, vacuum tubes are still used in some forms of hospital electronic equipment. The classic example of this is the Ritter CSV Bovie spark gap electrosurgical unit and older-model electrosurgical units that are still found in surgical suites in hospitals across the country. Also, as will be seen in Chapter 10, x-ray tubes are essentially vacuum-tube diodes and all cathode ray tubes (CRTs) used in ECG monitors and consoles, computer systems, and CAT scanners are vacuum tubes. Thus vacuum tubes are found in modern electronic equipment as well as in older types of equipment. Often a useful piece of medical electronic equipment is discarded on the assertion that it is "old," but the real reason may be that the clinical engineer does not have the appropriate background to deal with this type of electronic equipment. Since the clinical engineer may have to deal with vacuum tubes in one form or another, this section has been introduced to fill in the education gap that has been produced by "progress."

ELECTRON-TUBE CHARACTERISTICS

The term "vacuum tube" is somewhat descriptive of the physical appearance of this device, but the English had a better term for the device—"valve." This is a very appropriate term for the device, as an amplifier, either vacuum or solid state, is indeed acting as an electron valve in that the flow of electrons is controlled in a manner very much like a valve. It is precisely this control of the flow of electrons in a way that is proportional to a signal that is behind the concept of amplification.

A vacuum tube must have four basic characteristics if it is to be used as an amplifier:

1. It must have a source of electrons.
2. It must have a means to accelerate these electrons.
3. It must have a means to control the electron flow.
4. It must have a means of collecting these electrons.

As described in greater detail in Chapter 10, the source of electrons is the cathode, which is frequently a cylindrical metallic sleeve coated with a material which has the property that it emits a large number of electrons when heated to a high-enough temperature. When energy in some form, usually in the form of heat, is applied to the cathode, enough energy is imparted to the free electrons of the material making up the cathode that they escape from the surface of the material. Even though there are attractive forces which work to restrain the escape of free electrons from the surface of the metal, electrons with enough kinetic energy will leave the metal's surface. An analog to electron emission from a heated metal surface is the evaporation of water molecules from the surface of liquid water.

The number of electrons "evaporated" per unit area of an emitting surface is related to the absolute temperature T (°Kelvin) of the emitter. As described further in Chapter 10, the cathode emission current is given by the Richardson–Dushman equation:

$$I = aAT^2 e^{-b/T} \tag{2-1}$$

where a = a constant that is characteristic of the material

A = cathode emitter surface area

T = temperature, degrees Kelvin

b = measure of the work required for an electron to escape the emitter surface

The combination of the squared and exponential terms in the Richardson–Dushman emission equation makes the emission current in the tube very sensitive to small changes in temperature. For example, doubling the temperature of the emitting surface increases the electron emission by more than 10^6. This process of causing electron emission from an emitter by means of imparting kinetic energy to the electrons by means of heat transfer is called *thermionic emission*.

The most commonly used emitters in electron tubes are tungsten, heated to a temperature between 2200 and 3000°K; thoriated tungsten (thorium and carbon added to tungsten), heated to about 1900°K; and oxide-coated cathodes (oxides of barium and strontium), heated to between 1000 and 1150°K.

Electron emitters are heated electrically, either directly or indirectly. In the direct method, the electric current is flowing through a wire that also serves as an electron emitter. The I^2R energy loss in the wire is converted to heat, which increases the kinetic energy of the free electrons to the point where they can escape from the surface of the wire.

A second method of energy transfer to free electrons in the cathode is the indirect method. Here current is applied to a separate heating element, located in a cylindrical sleeve which acts as a cathode emitter. The cathode is heated indirectly through heat transfer from the heating element. Either direct or alternating current can be used for either method of heating the cathode emitter. Figure 2-1 illustrates the two methods of cathode heating.

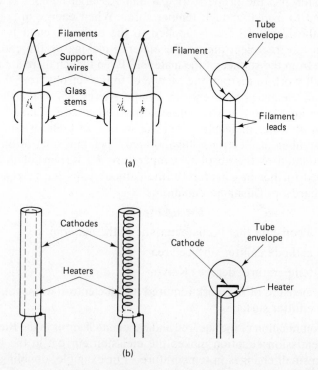

Figure 2-1 (a) Directly heated filament and schematic symbol; (b) indirectly heated cathodes and schematic symbol.

VACUUM-TUBE DIODES

The simplest combination of elements constituting an electron tube is the cathode, which is the electron emitter, and an anode, which acts as an electron gatherer. These two elements are enclosed in a glass or metal envelope which is evacuated so that the electrons do not collide with air molecules. The *anode* or *plate,* as it is sometimes called, is made of metal and is cylindrical in shape. It is usually made of nickel, molybdenum, graphite, tantalum, Monel, or iron. An electrical potential supplied by an external power source between the cathode and anode is used to accelerate the electrons from the cathode to the plate. Figure 2-2 illustrates the configuration of components that form the vacuum-tube diode. Since the vacuum-tube diode and its operating characteristics are discussed in detail in Chapter 10, we will move on to the next area in vacuum tubes, the triode.

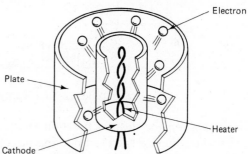

Figure 2-2   Elements of a vacuum-tube diode.

TRIODE VACUUM TUBES

A large step forward was made in 1907 when Lee DeForest added a third element, called a *control grid,* between the cathode and the plate of a diode, thus producing the first triode vacuum tube. This tube had the ability to amplify time-varying signals and led to the development of radio communications and the broadcasting and electronics industries in general as we know them today.

The grid usually consists of relatively fine wire wound onto two support rods and extends the full length of the cathode. The spaces between the turns are comparatively large so that the passage of electrons from cathode to plate is practically unobstructed by the grid, but when a voltage is placed on the grid, the electric field it produces has a profound effect on the electric field between the cathode and the anode (plate) and, hence, on the total electron flow.

The structure of the grid can take many forms. Different sizes and spacings of the control grid wires are employed, depending on the field configuration needed for the tube to perform its design function. Metals used for grids include Nichrome, molybdenum, iron, nickel, and tungsten.

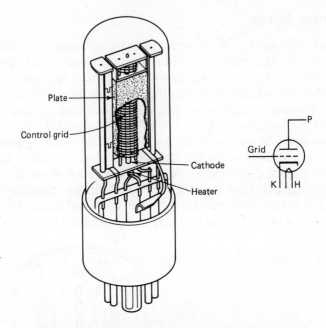

Figure 2-3 Triode construction with schematic symbol.

Depending on the desired function and power rating, the triode tube can differ widely in size and electrode spacing. Figure 2-3 shows the construction of a typical triode together with its schematic symbol.

Action of the Control Grid

Since the control grid is nearer to the cathode than to the plate, potentials placed on the grid produce a relatively high field intensity between the grid and the cathode as measured in volts per meter. Because of this fact, a potential placed on the grid has a much larger effect on the electric field between the cathode and the anode than does the anode potential, thus exerting great influence on the flow of electrons from the cathode to the plate. Therefore, small voltages placed on the control grid can cause large changes in the plate current. When the plate is positive relative to the cathode, and a steadily increasing negative direct-current (dc) grid voltage is applied, the plate is less able to attract electrons to it and the plate current decreases. When the grid is made less negative, the plate can attract more electrons and thus the plate current increases.

A triode requires three operating voltages, one for each electrode. The plate is normally connected to a large positive voltage, with respect to the cathode, called the *B+ voltage*. A relatively low voltage, ac or dc, usually under 10 volts (V), is used to heat the filament to bring the cathode up to proper emission temperature. The third voltage is placed on the grid to control the plate current. The grid voltage consists of two components, a bias voltage, which is usually a few volts negative with respect to the cathode, and the signal voltage, which swings about the bias voltage. The function of the bias voltage is basically the same as for the transistor, that is, to place the tube at the appropriate or quiescent operating point in its operating characteristics. This is the same concept as that for biasing a transistor to operate about a quiescent point in its operating characteristics.

The purpose of the grid voltage is to control the flow of plate current in faithful response to the grid or signal voltage. The plate current is an amplified replica of the signal voltage applied to the grid, as it is small variations of the grid voltage that produced the large variations of the plate current. If a resistive load is placed in series with the plate of the triode, the voltage drop produced across this resistance is a function of the plate current and hence is controlled by the grid voltage. Therefore, a small change in the grid voltage can cause a large change in the plate current and in the resulting voltage across the load resistance. We see, then, that the signal voltage appearing at the grid of the triode is amplified in the plate circuit of the tube. This amplification takes place without any grid current or power consumption in the grid circuit, as long as the grid voltage is negative with respect to the cathode.

Triode Characteristic Curves

Plate voltage, grid voltage, and plate current relationships in a triode can easily be summarized graphically. This graphical relationship is called the *triode characteristic curve*. If all of the triode parameters are plotted together, it would take a three-dimensional surface to show these relationships

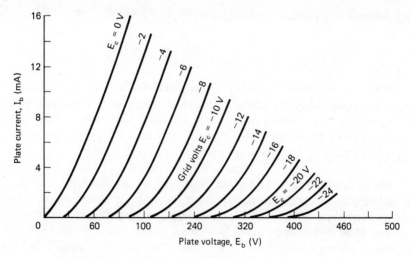

Figure 2-4   Static tube characteristics for 6J5 triode; $E_f = 6.5$ V.

properly all at the same time. For convenience, two-dimensional projections of the surface are used, by plotting any two of the three parameters while holding the third constant. Figure 2-4 is an example of triode characteristic curves plotting plate current (I_b) against plate voltage (E_b) while holding the grid voltage (E_c) at a fixed value. These characteristic curves are the ones that are typically given in vacuum-tube manuals.

Triode-Tube Constants

The families of tube characteristics that are related to tube performance are the product of careful design to make each tube behave in a prescribed way. These design factors are summarized by a series of numbers, called the *tube constants*. All major tube manufacturers publish tube manuals in which these constants are listed together with other ratings for each tube type. The three most important tube constants that we will take up here are:

1. The amplification factor, μ
2. The ac plate resistance, r_p
3. The transconductance, g_m

Amplification factor. The amplification factor of a triode is a measure of the relative effectiveness of the electric field set up by the control grid in altering the flow of electrons from the cathode to the plate. The amplification factor can be written

$$\text{amplification} = \frac{\text{small change in plate voltage}}{\text{small change in grid voltage}}$$

while holding the plate current, I_b, constant. Mathematically, this can be written

$$\mu = \frac{dV_b}{dV_c} \tag{2-2}$$

If the amplification factor, μ, of a triode is 25, a change in the grid voltage (or signal voltage) will be 25 times as effective in changing the plate current as the same change in the plate voltage. Thus the "valve" action of the electron tube is illustrated where small power levels at the input of the grid can control large power levels at the plate output.

Plate resistance. The plate resistance of an electron tube is the internal resistance to electron flow from the cathode to the plate. This is not the same type of resistance phenomenon as that found in current flow in wires. Plate resistance can be divided up into dc and ac plate resistances. Dc plate resistance is defined, using Ohm's law, as the ratio of the plate to cathode voltage divided by the plate current when each is not time varying: that is,

$$R_p = \frac{E_b}{I_b} \tag{2-3}$$

A more significant performance index for the electron tube is the ac plate resistance. This is more a figure of merit for the dynamic performance of the tube. The ac plate resistance is defined as the ratio of a small change in plate voltage to the change in plate current it produces when the grid voltage is held constant. Expressed in equation form, we have

$$r_p = \frac{\Delta v_b}{\Delta i_b} \qquad v_c = \text{constant}$$

As an example, consider that a 50-V change in the plate voltage produced a 50-milliampere (mA) current change in the plate. The ac plate resistance is computed using Ohm's law as

$$r_p = \frac{50}{0.005} = 10,000\Omega \tag{2-4}$$

Transconductance.   The third parameter used in describing the operating characteristics of an electron tube is the transconductance. This is sometimes referred to as the *mutual conductance* and is given the symbol g_m. It is a measure of the influence the control grid has on effecting changes in the plate current. It is a dynamic characteristic as opposed to a dc one and is defined as the ratio of a small change in plate current to a small change in the grid voltage that produced the plate current change. In equation form this becomes

$$g_m = \frac{\Delta i_b}{\Delta v_c} \qquad v_b = \text{constant} \tag{2-5}$$

Since g_m is a ratio of current to voltage, the unit for this parameter is that of conductance expressed in mhos. This is a rather large unit to be used easily, so the more practical unit "micromho" appears in the literature. In an electron tube where a 2-V change in grid voltage produces a 2-mA change in plate current, the transconductance of the tube is calculated as

$$g_m = \frac{0.002}{2} \times 10^6 = 1000 \text{ micromhos} \tag{2-6}$$

Relationship between tube parameters.   The tube parameters are interrelated, as we shall see in the following equations. For example, if we divide the amplification factor by the plate resistance, we have

$$\frac{\mu}{r_p} = \frac{\Delta v_b / \Delta v_c}{\Delta v_b / \Delta i_b} = g_m \tag{2-7}$$

From this relationship we see that

$$\mu = g_m r_p \tag{2-8}$$

From equation (2-8), if any two of the tube parameters are known, the third can be calculated.

As a general figure of merit for electron tubes, the transconductance is most often used when comparing tubes in the same general classification. Since transconductance is the ratio of the amplification to the plate resistance, a number of circuit design characteristics are reflected in this value. For example, if high amplification is desired to obtain maximum output voltage for a given input signal voltage applied to the grid of the triode, a tube with a large transconductance is used. On the other hand, when voltage amplification is not as important as power gain, a tube with a small plate resistance permits the flow of large plate current, resulting in large power output from the tube. Thus transconductance can be used as a firsthand measure of tube performance relative to voltage and power gain.

GRAPHICAL ANALYSIS OF ELECTRON-TUBE PERFORMANCE

Now that we have defined the tube parameters and shown how they are related mathematically, we will see how one can obtain these characteristic values from the characteristic curves provided by vacuum-tube manufacturers in their tube manuals. The use of these curves is very similar to the use of transistor characteristic curves in the design of electronic circuits.

Consider a set of characteristic curves for a triode (Figure 2-5). In these curves, plate current is plotted against plate voltage with the grid voltage held constant for each curve but curves shown for various grid voltages. From the definition of amplification, which is the change in plate voltage for a given change in grid voltage, we can determine from the characteristic curves the amplification factor for the tube around a particular operating point.

If A is the operating point for the tube, the grid voltage is at -8 V. The plate current can be read from the graph as 7 mA and the plate voltage is 235 V. To obtain the cause-and-effect relations between the tube parameters, a new point on the constant grid voltage curve is selected at point B. A horizontal line of constant plate current is drawn to the next, -10 V, curve of constant grid potential where it intersects point C. A vertical line is then drawn upward until it intersects the -8-V grid curve again at point D. From this construction we can make graphical evaluations of the tube's operating performance that are good enough for engineering design purposes and eliminate the need for test circuits to produce the same information.

Consider now the graphical computation of the amplification factor. The line BC is the change in grid voltage, 2 V, and the projections of the

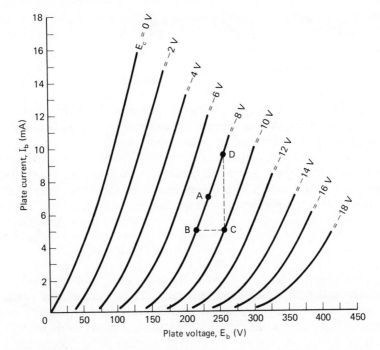

Figure 2-5 Determination of amplification factor and plate resistance from characteristic curves; type 6J5.

points *B* and *C* on the plate voltage axis show the corresponding change in plate voltage for the plate current remaining constant at 5 mA. The plate voltage corresponding to the projection of point *B* on the *x* axis is about 215 V. The projection of point *C* is about 255 on the *x* axis. The difference is

$$255 - 215 = 40 \text{ V}$$

By definition, then, the amplification of this triode is

$$\mu = \frac{40 \text{ V}}{2 \text{ V}} = 20$$

The calculation of the ac plate resistance, r_p, can be found graphically in the same easy manner as we did with the amplification. From the definition of the ac plate resistance, we wish to find the change in plate voltage produced by a change in plate current for constant grid voltage. From Figure 2-5 we see that this is the reciprocal of the slope of the grid curve around the operating point *A*. From the previous calculation, the change in plate voltage represented by the line *BC* is 40 V. The line *CD* represents the corresponding change in plate current for a constant grid voltage. This value, the difference in plate current, is defined by points *D* and C. The plate current corresponding to point D is about 9.5 mA, and that corresponding to point C is 5 mA. Thus the calculation of the r_p is given by

$$\frac{255 - 215}{9.5 - 5} = \frac{40 \text{ V}}{0.0045 \text{ A}} = 8888.89 \text{ ohms}$$

The calculated value of the r_p may differ from that published in tube manuals, as r_p is dependent on the operating point selected for the particular design, as is seen in the graphical solution obtained above.

Using the graphical techniques developed above, we can calculate the tube's transconductance, g_m. From Figure 2-5 we see that a change in grid voltage from -10 V to -8 V as determined from the points as represented by the line CD corresponds to a change in plate current from 5 mA to 9.5 mA. The transconductance, g_m, is easily calculated from the ratio

$$g_m = \frac{(9.5 - 5) \text{ mA}}{(10 - 8) \text{ V}} = 2250 \text{ mhos}$$

Besides the triode electron tube, there are other multielectrode tubes, the tetrode and pentode to name just two. The object of this portion of the chapter was simply to acquaint the clinical and biomedical engineer with the subject of vacuum tubes so as to deal better with them in the hospital or laboratory setting. It was not written to be a comprehensive section on vacuum-tube electronics and circuits. If the reader has an interest in pursuing the topic further, there are a myriad of older textbooks in electrical engineering that will cover the subject to any depth desired. This section on vacuum-tube characteristics, combined with the material in Chapter 10 on thermionic emission, represents a good start in an understanding of vacuum-tube electronics.

THE BIPOLAR TRANSISTOR

The solid-state counterpart of the electron tube is the transistor. The anatomy of the transistor with its corresponding currents and voltages of importance is as follows:

1. *Emitter:* emitter current, I_E; collector–emitter voltage, V_{CE}
2. *Base:* base current, I_B; collector–base voltage, V_{CB}
3. *Collector:* collector, I_C; emitter–base voltage, V_{EB}

The symbol for the transistor is shown in Figure 2-6. The arrow on the emitter lead specifies the direction of current flow when the emitter–base junction is forward biased. In either *NPN* or *PNP* cases, I_C, I_E, and I_B are considered positive when they flow into the transistor.

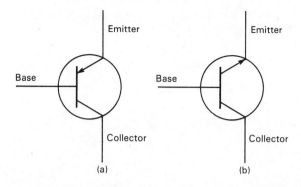

Figure 2-6   Schematic diagram of the *PNP* (a) and *NPN* (b) transistor.

Modes of Operation

There are four modes of operation for a transistor, which are defined on the basis of junction biasing.

1. *Active mode:* E-B junction forward biased, C-B junction reversed biased
2. *Reverse active mode:* E-B junction reversed biased, C-B junction forward biased
3. *Cutoff mode:* both junctions reversed biased
4. *Saturation:* both junctions forward biased

Transistor Configurations

There are three basic transistor configurations. They differ as to which terminals are the input and output terminals. These are the common-base, common-emitter, and common-collector transistor configurations. Figure 2-7 illustrates each of these configurations.

Analysis of common-base configuration.   In this configuration (Figure 2-8) the transistor is operated in its active region where the base–emitter junction is forward biased and the base–collector junction is reversed biased. This configuration is basically a current amplifier and is seldom used in electronic circuits.

The Common-Collector Configuration

This configuration provides power gain and current gain. The transistor is in its active region of operation when the collector junction is reversed biased and the base–emitter junction is forward biased. In the active region,

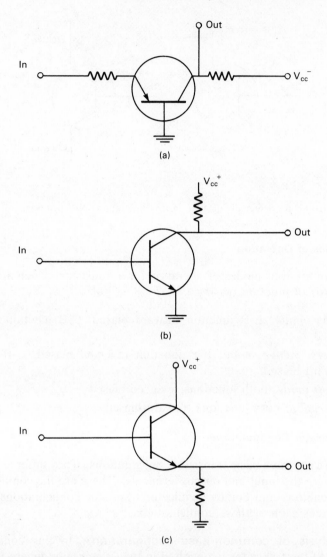

Figure 2-7   Circuit schematic diagram for the common-base (a), common-emitter (b), and common-collector (c) configurations.

the collector current is essentially independent of the collector voltage and depends only on the emitter current.

Common-emitter configuration.   The most widely used transistor circuit configuration is the common-emitter form, shown in Figure 2-7b. In this form the input current and output voltage are the independent

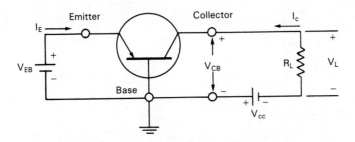

Figure 2-8 Common-base transistor configuration.

variables in the system and the input voltage and output current are the dependent variables in the system.

Consider the set of characteristic curves for a transistor in the common-emitter configuration shown in Figure 2-9a. From this graph we see a family of curves that display the relationship of the collector current to the collector–emitter voltage. The parabolic curve labeled P_c represents the maximum power dissipation at the collector for the different values of collector–emitter voltage and collector current. The curve P_c is called the *power parabola*. Any operation of the transistor at a point on the characteristic curves that is to the right of the power parabola exceeds the maximum collector power dissipation and should be avoided. Calculation of the transistor power dissipation can be approximated by the equation

$$P_c \simeq V_{CB}I_C \simeq V_{CE}I_C \qquad (2\text{-}9)$$

Load-line calculations. Consider the common-emitter transistor circuit given in Figure 2-9a. If we take the sum of the voltage around the output circuit, we have

$$V_{CE} = V_{CC} - I_C R_L \qquad (2\text{-}10)$$

This is the equation of a straight line when plotted on the output characteristic curves as shown in Figure 2-9b. This line is called the *load line* for a particular R_L. For any load, R_L, and for any base current input, I_B, that intersects the load line, the collector current and collector–emitter voltage are then known.

Example

Let $R_L = 500$ ohms (Ω) and $V_{CC} = 10$ V. Using equation (2-10), we have for $I_C = 0$

$$V_{CE} = V_{CC} = 10 \text{ V}$$

When $V_{CE} = 0$, we have

$$I_C = \frac{V_{CC}}{R_L} = \frac{10}{500} = 20 \text{ mA}$$

This locates two points on the load line and defines the load line as shown in Figure

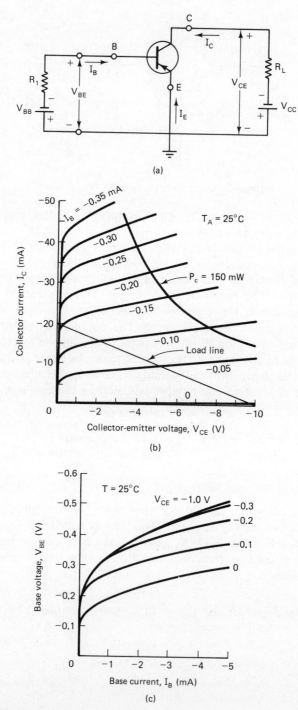

Figure 2-9 (a) Common-emitter transistor circuit; (b) output characteristics for the common-emitter transistor circuit; (c) base input characteristics for the common-emitter transistor circuit.

2-9b. Since the collector current is proportional to the base current for the common-emitter configuration, we can write this relationship in the following way:

$$I_C = \beta I_B \qquad (2\text{-}11)$$

where β is called the *current gain* of the transistor.

Example
Find the values for the bias resistor R_1 and the load resistor R_L and the quiescent operating point, Q point, for the circuit in Figure 2-9a for an input current of $i = 0.1 \sin \omega t$ milliamperes and a collector–emitter voltage, V_{CE}, of 12 V. The quiescent operating or Q point is the intersection of the load line with the base current with no signal applied.

Solution

$$R_1 = \frac{V_{BB}}{I_B} = \frac{3}{0.2 \times 10^{-3}} = 15 \text{ k}\Omega$$

$$R_L = \frac{V_{CC} - V_{CE}}{I_C} = \frac{20 - 12}{11 \times 10^{-3}} = 730 \ \Omega$$

The value 11×10^{-3} is obtained from the characteristic curves in Figure 2-9b.

Calculation of current gain. The current gain of the amplifier is defined at the change in collector current due to a given change in the base current, or

$$\text{current gain, } A_i = \frac{\Delta I_C}{\Delta I_B} \qquad (2\text{-}12)$$

When $\omega t = \pi/2$, the input current is 0.1 mA. The total base current is then equal to $0.1 + 0.2 = 0.3$ mA. From the characteristic curves we have that for this value of total base current the corresponding collector current is 15 mA. From this we can obtain the current gain as

$$A_i = \frac{\Delta I_C}{\Delta I_B} = \frac{15 - 11}{0.1} = 40$$

For this example we see how the characteristic curves in conjunction with the load line can be used for transistor circuit design.

Most transistor circuits are the common-emitter form as shown in Figure 2-9a. In this transistor configuration the input current and output voltages are independent variables and the input voltage and output current are dependent variables. For a fixed base current I_B, the collector current, I_C, is not a strong function of V_{CE}, the collector–emitter voltage. This is seen in Figure 2-10a, where the following equations describe the graph in the figure:

$$I_C = \beta I_B$$

$$I_C = f(\beta, i_B, V_{CE})$$

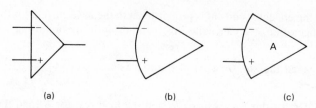

(a)   (b)   (c)

Figure 2-10   Circuit diagram symbols for the operational amplifier.

OPERATIONAL AMPLIFIER

The operational amplifier (op-amp) is a direct-coupled, high-gain amplifier that uses current feedback to control its performance. Internally, it consists of several series-connected transistor differential amplifiers and an output stage to convert the differential signal to a single-ended signal. Generally, we do not consider the inner workings of the operational amplifier but only its input/output characteristics in terms of three parameters: voltage, current, and impedance. Thus, as users of the op-amp in medical instrumentation we concern ourselves only with how the device performs at its terminals.

Characteristics of an Ideal Amplifier

To better understand the operational amplifier as a physical device we will first consider the characteristics of an ideal amplifier, which include the following:

1. Input impedance, $Z_i = \infty$, allows for any signal to be applied without loading effect.
2. Output impedance, $Z_o = 0$, provides for unlimited power out.
3. Voltage gain, $A = V_o/V_i = \infty$.
4. Bandwidth $= \infty$ (passes all frequencies the same from 0 to ∞).
5. $v_o = 0$ when $v_i = 0$.
6. Parameter changes with temperature $= 0$.
7. Internal noise $= 0$.
8. Common-mode voltage gain $= 0$.

None of these parameters are achieved by a real op-amp, but modern techniques have made some of the real parameters so close to the ideal that the difference is hardly discernible.

When selecting an op-amp for a specific application, the engineer usually will be looking for optimum performance in two to three parameters. After an op-amp is selected for these parameters, it may be found to be lacking in other areas; for example, an op-amp may be selected for low noise, but it may not have good bandwidth specifications.

Typical Applications for Operational Amplifiers

Some typical uses of operational amplifiers in medical equipment are as follows:

1. Linear amplifier
2. Nonlinear amplifier
3. Comparator
4. Active filter
5. Logarithmic application
6. Multivibrator
7. Wave shaping and generation
8. Voltage regulator

Operational Amplifiers Compared to Transistors

We look next at the advantages and disadvantages of op-amps compared with transistors. The transition from transistor to op-amp is easily done since we go from a device with many design difficulties to one that is relatively easy to use. Even though an op-amp may contain between 10 and 100 transistors in its integrated-circuit architecture, it is probably the easiest electronic device to work with when only its external parameters are considered.

	Transistor:	Op-amp:
Gain	Voltage Gain 0 to 100	0 to 10^6 Open Loop
Frequency response	Some as high as 5 to 10 GHz	1 Hz to 100 kHz (-3 dB); with feedback up to 100 MHz
Input Z	Bipolar a few kΩ; FET up to 10^{13} Ω	Typically 1 MΩ; some FET input types 10^{13} Ω; 3-pF-capacity input (same as FET since input is a FET)
Output response	From 1 to tens of kΩ	100 Ω or less open loop; below 1 Ω with feedback

Figure 2-10 shows typical notational representations of the operational amplifier and is used to symbolize the use of this amplifier.

Equivalent Circuits for Operational Amplifiers

Power supplies of both positive and negative voltages are used but are not shown in the circuit in Figure 2-11. These are necessary if the output voltage is to go positive and negative with respect to the input

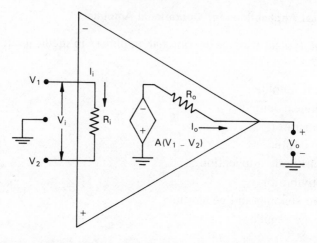

Figure 2-11   Schematic circuit model for the operational amplifier.

voltage. The pin locations for the application of the supply voltages are given in the specification sheet.

Basic Operational Amplifier Circuits

There are two basic circuits that are commonly used in medical instrumentation: inverting and noninverting amplifiers. Many of the more complex circuits are extensions of these two circuits.

Inverting amplifier. The analysis of the circuits shown in Figure 2-12 proceeds as follows. By applying Kirchhoff's voltage law at the input we have the following equations:

$$\text{(i) } V_s = I_1 R_1 + V_i$$

$$\text{(ii) } I_1 = \frac{V_s - V_i}{R_1}$$

$$\text{(iii) } V_o = -R_o I_o = A(V_1 - V_2)$$

But $V_2 = 0$. Therefore,

$$\text{(iv) } V_o = -R_o I_o - AV_1$$

For an ideal amplifier, $Z_i = \infty$ and $R_o = 0$. Then

$$\text{(v) } I_i = \frac{V_i}{R_i} = 0 \qquad R_o I_o = 0$$

Now using Kirchhoff's current law at the input node, we have

$$\text{(vi) } I_i = I_1 - I_f = \frac{V_s - V_i}{R_1} - \frac{V_i - V_o}{R_f} = 0$$

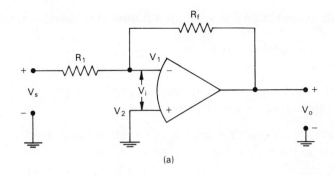

(a)

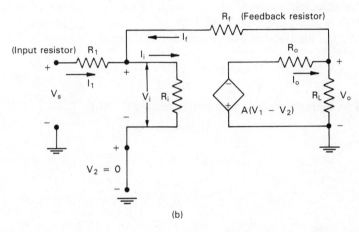

(b)

Figure 2-12   (a) Inverting amplifier ($v_2 = 0$); (b) circuit model for the inverting amplifier ($v_2 = 0$ and $v_1 = V_i$ with no load at the output. $R_L = \infty$.)

From equation (iv), $v_o = -AV_i = -AV_1$; thus

$$(\text{vii})\ \frac{V_s - (v_o/-A)}{R_1} - \frac{(v_o/-A) - v_o}{R_f} = 0$$

Once again for an ideal amplifier if we assume that $A = \infty$, then $V_o/A = 0$. Therefore,

$$(\text{viii})\ \frac{V_s}{R_1} - \frac{-v_o}{R_f} = 0$$

Solving for the gain (voltage transfer ratio V_o/V_s), we have

$$(\text{ix})\ \frac{v_o}{v_s} = -\frac{R_f}{R_1} \qquad (2\text{-}13)$$

For $Z_i = \infty$, the input current $I_i = 0$. Also, $A \rightarrow \infty$; then $V_i \rightarrow 0$ for a finite V_o. Thus V_i approaches ground potential or V_i is at a virtual ground.

Basic rules for design with operational amplifiers. Simplification of the analysis procedure used above provides a set of assumptions that allow for easier design and analysis of operational-amplifier circuits. The assumptions are:

1. Operational-amplifier input terminals draw no current.
2. Voltage across input terminals is zero.

These rules are adequate for most design work and analysis of circuits that the clinical engineer may do in the design or analysis of medical instrumentation circuits.

Noninverting amplifier. Consider now the analysis of the op-amp circuit shown in Figure 2-13. Making use of simplifying assumptions 1 and 2 above, the analysis of the circuit is made easier.

1. There is no current to either input. Therefore,

$$I_1 = I_f = I$$

2. The voltage across the input terminals is zero.

$$V_i = V_1 - V_2 = 0$$

Therefore

$$V_1 = V_2$$

$$V_1 = V_o \frac{R_1}{R_1 + R_f} = V_2$$

$$\text{voltage gain} = \frac{V_o}{V_2} = \frac{R_1 + R_f}{R_1} = 1 + \frac{R_f}{R_1} \qquad (2\text{-}14)$$

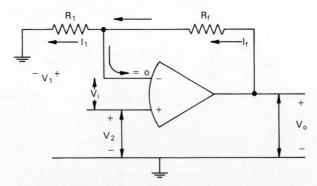

Figure 2-13   Noninverting amplifier.

Voltage-follower circuit. This circuit is used as a buffer or isolation stage when voltage gain is not needed in the circuit. Consider the operational-amplifier circuit given in Figure 2-14a. Let $R_1 \to \infty$ and $R_f \to 0$. From

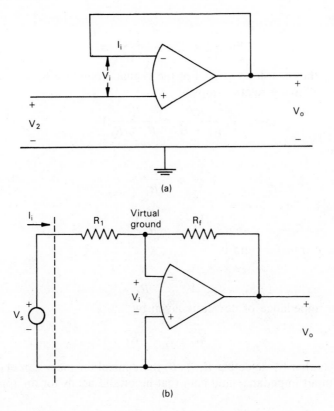

Figure 2-14 (a) Voltage follower; (b) noninverting amplifier showing point of virtual ground.

equation (2-14), we can compute the gain of the amplifier by taking the limit of equation (2-14) as $R_1 \to \infty$ and $R_f \to 0$.

$$\lim_{\substack{R_1 \to \infty \\ R_f \to 0}} \frac{V_o}{V_o} = \lim_{\substack{R_1 \to \infty \\ R_f \to 0}} \left(1 + \frac{R_f}{R_1}\right) = 1$$

Analysis Using Virtual Ground Concept

Kirchhoff's voltage law at the input of the operational amplifier produces

$$V_2 + V_i = 0$$

But $V_o = -AV_i$. Therefore,

$$V_2 = V_o\left(1 + \frac{1}{A}\right)$$

$$\frac{V_o}{V_2} = \frac{1}{1 + 1/A}$$

Ideal case:

$$\lim_{A \to \infty} \frac{V_o}{V_2} = \lim_{A \to \infty} \frac{1}{1 + 1/A} = 1$$

where A is the open-loop gain. For the virtual ground we have from Figure 2-14b the following relationships:

$$V_s \frac{R_f}{R_1 + R_f} = \frac{R_1}{R_1 + R_f} V_o$$

Therefore,

$$\frac{V_o}{V_s} = -\frac{R_f}{R_1}$$

$$V_o = -\frac{R_f}{R_1} V_s$$

The voltage transfer ratio is

$$\frac{V_o}{V_s} = -\frac{R_f}{R_1}$$

The input impedance of device is given by

$$Z_{\text{in}} = \frac{V_s}{I_1} = R_1$$

This shows that by selecting $R_1 + R_f$ the clinical engineer can choose any value of input impedance and gain that he or she needs for an application.

Summing amplifier (adder).   Applying Kirchhoff's current law at the input node of the operational amplifier in Figure 2-15, we have

$$\sum_{M=1}^{4+f} I_M = I_1 + I_2 + I_3 + I_4 + I_f = 0$$

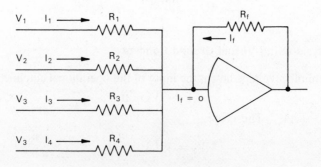

Figure 2-15 Summing amplifier.

Using Ohm's law yields the equation

$$-\frac{V_o}{R_f} = \frac{V_1}{R_1} + \frac{V_2}{R_2} + \frac{V_3}{R_3} + \frac{V_4}{R_4}$$

Solving for the output voltage V_o, we have

$$V_o = -\left(\frac{R_f}{R_1} V_1 + \frac{R_f}{R_2} V_2 + \frac{R_f}{R_3} V_3 + \frac{R_f}{R_4} V_4\right) \qquad (2\text{-}15)$$

If $R_1 = R_2 = R_3 = R_4$, then $-V_o = V_1 + V_2 + V_3 + V_4$, which shows that the output voltage is the negative sum of all the input voltages.

Differential input amplifier. In many applications, it is desirable to obtain the difference between two voltages as opposed to their actual value. Since we have two voltages acting, V_o is the result of the effects of V_{s1} and V_{s2} (see Figure 2-16). Thus, using the principle of superposition, we can let $v_{s1} = 0$ (ground point 1) which converts the circuit to a noninverting amplifier. Then

$$\frac{V_o}{V_{in}} = 1 + \frac{R_f}{R_1}$$

or

$$V_o = V_{in}\left(1 + \frac{R_f}{R_1}\right)$$

$$V_{in} = \frac{R_f}{R_1 + R_f} V_2$$

Therefore,

$$V_{2_0} = \frac{R_1 + R_f}{R_1}\left(\frac{R_f}{R_1 + R_f}\right) V_2 = \frac{R_f}{R_1} V_2$$

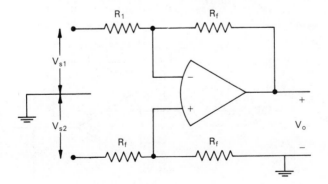

Figure 2-16 Differential input amplifier.

With $V_2 = 0$ there is no current flowing in the (+) input, and

$$V_{1_0} = \frac{R_f}{R_1} V_1$$

$$V_o = V_{1_0} + Z_{1_0} = \frac{R_f}{R_1} (V_2 - V_1) \qquad (2\text{-}16)$$

Example: Ideal Amplifier

Design an amplifier of gain of -100 with an input resistance of 1000 Ω (see Figure 2-17).

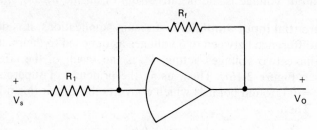

Figure 2-17

Solution. Since the node is at a virtual ground, the input impedance is R_1. Therefore, $R_1 = 1$ kΩ by design selection.

$$\frac{V_o}{V_s} = -\frac{R_f}{R_1} = -100$$

Thus $R_f = 100R_1 = 100(10^3) = 100$ kΩ.

Example: Integrator

Let $V_s(t) = vu(t)$, $u(t)$ = unit step function. Find $V_o(t)$, the output voltage in the circuit shown in Figure 2-18.

Solution. Because there is a virtual ground at node 1 we have

$$V_s = L \frac{di(t)}{dt} + R_i(t) \qquad \text{at } t = 0, i(0) = 0$$

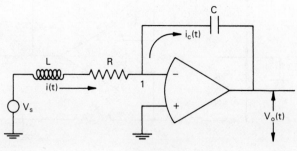

Figure 2-18

Solving for $i(t)$, we have

$$i(t) = \frac{V_s(t)}{R}(1 - e^{-(R/L)t}) = \frac{V}{R}(1 - e^{-(R/L)t})$$

The voltage across capacitor C is $V_o(t)$. Therefore,

$$V_o(t) = \frac{1}{C}\int_0^t i_c(t)\, dt$$

Since the amplifier draws no current, $i_c(t) = i(t)$, and thus

$$V_o = \frac{V}{RL}\int_0^t (1 - e^{-(R/L)t})\, dt$$

$$V_o(t) = \frac{V}{RC}\left[t + \frac{L}{R}(e^{-(R/L)t} - 1)\right]$$

Offset Error Voltages and Currents (dc)

For an ideal operational amplifier we have the output voltage, V_o, as zero when the input voltages to both input ports are equal (i.e., when $V_1 = V_2$). A real operational amplifier exhibits an unbalance caused by mismatch of the input transistors. We now consider dc error voltages and currents that can be measured at the input and output terminals.

Input bias current: This is the bias current required to bias the input transistor of the op-amp into the active region when $V_o = 0$. If I_{B1} is the bias current into terminal 1 and I_{B2} is the bias current into terminal 2, then

$$I_B = \frac{I_{B1} + I_{B2}}{2}$$

when the output voltage V_o is zero.

Input offset current: The input offset current I_{io} is the difference between the separate currents entering the input terminals of a balanced amplifier. As shown in Figure 2-19, we have $I_{io} = I_{B1} - I_{B2}$ when $V_o = 0$.

Input offset current drift: The input offset current drift $\Delta I_{io}/\Delta T$ is the ratio of the change of input offset current to the change of temperature.

Input offset voltage: The input offset voltage V_{io} is that voltage which must be applied between the input terminals to balance the amplifier for $V_o = 0$.

Input offset voltage drift: The input offset voltage drift $\Delta V_{io}/\Delta T$ is the ratio of the change of input offset voltage to the change in temperature.

Output offset voltage: The output offset voltage is the difference between the dc voltages present at the two output terminals (or at the output terminal and ground for an amplifier with one output) when the two input terminals are grounded.

Power supply rejection ratio: The power supply rejection ratio (PSRR), $\mu V/V_1$, is the ratio of the change in input offset voltage to the corresponding change in one power supply voltage, with all remaining power supply voltages held constant.

Slew rate: The slew rate is the time rate of change of the closed-loop amplifier output voltage under large-signal conditions. It is measured in volts per microsecond. Figure 2-19 is the model of an op-amp taking offset voltage and bias currents into account.

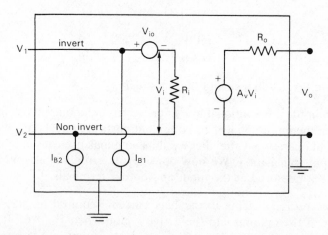

Figure 2-19 Circuit model for use in common-mode rejection analysis.

Common-mode rejection: This term describes the ability of the amplifier to reject a common-mode input voltage while amplifying a differential input voltage, expressed in decibels. This topic is covered in more detail in Chapter 5.

    The use of operational amplifiers has considerably simplified the work of the electronics engineer in the design of electronic circuits. The subject of operational-amplifier circuits is far more extensive than can be covered in an introductory chapter such as this and the reader is referred to more complete texts on the subject if more information on this topic is required. It is the purpose of this chapter to introduce the reader to the basic aspects of electronics that he or she needs as a clinical engineer in the management of medical electronics.

PROBLEMS

1. The constant b in equation (2-1) can be written as $b = W_w/k$, where W_w is the amount of energy an electron must have in order to escape from the surface of a cathode. The constant k is Boltzmann's constant, which is equal to 1.380

$\times\ 10^2$ joules per degree Kelvin. Calculate the saturation current for an oxide-coated cathode with an area of 2 cm^2 and a temperature of 300°K (room temperature). Repeat for a temperature of 1100°K. The constants for the material are as follows:

$$a = 60 \times 10^4$$
$$W_w = 3.40$$

2. Calculate the saturation current for a 2-cm^2 emitter made of thorium at 2000°K.

$$a = 60 \times 10^4$$
$$W_w = 3.40$$

3. A tungsten wire filament 0.02 cm in diameter and 5 cm long is the cathode of a vacuum-tube diode. The plate is a cylinder concentric with the filament. What is the temperature-limited current produced by the cathode at a temperature of 2500°K where

$$a = 60 \times 10^4$$
$$W_w = 4.52$$

for tungsten?

4. A vacuum-tube diode circuit is connected as shown in Figure 2-20.
   (a) If $v_i(t)$ = instantaneous input voltage
       $i_p(t)$ = instantaneous plate current
       $v_p(t)$ = plate voltage (voltage across the diode)
       R_L = load resistance

   write an equation that expresses the plate voltage as a function of the remaining circuit variables.
   (b) The characteristic curve for this diode is given in Figure 2-21, which plots the plate current as a function of plate voltage. On this same graph, plot your equation from part (a) for $v_i(t)$ = 50 V and 100 V. What is the significance of the intersection of the graph of the equation developed in part (a) with the characteristic curve?
   (c) If the input voltage is v_i = 100 sin ωt, draw the output voltage $v_o(t)$ as a function of time for this circuit.

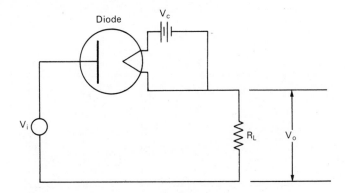

Figure 2-20

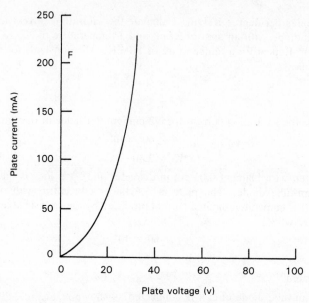

Figure 2-21

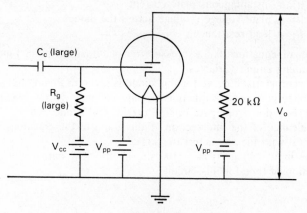

Figure 2-22

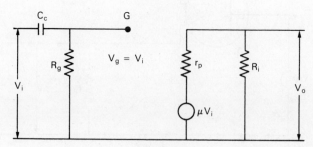

Figure 2-23

5. A 6J5 triode is used in the circuit configuration shown in Figure 2-22. Its equivalent circuit is given in Figure 2-23. The associated plate characteristics for this tube are given in Figure 2-24. If the plate voltage, $V_{pp} = -6$ V and the load resistance, $R_1 = 20$ kΩ, find the voltage gain for the amplifier circuit.

6. The transistor circuit in Figure 2-25 is a basic common-emitter amplifier. In the design of this amplifier, the engineer must select R_1, R_2, R_L, and R_e together

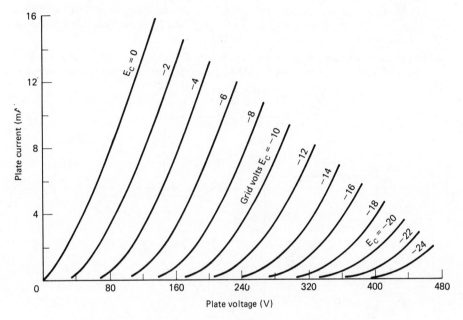

Figure 2-24   Average plate characteristics, type 6J5, $E_f = 6.3$ V.

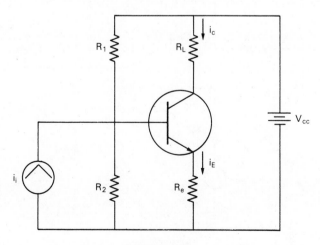

Figure 2-25

with the supply voltage V_{cc} so that the transistor operates linearly and with maximum collector current i_c.

(a) Write an expression for V_{BB}, the base bias voltage, in terms of R_1, R_2, and V_{cc}.

(b) Write an expression for the base bias resistance, R_b, in terms of R_1 and R_2.

(c) Write an expression for R_1 and R_2 as a function of R_b, V_{bb}, and V_{cc}.

(d) Draw a Thévenin equivalent circuit for the bias circuit of this amplifier, with V_{bb} as the Thévenin equivalent voltage and R_b as the Thévenin equivalent resistance.

7. In the transistor circuit of problem 6, Figure 2-25, let $V_{cc} = 15$ V, $R_L = 1$ kΩ, and $R_e = 500$ Ω. Determine the maximum symmetrical swing of the collector

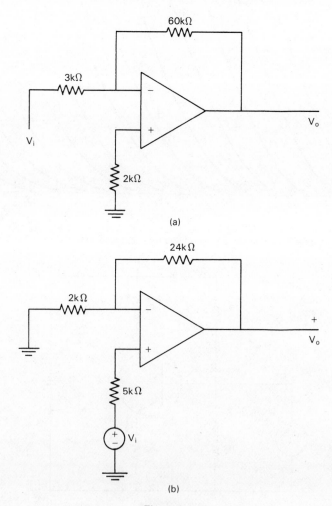

(a)

(b)

Figure 2-26

current and the quiescent operating point, Q, for the circuit. (*Hint:* Use a load-line construction.)

8. Calculate the gain for an ideal inverting amplifier circuit using an operational amplifier where $R_1 = 1$ kΩ, $Z_i = \infty$, and the gain of the operational amplifier is $A = 100,000$ for the following values of R_f:
   (a) $R_f = 75$ kΩ
   (b) $R_f = 480$ kΩ
9. In each of the circuits shown in Figure 2-26, find the output voltage, V_o. Assume that all the operational amplifiers are ideal.
10. Given the operational-amplifier circuit in Figure 2-27:
   (a) Find the voltage transfer ratio (circuit gain) as a function of $j\omega$.

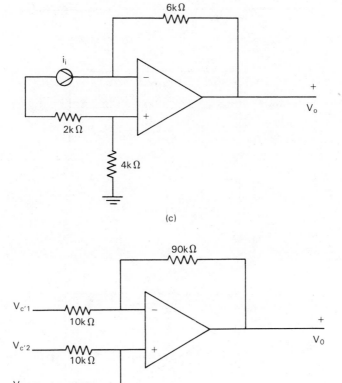

(c)

(d)

Figure 2-26 (*cont.*)

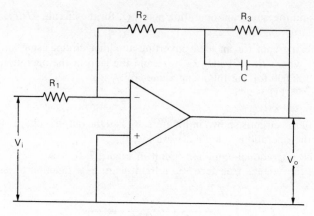

Figure 2-27

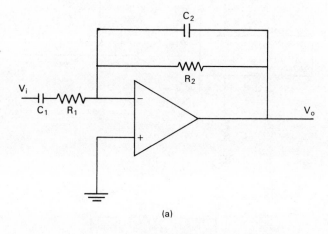

(a)

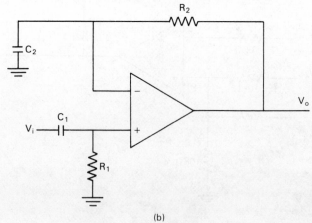

(b)

Figure 2-28

(b) If the input voltage is $v_i(t)$, show that the output voltage is determined by the following differential equation:

$$C \frac{dv_o(t)}{dt} + \frac{V_o(t)}{R_3} + \frac{v_i(t)}{R_1}\left(1 + \frac{R_2}{R_3}\right) = 0$$

11. Determine the voltage gain for each of the amplifier circuits shown in Figure 2-28.

12. For the amplifier circuit shown in Figure 2-29, sketch the output voltage, $v_o(t)$, for the input voltage, $v_i(t)$, given in Figure 2-30. Assume that the operational amplifier and the diode in the circuit are ideal.

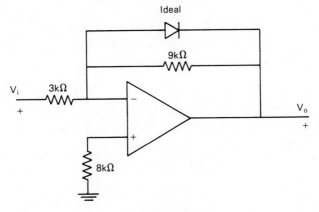

Figure 2-29

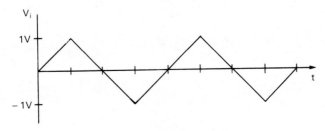

Figure 2-30

REFERENCES

1. Brown, Paul F., Gunter N. Franz, and Howard Moraff, *Electronics for the Modern Scientist* (New York: Elsevier, 1982).

2. Dow, William G., *Fundamentals of Engineering Electronics,* 2nd ed. (New York: Wiley, 1952).

3. Milman, Jacob, *Micro-electronics: Digital and Analog Circuits and Systems* (New York: McGraw-Hill, 1979).

4. Stout, David F., and Milton Kaufman (eds.), *Handbook of Operational Amplifier Circuit Design* (New York: McGraw-Hill, 1976).

5. Vassos, Basil H., and Galen W. Ewing, *Analog and Digital Electronics for Scientists* (New York: Wiley-Interscience, 1972).

3

FUNDAMENTALS
OF MEDICAL
INSTRUMENTATION

The primary purpose of medical instrumentation is to extend the human senses in order to obtain measures which provide the data needed for more accurate diagnosis and treatment of the sick or injured. Since a large portion of what is called "medical electronic devices" comprises in essence different forms of instrumentation systems designed for medical use, some common aspects of instrumentation will be taken up in light of their utilization in the medical field. This is in contradistinction to other forms of medical electronics, such as defibrillators and x-ray machines, which are not instrumentation devices.

In a medical instrumentation system, the primary function is to measure or determine the presence of some physical quantity that is in some way useful to medical personnel. Many types of instrumentation systems are used in hospitals and physician's offices. The majority are electrical or electronic systems, although mechanical systems, such as the water-filled spirometer (a device used in pulmonary function testing) are also used. Because of the extremely large number of electronic instrumentation systems that the clinical engineer must deal with in the practice of his or her profession, this chapter will take up the concepts of the electronic medical systems, although many of these concepts can be applied to nonelectronic medical instrumentation, for example, the mercury thermometer, as well.

THE INSTRUMENTATION SYSTEM

Certain characteristic features are common to most instrumentation systems, including medical electronic instrumentation systems. We can consider that the instrumentation system consists of three major functional blocks (see Figure 3-1):

1. transducers or sensors,
2. signal conditioning,
3. display or recording.

Transducer or Sensor

A transducer or sensor is a device that converts energy in one form to energy in another form. In most cases, this is electrical energy due to the ease of transmission of the electrical signal produced by the transducer. This is not always the case, however, as in the case of an electromechanical ammeter, which converts electrical current into mechanical motion. Transducers can be either active or passive. An *active transducer* directly converts one form of energy into a second form, whereas a *passive transducer* requires energy to be put into it for the transducer to produce the required transduction. In general, we assume that the output signal from the transducer is directly related to or even proportional to the quantity we wish to measure.

The justification for this predominant use of medical electronic instrumentation is the ease with which information in electrical form may be processed and transmitted. This may change in the future with the use and development of fiber-optic transmission systems and optical couplers. Then we may see that electro-optical systems will prevail over totally electronic medical instrumentation systems.

Another way to consider a transducer is as a device which provides a usable output in responses to a specific physical quantity, property, or possibly a condition—to the presence or absence of a particular quantity. This physical quantity, property, or condition to which the transducer is responding is defined as the *measurand*.

Signal Conditioning

The main purpose of the signal conditioning section of an instrumentation system is to modify the signal coming from the transducer into a form that is usable by the final stages of the instrumentation system. The signal conditioner is involved with transmission and transformation of sensor or transducer response. Signal conditioning can take any of the following forms:

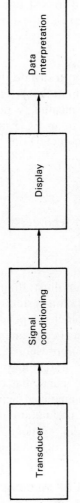

Figure 3-1 Block diagram of an instrumentation system.

1. Amplification
2. Filtering, analog or digital
3. Analog-to-digital and digital-to-analog conversion
4. Radiated in radio transmission

Signal conditioning is not confined to just one of the foregoing processes but may be involved in one or more of these processes.

Display

The object of the display is to translate the output of the signal conditioning section of the instrumentation system into a form that can be read, heard, or otherwise communicated to our senses. Some examples are the cathode ray tube, as used in an ECG monitor; a digital display, as in a cardiotachometer (heart rate meter); or a meter deflection. Other typical forms are:

1. Scales (manometer)
2. Meters, digital or analog
3. Graphical (X-Y or strip-chart recorders)
4. Video or cathode ray output
5. Printed computer output

Figure 3-1 is a simple block diagram illustrating each of the basic elements that comprise a medical instrumentation system. Each of these blocks could be broken down into smaller subblocks depending on the type of system. The transducer constitutes the critical component of each measuring system in which it is used. The portion represented by the signal conditioning block of a medical instrumentation system may vary in its complexity from a simple resistance or resistance–capacitance (RC) filter network and impedance-matching network, to an amplifier, active analog or digital filtering, or analog-to-digital converter, to modulation–demodulation systems used in electrocardiographic transmission. Among the many types of display devices the display itself can be any combination of the items listed above.

PERFORMANCE CHARACTERISTICS OF MEDICAL INSTRUMENTATION SYSTEMS

The performance characteristics of a medical instrumentation system are both static and dynamic in nature. A transducer is normally designed to sense a specific quantity to be measured, the measurand. The characteristics of a medical transducer to perform a specific detection and conversion process are determined during a calibration cycle by applying the measurand in discrete amplitude intervals and allowing the system to reach equilibrium.

This eliminates any frequency-dependent effects that may occur. The input/output characteristics obtained in this way are called the *static characteristics*.

Error in Measurement Systems

Since no transducer can be fabricated ideally, the output of an actual medical transducer causes the indicated measurand value to deviate from the true value.

Accuracy: This term describes the algebraic difference between the indicated value and the true or theoretical value of the measurand. This deviation of the indicated value from the true value is usually expressed as a percent of full-scale output. This is a measure of total error without regard to its source.

Reproducibility (repeatability): This is a measure of the ability of a transducer to reproduce output readings when the same measurand value is applied to it repeatedly, under the same conditions and in the same direction.

Precision: Stated simply, precision refers to the degree of reproducibility of a measurement. It should be noted that precision in a transducer or instrumentation system does not imply accuracy and is sometimes a point of confusion. Since precision is a measure of reproducibility, an instrumentation system with an uncompensated offset voltage in an operational amplifier may give very reproducible results that are not accurate.

Static sensitivity: This is defined for a transducer or instrumentation system as the ratio of the incremental output quantity to the incremental change in the measurand or other output quantity to the incremental change in the measurand or other input quantity. The static sensitivity for an instrumentation system may be constant for only part of the normal operating range of the instrument. Figure 3-2 shows a calibration curve for an instrumentation system and its use in defining sensitivity.

Threshold: The threshold of the transducer is the smallest change in the measurand that will result in a measurable change in the transducer output. This specification sets a lower limit on the measurement capability of a particular transducer that is to be used for medical instrumentation systems.

Resolution: The resolution of a transducer is the smallest incremental quantity of the measurand that can be measured with certainty.

Zero drift: This occurs when all output values increase or decrease by the same absolute amount. Thus the sensitivity of the device does not change; the calibration curve is shifted up or down. Factors that contribute to zero drift are:

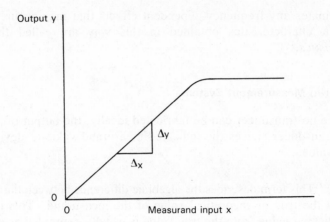

Figure 3-2 Input/output relationship for an instrumentation system. Sensitivity $s = \Delta y / \Delta x$.

1. Manufacturing misalignment
2. Sensitivity to ambient temperature variations
3. Amplifier offset

Slow changes in dc offset voltage do not cause a big problem in medical electronic instruments since most amplifiers are ac coupled.

Sensitivity drift: When the slope of the calibration curve changes, we have sensitivity drift. This causes errors that are proportional to the input. This drift can be caused by:

1. Power supply variations
2. Ambient temperature changes
3. Ambient pressure changes
4. Component changes

LINEARITY IN MEDICAL TRANSDUCER AND INSTRUMENTATION SYSTEMS

An instrumentation system or biomedical transducer as pictured in Figure 3-3 is linear if an input stimulus $x_1(t)$ produces a response $y_1(t)$ and a second stimulus $x_2(t)$ produces a response $y_2(t)$. Then the particular instrumentation system or transducer is linear if a stimulus of $x_1(t) + x_2(t)$ produces a response of $y_1(t) + y_2(t)$. Symbolically, we have that if

$$x_1(t) \longrightarrow y_1(t) \tag{3-1}$$

$$x_2(t) \longrightarrow y_2(t) \tag{3-2}$$

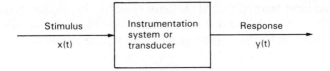

Figure 3-3   System block diagram.

then a medical instrumentation system, transducer, or component is linear
if

$$x_1(t) + x_2(t) \longrightarrow y_1(t) + y_2(t) \qquad (3\text{-}3)$$

This is nothing more than a statement of the superposition principle and
is a necessary condition for any system to be linear. Basically, this says
that the output is in some way proportional to the input. Graphically, this
can be represented as in Figure 3-4. Thus, if the variables $y(t)$ and $x(t)$ are
linearly related, they are proportional to one another and the transfer characteristic between the input and the output is a straight line through the
origin.

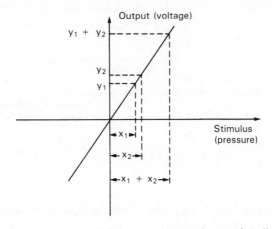

Figure 3-4   Transfer characteristic between input and output for a linear system.

NONLINEAR STATIC CHARACTERISTICS

Since there are many different ways to describe the nonlinear behavior of
a medical instrument, it is simpler to contrast this state with that of linear
performance with the statement that nonlinear performance is any deviation
from the definition of linearity as previously described. Since the design
of instrumentation systems is based on the concept of linearity and the
relatively straightforward mathematical analysis associated with linear systems, the nonlinear behavior of a transducer or instrumentation system
negates the use of linear mathematics, and alternative methods and ap-

proximation must be used to analyze these systems. In practice, no instrument is perfectly linear. Several fundamental types of nonlinear input/output characteristics will be discussed together with typical causes for the nonlinear behavior.

Saturation

All real instruments exhibit saturation when input magnitudes of the measurand are too large. For any physical transducer or instrumentation system, there is a range where the output is proportional to the input for linear operation. As the amplitude of the measurand increases, a point is reached where the response of the transducer is no longer proportional to the measurand. If the measurand continues to increase, positively or negatively, a point is reached where the transducer will no longer increase its output for increased input. A graph of this nonlinear relationship is given in Figure 3-5.

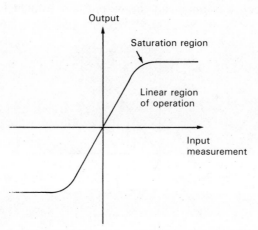

Figure 3-5 Transfer characteristic between input and output for an instrumentation system indicating system saturation.

Hysteresis

Hysteresis is observed when the input/output characteristics for a transducer are different for increasing inputs than for decreasing inputs. Thus, when a given measurand level is approached with increasing values of the measurand, then with decreasing measurand, the effect of hysteresis is to obtain different output levels for the same value of measurand. Hysteresis results when some of the energy applied to the instrument for increasing inputs is not recovered when the input decreases. The lost energy is usually dissipated as heat. Actually, all instruments have some hysteresis because the second law of thermodynamics prohibits perfect reversibility. The measure of how much hysteresis a transducer has is to look at the maximum difference in any pair of output readings during any one calibration cycle. This difference is usually expressed as a percent of full-scale output (see Figure 3-6).

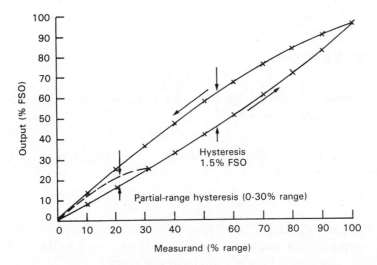

Figure 3-6 Transfer characteristic for an instrumentation system with hysteresis (scale of errors, 10:1). (From Harry N. Norton, *Sensor and Analysis Handbook*, Prentice-Hall, Inc., Englewood Cliffs, N.J., 1982, by permission.)

DYNAMIC RESPONSE OF INSTRUMENTS

Nonlinear dynamic systems are a study by themselves. The mathematics and graphical technique required to solve for a nonlinear system's performance is usually tailored to that system for a specific input. The response of some nonlinear systems can change totally as the amplitude of the input signal is changed. Few generalities can be drawn about a nonlinear system, and this fact makes them difficult to use in the medical instrumentation area. Therefore, linear systems are by far the most useful of medical instrumentation systems and represent the majority of medical instrumentation systems designed. Because of this fact, we will review only the linear system response here.

Any medical instrumentation system represents the conversion of energy from one form to another. The differential equations that describe this transformation of energy within the system and their solutions characterize the medical instrumentation system. Since there are a large number of different types of medical instrumentation—mechanical, electrical or electronic, electromechanical, hydraulic, and so on—we will take up only the general characteristics of the instrument response here. Common transducer types and associated systems will be discussed later.

For any dynamic system, the order of the differential equation that describes the system is called the *order* of the system. Most medical instrumentation systems can be classified into zero-, first-, second-, and higher-order systems. In the main, the majority of medical instrumentation systems fall into one of the first three classifications.

ZERO-ORDER SYSTEMS

Let the measurand $x(t)$ be the input and let the instrumentation system or transducer response be $y(t)$. Then the relationship between input and output in the time domain is given by

$$a_0 y(t) = b_0 x(t) \qquad (3\text{-}4)$$

which is a zero-order differential equation. The transfer function in the frequency domain is given by

$$F(j\omega) = \frac{y(j\omega)}{x(j\omega)} = \frac{b_0}{a_0} = K \qquad (3\text{-}5)$$

where K is the static sensitivity of the system. This simple type of instrument or transducer has an output that is directly proportional to the input and there is no phase shift between the input and output.

The potentiometer used as a displacement transducer, at frequencies or acceleration where inertial effects are small, can be considered as a zero-order system (Figure 3-7).

$$V_o = V \frac{R_i}{R_t}$$

$$R_i = \alpha d$$

Therefore,

$$V_o = \frac{V}{R_t}(\alpha d) \qquad (3\text{-}6)$$

where α = proportionality constant

d = displacement

Equation (3-6) shows that the voltage output is proportional to the displacement, d. Thus

$$y(t) = V_o(t)$$

$$x(t) = d(t)$$

$$\frac{y(j\omega)}{x(j\omega)} = \frac{\alpha V}{R_t} = K$$

for this system and represents a system of order zero.

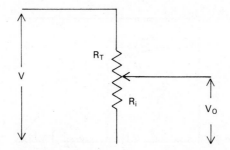

Figure 3-7 Schematic diagram of a potentiometer.

FIRST-ORDER SYSTEMS

The first-order transducers or medical instrumentation systems are characterized by a first-order differential equation of the form

$$a_1 \frac{dy(t)}{dt} + a_0 y(t) = b_0 x(t) \tag{3-7}$$

where $x(t)$ and $y(t)$ are as defined previously. The transfer function in the frequency domain is given by

$$F(j\omega) = \frac{y(j\omega)}{x(j\omega)} = \frac{b_0/a_0}{1 + (a_1/a_0)j\omega} = \frac{K}{1 + j\omega\tau}$$

The magnitude of this input/output relationship is given by

$$|F(j\omega)| = \frac{K}{\sqrt{1 + \omega^2 \tau^2}} \tan^{-1}(-\omega\tau)$$

where $K = b_0/a_0$ = static sensitivity

$\tau = a_1/a_0$ = time constant of the transducer or instrument response

The plot of the magnitude and phase characteristics of the system as a function of frequency is shown in Figure 3-8.

In the time domain, the transient response to a unit step input of the measurand where $x(t) = u(t)$ [$u(t)$ is the unit step input] is given by the solution of equation (3-7). Thus the transient response is given by

$$y(t) = K(1 - e^{-t/\tau})u(t) \tag{3-8}$$

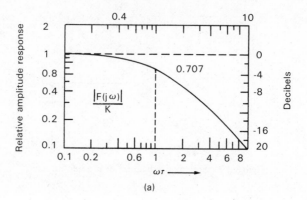

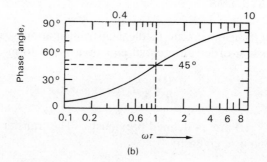

Figure 3-8 Amplitude (a) and phase (b) characteristics of a first-order system.

where $K = b_0/a_0 =$ static sensitivity

$\tau = a_1/a_0 =$ system time constant

A simple example of a first-order medical instrumentation system is the mercury thermometer used to take a patient's temperature. When the thermometer is placed in the patient's mouth, we can consider this to be a step input of temperature into the system. Heat transfer takes place through the glass bulb, which causes the mercury inside to expand. In this case, the system response $y(t)$ is the indicated temperature change $\Delta T(t)$ and the input $x(t)$ is $T_p u(t)$, where T_p is the body temperature of the patient. The change in temperature response of the thermometer is then given by

$$\Delta T(t) = \Delta T(1 - e^{-t/\tau})u(t)$$

where $\Delta T = T_0 - T_p$

$T_0 =$ initial temperature of the mercury thermometer

$T_p =$ patient body temperature

A typical plot of the relative time response of the system is shown in Figure 3-9. When $t = \tau$, the time constant of the system, the expression becomes

$$1 - e^{-t/\tau} = 1 - e^{-1} = 0.63$$

Thus a first-order system reaches 63% of its maximum value in one time constant.

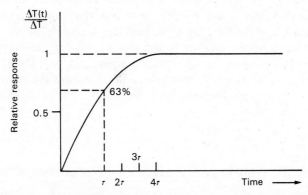

Figure 3-9   Transient response to a step input for a first-order system.

SECOND-ORDER SYSTEMS

A transducer or measurement system that is described by a second-order differential equation of the form

$$a_2 \frac{d^2y(t)}{dt^2} + a_1 \frac{dy(t)}{dt} + a_0 y(t) = b_0 x(t) \tag{3-9}$$

is a second-order system. Rewriting this equation to put it in a standard form, we have

$$\frac{d^2y(t)}{dt^2} + \frac{a_1}{a_2} \frac{dy(t)}{dt} + \frac{a_0}{a_2} y(t) = \frac{b_0}{a_2} x(t) \tag{3-10}$$

We define the coefficients in the following manner:

$$K = \frac{b_0}{a_0} = \text{static sensitivity}$$

$$\omega_n = \sqrt{\frac{a_0}{a_2}} = \text{undamped natural frequency}$$

$$\rho = \frac{a_1}{2\sqrt{a_0 a_2}} = \text{damping ratio}$$

$$\omega_0 = \omega_n \sqrt{1 - \rho^2} = \text{damped frequency}$$

Thus the second-order equation takes the form

$$\frac{d^2y(t)}{dt^2} + 2\rho\omega_n \frac{d_g(t)}{dt} + \omega_n^2 y(t) = K\omega_n^2 x(t) \tag{3-11}$$

which is a standard form for a system described by a second-order differential equation. The solution to the differential equation is composed of two components: the homogeneous solution, that is, the solution to the differential equation when $x(t) = 0$ for all t, and the nonhomogeneous solution, where $x(t)$ is some nonzero function of time.

The nonhomogeneous solution is a function of the form of the measurand $x(t)$. Our discussion of the various solutions that are dependent on the input will include only the step input and the sinusoidal input.

The homogeneous solution to the differential equation is the solution to the second-order differential equation when $x(t) = 0$ for all t. This is also referred to as the *transient response* of the system or the *initial conditions response*. The systems response to a unit step input of the measurand is given by the equation

$$y(t) = 1 - \frac{e^{-\rho\omega_n t}}{\sqrt{1 - \rho^2}} \sin(\omega_n \sqrt{1 - \rho^2}\, t + \cos^{-1}\rho)$$

The form of the solution is dependent on the magnitude of the damping ratio ρ:

$\rho = 0$: $y(t)$ is a pure sinusoid.

$\rho < 1$: $y(t)$ is an exponentially decaying sinusoid and is underdamped.

$\rho = 1$: $y(t)$ is nonoscillatory but has the fastest rise with no overshoot. In this case the system is critically damped.

$\rho > 1$: $y(t)$ is nonoscillatory and is overdamped.

Figure 3-10 illustrates the response of a second-order medical instrumentation system as a function of the damping ratio.

The transfer function in the frequency domain is given by the equation

$$\frac{y(j\omega)}{x(j\omega)} = \frac{K}{(j\omega/\omega_n)^2 + (2\rho/\omega_n)j\omega + 1} \tag{3-12}$$

The magnitude and phase of the transfer function are given by

$$\left|\frac{y(j\omega)}{x(j\omega)}\right| = \frac{K}{\{[1 - (\omega/\omega_n)^2]^2 + 4\rho^2\omega^2/\omega_n^2\}^{1/2}}$$

$$\underline{/\frac{y(j\omega)}{x(j\omega)}} = \tan^{-1}\frac{2\rho\omega/\omega_n}{1 - \omega^2/\omega_n^2}$$

The magnitude and phase plots are shown in Figures 3-11 and 3-12.

If a system is of higher order than two, in most cases it will respond in a manner that a second-order description can be used to describe the higher-order system.

To conclude this section on medical instrumentation system response it is necessary to take up the concept of time-response specification where the desired performance characteristics of the medical instrumentation system

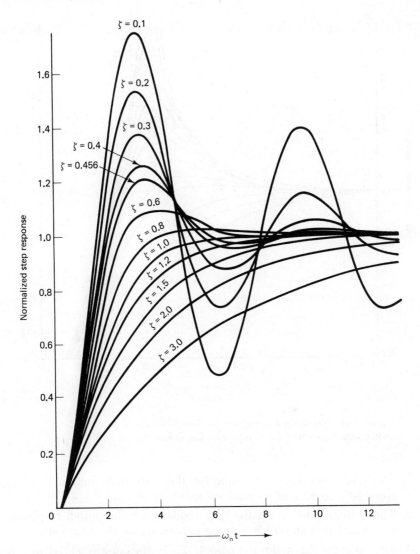

Figure 3-10 Normalized transient response of a second-order system to a step function shown as a function of the damping ratio.

of any order are specified in terms of the transient response to a unit step input of the measurand. This sets up a set of dynamic specifications that complement the static performance characteristics. The dynamic performance of a medical instrumentation system may be evaluated in terms of the following quantities, as illustrated in Figure 3-13.

1. The maximum overshoot, y_m, is the magnitude of the first overshoot. Sometimes this is expressed as a percentage of the final value.

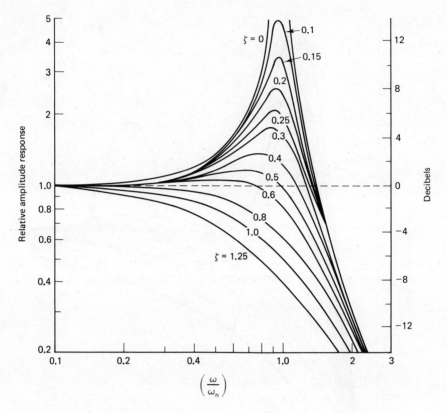

Figure 3-11 Normalized amplitude characteristics for a second-order system shown as a function of the system damping ratio.

2. The rise time, T_R, is the time for the instrument output to change from 10% to 90% of its final value.
3. The settling time, T_s, is the time required for the output response to reach and remain within a given percentage of the final value.
4. The time to maximum overshoot, T_p, is the time required to reach the maximum overshoot.

TRANSDUCERS

We now move from the more general description of instrumentation system types to the specific of transducer types and measurement circuits. We begin by defining the following transducer types.

1. *Passive:* a transducer that requires energy to be put into it in order to translate changes due to the measurand

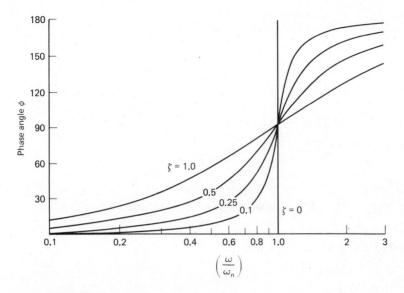

Figure 3-12 Normalized phase characteristics for a second-order system shown as a function of the system damping ratio.

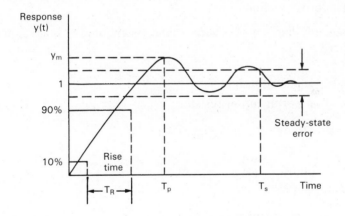

Figure 3-13 Dynamic response of an instrumentation system to a step input.

2. *Active:* a transducer that converts one form of energy directly into another

Passive Transducer—Variable-Resistance Type

The potentiometer is a transducer in which the voltage pickoff arm is actuated by some mechanical input mechanism, such as displacement, pressure, or acceleration. The principles of its operation are given below.

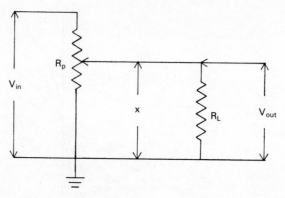

Figure 3-14   Schematic diagram of a wire-wound potentiometer.

The *wire-wound potentiometer* makes use of the resistance property of wire given by the equation

$$R = \frac{\rho L}{A} \tag{3-13}$$

where ρ = resistivity of wire
  L = length of wire
  A = cross-sectional area

The schematic diagram of a wire-wound potentiometer is given in Figure 3-14, where

$$0 \leqslant x \leqslant 1$$

$x = 0$: wiper at ground
$x = 1$: wiper at top of resistor

Case 1. $R_L = \infty$.   See Figure 3-15.

$$V_{out} = \frac{xR_p}{R_p} V_{in} = xV_{in}$$

If $x = 0.4$, then $V_{out} = 0.4V_{in}$.

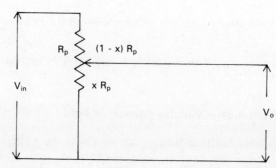

Figure 3-15   Schematic diagram of a potentiometer with no load.

Note: This is a zeroth-order system. The output is proportional to the setting and is independent of the value R_p.

Case 2. Potentiometer with a load resistance R_L.   The simple potentiometer calculation shown above does not hold for the case of a finite load resistance on the potentiometer (see Figure 3-16). Let $R_L = \alpha R_p$, where α is a proportionality constant. Then

$$\frac{V_{out}}{V_{in}} = \frac{\alpha x}{\alpha + x(1 - x)}$$

(x is defined as above.) If the true value is x, the error between the output voltage for this case and when $R_L = \infty$ is given by the equation

$$\text{error} = \frac{x^2(1 - x)}{\alpha + x(1 - x)}$$

Maximum error occurs as $d(\text{error})/dx = 0$ or when $x = 0.5$. For this value of x the maximum error is given by the equation

$$\text{max. error} = -\frac{1}{4\alpha + 1} \qquad (3\text{-}14)$$

Example
Let $R_p = 1\ M\Omega$, $R_L = 2\ M\Omega$. Therefore, $\alpha = 2$. What is the maximum error?
Solution

$$\text{max. error} = -\frac{1}{4(2) + 1} = \frac{-1}{9} = -11.11\%$$

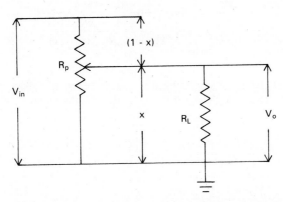

Figure 3-16   Schematic diagram of a potentiometer with load.

Error Due to High-Frequency Effect of Capacitance

In the wire-wound potentiometer used in high-frequency measurements we must account for inductive and capacitive effects, which we now present (Figure 3-17). C_1 and C_2 represent the distributed capacitance of the windings.

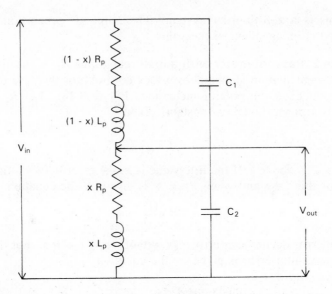

Figure 3-17 Schematic diagram of a high-frequency model of a wire-wound potentiometer.

It is assumed that the resistance R_P and the distributed inductance L_p are partitioned proportionately to x. If the inductance can be neglected for simplicity, the voltage transfer function can be written as in the equation.

$$\frac{V_{out}}{V_{in}} = \frac{1}{1 + \dfrac{(1 - x)(1 + xR_pC_2j\omega)}{x[1 + j\omega R_pC_1(1 - x)]}} \tag{3-15}$$

Example

Let $R_p = 1$ MΩ (10^6 Ω) at $f = 159{,}200$ Hz, $\omega = 2\pi f = 10^6$, stray capacitances C_1 and C_2 at 2 pF. Then $\omega R_pC_1 = \omega R_pC_2 = 2$. What is the voltage transfer function?

Solution

$$\frac{V_o}{V_i} = \frac{x[1 + j2(1 - x)]}{1 + j4(1 - x)} \qquad \left[\left|\frac{V_o}{V_i}\right| = x\left\{\frac{1 + [2(1 - x)]^2}{1 + [4x(1 - x)]^2}\right\}^{1/2}\right]$$

$$\text{error} = \left(\frac{1 + [2(1 - x)]^2}{1 + [4x(1 - x)]^2}\right)^{1/2} - 1$$

As $x \to 0$, error $\to \sqrt{5} - 1$ or 123%. In general, under the assumption that stray capacitance introduces large linearity error, one must be careful not to use a wire-wound potentiometer with nonsinusoidal inputs that are rich in harmonics.

MEASUREMENT DUE TO CHANGE IN RESISTIVITY

Classic examples of transducers that make use of resistance changes as a function of changes in the measurand are:

1. The strain gauge
2. The thermistor

Strain Gauge

Here we have a mechanical deformation of the wire which causes a change in the resistance of the wire in the strain gauge (Figure 3-18). The strain gauge is an isometric device permitting only small displacement to be measured. The extension permissible before permanent deformation of the strain gauge results in dependency on the modulus of elasticity of the material used. Tension in the strain gauge causes a change in length ΔL and changes in diameter ΔD. The resistance of a piece of wire before it is subjected to tensil forces is given in the equation

$$R_1 = \rho_1 \frac{L}{A} = \rho_1 \frac{L}{(\pi/4)D^2} \tag{3-16}$$

Under tension the new resistance, R_2, becomes

$$R_2 = \rho_2 \frac{L + \Delta L}{(\pi/4)(D + \Delta D)^2} \approx \frac{\rho_2 L(1 + \Delta L/L)}{(\pi/4)D^2[1 - 2(\Delta D/D)]} \tag{3-17}$$

But

$$\frac{\Delta D/D}{\Delta L/L} = \mu$$

which is known as *Poisson's ratio* and relates the change in diameter to change in the length of the wire. The equation for R_2 written in terms of Poisson's ratio is given in the equation

$$R_2 = \frac{\rho_2}{\rho_1} R_1 \left[1 + (1 + 2\mu)\frac{\Delta L}{L} \right] \tag{3-18a}$$

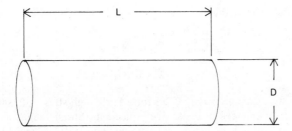

Figure 3-18 Segment of wire in a wire strain gauge.

If we assume that $\rho_2 = \rho_1$, that is, there is no change in resistivity of the wire, then the expression for the resistance R_2 is given by

$$R_2 = R_1 + \Delta R = R_1 \left[1 + (1 + 2\mu)\frac{\Delta L}{L} \right] \qquad (3\text{-}18b)$$

Solving for $(\Delta R/R)/(\Delta L/L) = 1 + 2\mu$. This is called the *strain sensitive factor*. Generally, the strain gauge is made of resistance wire, foil, or a helix with multiple turns of wire. In this case the lumped quantity is called the *gauge factor*, G:

$$\text{gauge factor} = \frac{\Delta R/R}{\Delta L/L} = \frac{\text{fractional change in resistance}}{\text{fractional change in length}} \qquad (3\text{-}19)$$

where ΔR = change in resistance
    R = unstrained resistance
    ΔL = change in element length
    L = unstrained length

This is somewhat less than for a straight wire. Basing the gauge factor on $(\Delta R/R)/(\Delta L/L)$, strain can be found directly from observed resistance changes.

$$\frac{\Delta L}{L} = \frac{S}{E}$$

where E = modulus of elasticity
    S = stress, psi

It is assumed that $\Delta L/L$ for the stressed object is the same for the gauge.

Example

A steel member is stressed at 30,000 psi, gauge factor = 2. Find the change in resistance of the strain gauge.

Solution

$$E_{\text{steel}} = 30 \times 10^6 \, 1/\text{in}^2 \text{ (psi)}$$

Therefore

$$\frac{\Delta L}{L} = \frac{3 \times 10^4}{3 \times 10^7} = \frac{1}{10^3} = 0.1\%$$

If the gauge factor is 2, then

$$\frac{\Delta R/R}{\Delta L/L} = 2$$

$$\frac{\Delta R}{R} = 2 \times 0.1\% = 0.2\%$$

Since the resistive change is only 0.2% for a fairly high stress, 30,000 psi, a circuit sensitive to small changes in resistance is needed. Strain gauges have been used to measure the strength of heart contraction, bone stresses, and blood pressure. Generally, the gauge factor, G, and the resistance, R, of the strain gauge are given by the manufacturer in the specification sheet that comes with the strain gauge.

Material	Gauge Factor, G	Temperature Coefficient (°C^{-1})
Constantan	2.1	0.00002 (25°C)
Nichrome	2.5	0.004
Nickel	12.1 to -20	0.006
Platinum	6.0	0.003
Silicon	120	0.005 to 0.007 (25°C)

General Bridge Circuits

A circuit that is ideally suited to the task of sensing changes in resistance and producing voltages proportional to these changes is the four-arm bridge network. A general form of a four-arm bridge network is shown in Figure 3-19. The detector can be any device that can respond to a voltage differential. Bridge balance is obtained when the voltage between points A and C is zero. Z_1, Z_2, Z_3, and Z_4 are the impedance of each arm and $\mathbf{I}_1$ and $\mathbf{I}_2$ are the phasor currents in each arm. For a null balance we have the condition that the voltage difference between A and C is zero. Thus there is no current flowing in the detector. Under this condition we have the relation given in equation (3-20) in phasor notation.

$$\mathbf{I}_1\,\mathbf{Z}_1 = \mathbf{I}_2\,\mathbf{Z}_2 \tag{3-20}$$

$$\mathbf{I}_1 = \frac{\mathbf{V}}{\mathbf{Z}_1 + \mathbf{Z}_3} \tag{3-21}$$

$$\mathbf{I}_2 = \frac{\mathbf{V}}{\mathbf{Z}_2 + \mathbf{Z}_4} \tag{3-22}$$

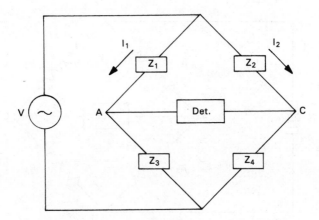

Figure 3-19 General form of a four-arm bridge network.

Substituting equations (3-21) and (3-22) into (3-20), we have for balance of the bridge network that

$$Z_1 Z_4 = Z_2 Z_3 \qquad (3\text{-}23)$$

or

$$\frac{Z_1}{Z_2} = \frac{Z_3}{Z_4} \qquad (3\text{-}24)$$

If we write this in terms of admittances, this gives rise to the equation

$$Y_1 Y_4 = Y_2 Y_3 \qquad (3\text{-}25)$$

Equation (3-23) or (3-25) represent the general condition for balance of a four-arm bridge. From these equations we can derive the conditions for the balance of the bridge in terms of the magnitude and phase angles of the impedances. Writing these impedances in polar form, we have

$$\mathbf{Z}_1 = Z_1 \,\underline{/\phi_1} \qquad \mathbf{Z}_2 = Z_2 \,\underline{/\phi_2} \qquad \mathbf{Z}_3 = Z_3 \,\underline{/\phi_3} \qquad \mathbf{Z}_4 = Z_4 \,\underline{/\phi_4}$$

By using the polar form of the impedance, equation (3-23) can be written in the form

$$Z_1 Z_4 \,\underline{/\phi_1 + \phi_4} = Z_2 Z_3 \,\underline{/\phi_2 + \phi_3} \qquad (3\text{-}26)$$

Equation (3-26) then gives rise to both the magnitude and phase conditions for the balance of a bridge network as given in equations (3-27) and (3-28):

$$Z_1 Z_4 = Z_2 Z_3 \qquad (3\text{-}27)$$

$$\phi_1 + \phi_4 = \phi_2 + \phi_3 \qquad (3\text{-}28)$$

The Wheatstone bridge.   A dc version of the general bridge circuit is the Wheatstone bridge (Figure 3-20). It consists of four resistance arms as illustrated in the figure. At balance there is no current through the branch AC. The condition for balance is obtained from equations (3-27)

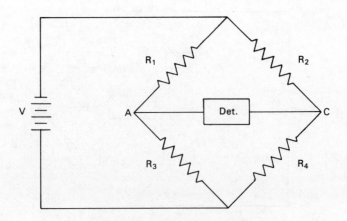

Figure 3-20 Wheatstone bridge.

and (3-28), where

$$\phi_1 = \phi_2 = \phi_3 = \phi_4 = 0$$

$$Z_1 = R_1 \qquad Z_2 = R_2 \qquad Z_3 = R_3 \qquad Z_4 = R_4$$

Therefore,

$$R_1 R_4 = R_2 R_3$$

A Wheatstone bridge can be used to measure unknown resistances, or in the case of medical instrumentation it can be used to respond to changes in resistance of any of the arms of the bridge network. This is its primary function when it is used in many medical devices.

Analysis of Wheatstone Bridge.   When dealing with medical instrument design and analysis which makes use of a bridge network, the circuit analysis and computations using mesh and node methods are tedious when dealing with the unbalanced bridge condition. It is better to make use of a Thévenin equivalent circuit because using a Thévenin equivalent circuit is directed toward finding the detector response only, and this is the major virtue of the method. Detector response is the item of interest most of the time, and the use of the Thévenin equivalent mode of analysis finds it easily and directly and is superior to a formula approach. Thus the voltage input to a detector or amplifier from a bridge network is represented by the Thévenin voltage at that point AC.

Calculation of voltage output for an unbalanced bridge.   Assume that

$$Z_1 = R + \Delta R$$

$$Z_2 = Z_3 = Z_4 = R$$

$$V_{Th} = \frac{V_{in} R}{2R} - \frac{V_{in} R}{2R + \Delta R}$$

$$V_{Th} = V_{in} R \left(\frac{1}{2R} - \frac{1}{2R + \Delta R} \right)$$

$$V_{Th} = V_{in} R \frac{2R + \Delta R - 2R}{2R + 2R + \Delta R}$$

$$V_{Th} = \frac{\Delta R}{2(2R + \Delta R)}$$

If $\Delta R \ll R$, then

$$V_{Th} \approx V_{in} \frac{\Delta R}{4R}$$

The voltage gain of the bridge network is given by the equation

$$\frac{V_{Th}}{V_{in}} = \frac{\Delta R}{4R} \tag{3-29}$$

The input or Thévenin equivalent impedance of the bridge network is

$$Z_i = \frac{R(4 + 3\Delta)}{2(2 + \Delta)} \approx R \qquad \text{for } \Delta \text{ small} \tag{3-30}$$

Example

Given the Wheatstone bridge $R_1 = R_2 = R_3 = R_4 = 4000 \ \Omega$ (see Figure 3-21). Let $\Delta R = 10 \ \Omega$ in R_3. What is the output voltage?

Solution

$$V_{Th} \approx V_{in} \frac{\Delta R}{4R}$$

$$= 4 \left(\frac{10}{16,000} \right) = 2.5 \text{ mV}$$

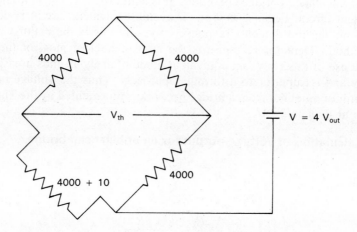

Figure 3-21 Bridge network with strain gauge in one arm.

Bridge Amplifier

It is possible to use a differential amplifier in a bridge configuration to amplify the output of a thermistor or strain gauge bridge network. A typical configuration is shown in Figure 3-22. Assume that the input impedance of the differential amplifier is infinite (i.e., $Z_i = \infty$); the differential amplifier gain $= A_d$. By voltage division

$$V_2 = \frac{V_{in}R}{R + R + \Delta R} = \frac{V_{in}R}{2R + \Delta R}$$

$$V_1 = \frac{V_{in}R}{R + R} = \frac{V}{2}$$

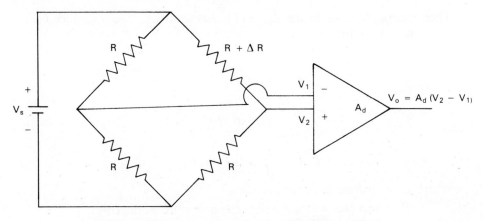

Figure 3-22 Bridge network used with a differential amplifier.

The differential voltage

$$V_2 - V_1 = \frac{V_{in}R}{2R + \Delta R} - \frac{V_{in}}{2}$$

$$= -\frac{V_{in}}{4} \frac{\Delta R/R}{1 + (\Delta R/R)/2}$$

Let $\delta = \Delta R/R$. Then

$$V_o = A_d(V_2 - V_1) = \frac{-A_d V_{in}}{4} \frac{\delta}{1 + \delta/2}$$

which holds for large changes in the resistance transducer. For $\Delta R/R \ll 1$, then

$$V_o = \frac{-A_d V_{in}}{4} \delta$$

is good for small changes in the reactance transducer.

Thermistor

Thermistors are temperature-sensitive resistors made of semiconductor material. They have a negative coefficient of resistivity as a function of temperature. The general characteristics of modern thermistors are as follows:

1. Small
2. Rugged
3. High sensitivity
4. Made of oxides of manganese, nickel, and cobalt
5. May suffer drift as a function of time
6. Nonlinear

Thermistors are made in the shape of beads or disks. The resistance characteristics as a function of temperature are described by the following equations:

$$R_T = R_0 \exp \left[\beta \left(\frac{1}{T} - \frac{1}{T_0} \right) \right]$$

$$= R_0 \exp \left(\beta \frac{T_0 - T}{TT_0} \right) \qquad (3\text{-}31)$$

$$L_n R_T = \frac{\beta}{T} - \frac{\beta}{T_0} + \ln R_0 \qquad (3\text{-}32)$$

where R_T = resistance at temperature T

R_0 = resistance at a reference temperature T_0 (usually 298°K = 25°C)

β = thermistor material constant (typically, 4000°K)

T is in degrees Kelvin (°K)

Example

Consider the bridge circuit shown in Figure 3-23, where one of the impedance arms of the bridge is a thermistor. The thermistor is to be used for measurement of a temperature differential of between 0 and 100°C. Calculate the meter voltage and current at 100°C. R_s = meter internal resistance = 10 kΩ and V_{in} = 4 V.

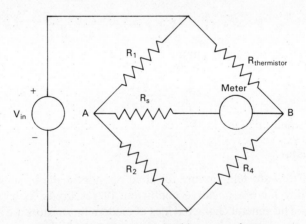

Figure 3-23 Bridge network with a meter detector.

Solution. Let

$$R_{Ti}(50°C) = 5 \text{ kΩ} \qquad R_T(0°C) = 50 \text{ kΩ} \qquad R_T(100°C) = 1 \text{ kΩ}$$

$R_2 = R_4 = R_{\text{therm}}(50°C) = 5$ kΩ, the midscale temperature.

Let $R_1 = 50$ kΩ for bridge balance at 0°C. For the purpose of analyzing, we can replace the bridge network by its Thévenin equivalent. The voltage between the nodes AB at 100°C is given by equation (3-33a), which is the Thévenin voltage.

$$V_{AB}(100°C) = V_{Th} = V_{in}\frac{R_iR_4 - R_2R_{therm}}{(R_1 + R_2)(R_{therm} + R_4)} \qquad (3\text{-}33a)$$

The Thévenin equivalent resistance of the bridge network at 100°C is given by

$$R_{AB}(0°C) = R_{Th} = \frac{R_1R_2}{R_1 + R_2} + \frac{R_{therm}R_4}{R_{therm} + R_4} \qquad (3\text{-}33b)$$

The Thévenin equivalent circuit, which is what the meter sees looking into the terminals AC, is given in Figure 3-24.

$$V_{Th} = V_{in}\left[\frac{50.5 - 5.1}{(50 + 5)(1 + 5)}\right] = V_{in}\left[\frac{250 - 5}{(55)(5)}\right]$$
$$= V_{in}(0.89)$$

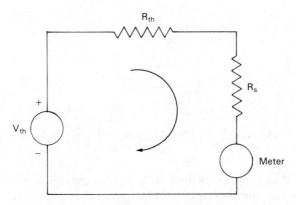

Figure 3-24   Thévenin equivalent circuit for bridge circuit.

If $V_{in} = 4$ V, then $V_{Th} = 3.56$ V. Thus the voltage that the meter sees is 3.56 V when the thermistor is at 100°C.

$$R_{Th} = \frac{(50)(5)}{50 + 5} + \frac{(1)(5)}{1 + 5} = \frac{250}{55} + \frac{5}{6}$$
$$= 5.38 \text{ k}\Omega$$

The current through the meter circuit is given by

$$I_m = \frac{V_{Th}}{R_{Th} + R_s} = \frac{3.56}{5.38 \text{ k}\Omega + 10 \text{ k}\Omega} = 231.47 \text{ }\mu A$$

Thus the thermistor is a very useful transducer when integrated into the appropriate measuring circuit. It can be used to convert temperature changes into corresponding electrical signals.

Inductance

Changes in inductance can be used to measure displacement by varying any of the three coil parameters as given in equation (3-34) for the inductance of a coil.

$$L = n^2 G \mu \tag{3-34}$$

where n = number of turns in a coil

   G = geometric form factor

   μ = effective permeability of the medium

   One of the most common transducers making use of changes in inductance is the *differential transformer*, which measures displacement (Figure 3-25). When used in conjunction with diaphragms or springs, the differential transformer can be used to measure pressure and force. This device is commonly known as a *linear variable differential transformer* (LVDT).

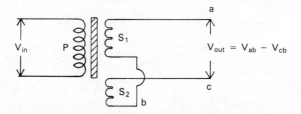

Figure 3-25   Diagram for a linear variable differential transformer (LVDT).

   The primary coil is excited with a frequency between 60 Hz and 20 kHz. When the slug is placed symmetrically, the secondary voltages are symmetrical and of opposite phase, causing them to cancel each other out. Therefore, $V_{out} = 0$ under this condition.

   The characteristics shown in Figure 3-26 illustrate the following aspects of the differential transformer:

1. Linearity over a large range of displacement
2. Change in phase when core passes through the center position

   Sensitivity of linear variable differential transformer (LVDT). Since the output voltage of this transformer is proportional to the excitation voltage, it is convenient to specify the sensitivity for a 1-V excitation. Commercial devices typically have a sensitivity of 0.5 to 2.0 mV per 0.001 cm displacement for a 1-V input. Full-scale displacements of 0.001 to 25 cm with a linearity of $\pm 0.25\%$ are available.

CAPACITIVE TRANSDUCER

Given two parallel plates separated by a distance X (cm) of area A (cm^2) and a dielectric permittivity of the medium, the capacitance between the plates of a capacitive transducer for a fixed place separation E is given by

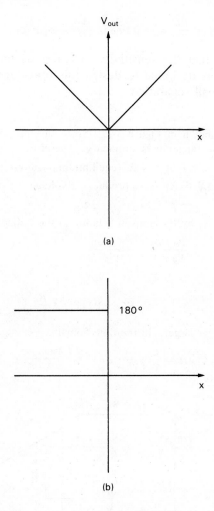

Figure 3-26 (a) Voltage-output characteristics for LVDT; (b) phase characteristics for the LVDT.

the equation

$$C = 0.0885\varepsilon \frac{A}{X} \qquad pF \qquad\qquad (3\text{-}35)$$

In principle, C can be changed by changing either A, X, or ε, but changing X is the most common.

Sensitivity of a Capacitive Transducer

The sensitivity of a capacitive transducer is defined as the change in capacitance with respect to plate separation and is given by the equation

$$S = \frac{dc}{dX} = -0.0885\varepsilon\frac{A}{X^2} \qquad \text{pF/cm} \qquad (3\text{-}36)$$

It should be noted that the sensitivity increases as the plate separation decreases. Thus it is desirable to design this transducer to have a large plate area with a small separation.

Example

Consider the circuit shown in Figure 3-27. Since the input voltage E is dc, no current flows when the capacitor is stationary. Therefore,

$$e_1 = E \text{ at } X = X_0 \text{ (equilibrium position)}$$

A change in position $\Delta X = X_1 - X_0$ produces a voltage

$$e_0 = e_1 - E$$

where e_0 is related to X_1 by the transfer relation in the frequency domain:

$$\frac{e_0(j\omega)}{X_1(j\omega)} = \frac{E}{X_0}\frac{j\omega\tau}{1 + j\omega\tau}$$

where

$$\tau = RC = 0.0885EA\frac{R}{X_0}$$

Typically, $R = 10^6 \ \Omega$ or larger. In the time domain,

$$e_0(t) = \frac{E}{X_0}\frac{j\omega\tau}{1Xj\omega\tau}Xe^{j\omega\tau} \approx \frac{EX\sin\omega\tau}{X_0}$$

for $\omega\tau \gg 1$.

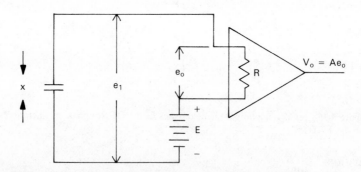

Figure 3-27   Capacitive transducer and amplifier.

Differential Capacitor

Differential capacitor systems (Figure 3-28) have the advantage that more accurate measurements of displacement are possible than with the standard capacitance transducer. The chief advantage of the differential parallel-plate capacitor is its linearity of the fractional difference with displacement. When the movable plate is displaced a distance X from the

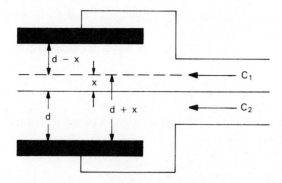

Figure 3-28   Differential capacitor.

equilibrium separation d, then at equilibrium we have a separation of the plates of $d - X$ on one side of the equilibrium position and $d + X$ on the other side. Thus the capacitances C_1 and C_2 are given by the relations

$$C_1 = \frac{\varepsilon A(0.0885)}{d - X}(1) \qquad \text{pF} \tag{3-37}$$

$$C_2 = \frac{\varepsilon A}{d + X}(0.0885) \qquad \text{pF} \tag{3-38}$$

Forming the fractional displacement we have, by solving for the ratio X/d, the equation

$$\frac{X}{d} = \frac{C_1 - C_2}{C_1 + C_2} \tag{3-39}$$

Consider the bridge network in Figure 3-29, where the differential capacitor is used as two arms of the capacitance bridge shown in Figure 3-29. C_1 and C_2 are the capacitances of the differential capacitor. For the bridge

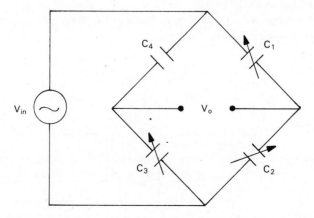

Figure 3-29   Bridge network with differential capacitor in two arms.

to balance at equilibrium, we have $C_1 = C_2$, and C_3 is adjusted to equal C_4. Under this condition, there is no output voltage V_o.

The output voltage for the bridge for a displacement of X is given by the equation

$$V_o = \frac{V_{in}}{2} \frac{C_1 - C_2}{C_1 + C_2} \tag{3-40}$$

The output voltage of the bridge as a function of the fractional displacement is given by the equation

$$V_o = \frac{V_{in}}{2} \frac{x}{d} \tag{3-41}$$

It should be noted that the output voltage of the bridge is independent of the area of the plates making up the differential capacitor and the permittivity of the medium between the plates.

PROBLEMS

1. A system is devised to measure the liquid level in a tank where the liquid is electrically conductive. The system is given in Figure 3-30. Figure 3-31 illustrates the variation of conductance with height of the tank. The conductance between the wire and a cylinder plate in the tank is found to vary directly with the height of the liquid, as shown in the figure. Measurements will be made in terms of the terminal voltage, V, and current, I.

   (a) If a linear output electrical quantity versus height is desired, should a current source be used and the voltage measured at the terminals, or should a voltage source be used with the current measured at the terminals? Explain your answer.

   (b) What is the sensitivity of the transducer if V is 5 V and I is the output variable? See Figure 3-31.

   (c) What is the order of this measurement system? Derive your relationship and explain your answer.

2. A modified form of the instrument in Problem 1 is to be used as a urine output meter, as shown in Figure 3-32.

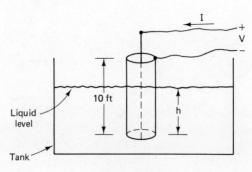

Figure 3-30

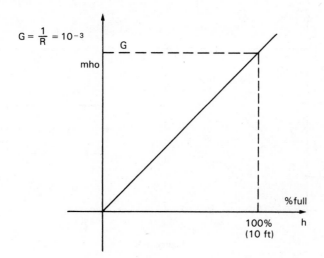

$$G = \frac{1}{R} = 10^{-3}$$

Figure 3-31

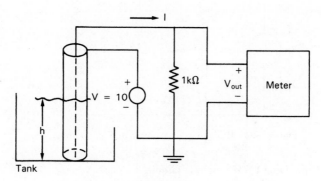

Figure 3-32

(a) If the meter can be read to the nearest tenth of a volt, what is the precision of the transducer in terms of the input urine height? Does your answer depend on the height h?

(b) If the accuracy of the measurement system is to be within 5% of the true value, what would you specify as limits in the source voltage and on the resistance of the meter?

3. A blood pressure transducer has the transfer characteristic approximated by Figure 3-33, where the input variable is pressure and the output variable is resistance.

(a) What is the sensitivity of this transducer?

(b) What is the dynamic range of this transducer for linear operation?

(c) What is the precision of this transducer?

4. Matching strain gauges are used in a bridge circuit as part of the blood pressure transducer system shown in Figure 3-34.

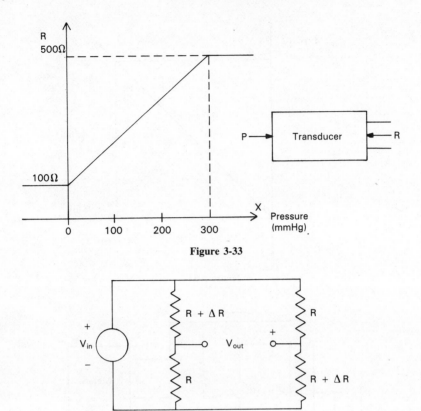

Figure 3-33

Figure 3-34 Bridge circuit using matching strain gauges.

    (a) Find the gain, V_o/V_i, for this circuit.

    (b) What is the difference in voltage gain between this circuit configuration and one using only one strain gauge in the bridge circuit?

5. For the circuit in Figure 3-35, all the resistors are at 20°C except for a thermistor, R_t, which is used as a patient temperature sensor. R_t varies with temperature according to the relationship

$$R_t = R_{20°C}[1 + 0.05(T - 20)]$$

T is given in degrees Celsius and the thermistor resistance at 20°C is 10 kΩ. If V_{out} is to change by 1 V per degree C change in temperature of R_t, what voltage should be used for V_{in}?

6. Consider the thermistor bridge circuit shown in Figure 3-36.

    (a) If the meter reads 50 μA, what is the resistance of thermistor R_3?

    (b) What is your estimate of the thermistor temperature when $I = 50$ μA?

7. An interarterial system for measuring blood pressure consists of a saline-filled catheter that acts as a hydraulic transmission line between the artery and a pressure transducer located outside the body. A simple model of the arterial line-pressure transducer system is given in Figure 3-37. For this simple model,

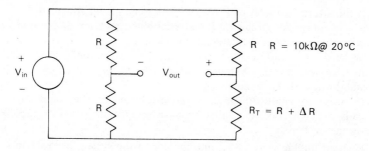

Figure 3-35

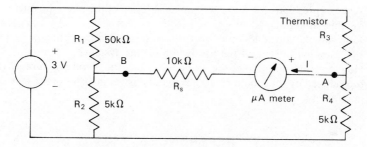

Figure 3-36   Thermistor bridge for measuring temperatures between 0 and 100°C.

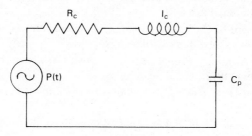

Figure 3-37

the catheter is assumed to be noncompliant. The parameters of the circuit are as follows:

$P(t)$ = arterial pressure as a function of time

R_c = fluid resistance of the catheter

I_c = inertia of the fluid in the catheter

C_p = mechanical compliance of the diaphragm in the pressure transducer

(a) Write a differential equation for this system in terms of the system parameters. What order of system is this blood pressure monitoring system?

(b) Write expressions for the static sensitivity, undamped natural frequency, damping ratio, and the damped frequency in terms of the system parameters.

8. (a) If the arterial catheter of Problem 7 has some lateral compliance, C_c, draw a new equivalent circuit for the arterial blood pressure monitoring system, taking into account the catheter compliance.
   (b) Write the new differential equation for this circuit configuration.
   (c) In what way are the system parameters K, ω_n, ρ, and ω_0 affected by this additional catheter compliance?
   (d) How is the bandwidth of this system affected by the compliance of the catheter? What does the catheter compliance do to the waveform reproduction by the system?

9. (a) If an air bubble were to get into the line of the system of Problem 7, draw a new equivalent circuit for this situation.
   (b) Write the differential equation for this equivalent system.
   (c) Write expressions for K, ω_n, ρ, and ω_0 in terms of the new equivalent circuit for this situation.
   (d) How has the introduction of an air bubble changed the resonant frequency, the damping ratio, and the bandwidth of this system?
   (e) How does the addition of the air bubble affect the blood pressure waveform reproduced by the monitor? Explain your answer in terms of the new system parameters.

REFERENCES

1. Busser, John H., and Barry N. Feinberg, "Measurements," in *CRC Handbook of Engineering in Medicine and Biology,* David G. Fleming and Barry N. Feinberg (eds.) (Boca Raton, Fla.: CRC Press, 1976).

2. Cobbold, Richard S. C., *Transducers for Biomedical Measurements: Principles and Applications* (New York: Wiley, 1974).

3. Diefenderfer, A. James, *Principles of Electronic Instrumentation* (Philadelphia: W. B. Saunders, 1972).

4. Ferris, Clifford D., *Introduction to Bioinstrumentation* (Clifton, N.J.: Humana Press, 1978).

5. Geddes, L. A., and L. E. Baker, *Applied Biomedical Instrumentation,* 2nd ed. (New York: Wiley-Interscience, 1975).

6. Glass, Carter M., *Linear Systems with Applications and Discrete Analysis* (St. Paul, Minn.: West, 1976).

7. Norton, Harry N., *Handbook of Transducers for Electronic Measuring Systems* (Englewood Cliffs, N.J.: Prentice-Hall, 1969).

8. Norton, Harry N., *Sensor and Analyzer Handbook* (Englewood Cliffs, N.J.: Prentice-Hall, 1982).

9. Welkowitz, Walter, and Sid Deutch, *Biomedical Instruments: Theory and Design* (New York: Academic Press, 1976).

10. Wobschall, Darold, *Circuit Design for Electronic Instrumentation* (New York: McGraw-Hill, 1979).

4

BIOELECTRIC
POTENTIALS AND
ELECTROCARDIOGRAPHY

THE CELL AS A BIOELECTRIC GENERATOR

Certain cells in the human body comprise the origin of all of the body's bioelectric potentials, which are potentials between the inside and the outside of a cell wall or membrane. The source of these potentials is strictly ionic in nature. Such cells, as nerve or muscle cells, have a cell wall that is semipermeable to certain ions and as such act as a selective ionic filter to these ions. What this means is that some ions can pass through the membrane freely whereas other ions cannot. The exact structure of the membrane and the exact mechanism of its semipermeability are not known. Experiments demonstrate only that it takes place in the lipid membrane of the cell.

Surrounding the cells of the body are body fluids. These fluids are ionic and represent a conducting medium for electric potentials. The principal ions involved with the mechanism of producing cell potentials are sodium (Na^+), potassium (K^+), and the chloride ion (Cl^-). The membranes of excitable cells readily permit entry of K^+ and Cl^- but impede the flow of Na^+ even though there may be a very high concentration gradient of sodium across the cell membrane. The effect of this is that the concentration of sodium ion inside the cell is less than it is on the outside of the cell membrane. Since sodium is a positive ion, the outside of the cell becomes more positive than the inside of the cell.

At all times, the ions seek an equilibrium under the two major forces that act on them. These forces are those produced by concentration gradients and electric field gradients produced by differences in concentration of the ions and their charge distributions. Because there is a higher concentration of the positive sodium ion outside the cell, an electric field gradient is formed that will tend to move positive ions inside the cell. Thus potassium (K^+) tends to move into the cell under the influence of both concentration and electric field gradients, and the concentration of potassium inside the cell becomes greater than that on the outside. Chloride ion (Cl^-) moves with much less impedance than sodium or potassium across the cell membrane and thus is influenced mainly by the electric field gradient.

At equilibrium between concentration and charge gradients, there is a resulting difference of potential of between -50 and -100 mV across the cell membrane (see Figure 4-1). When equilibrium is reached, the resulting potential across the cell membrane is called the *resting membrane potential*. When this type of potential difference exists across the cell membrane, the cell is said to be *polarized*. A decrease in this resting membrane potential difference is called *depolarization*. On the other hand, any increase in the potential difference across the cell membrane is called *hyperpolarization*.

This static or resting potential, as it is called, remains constant until the cell is disturbed or stimulated by some other action. The fact that this membrane potential is maintained in the resting state even though there are concentration gradients of potassium and sodium across the cell membrane which are not reduced in time gave rise to the concept of active transport across the cell membrane.

Active transport is the ability of the cell to move ions against a concentration and electric field gradient; thus this process requires the input of energy. Because of this mechanism, the concentration of sodium is greater on the outside of the cell than on the inside, whereas the concentration of potassium on the inside of the cell is greater than it is on the outside. Little is known about the mechanism of this active transport, but one theory is that the ion is complexed with an enzyme at the surface of the cell membrane and carried through the membrane in this form. Upon reaching

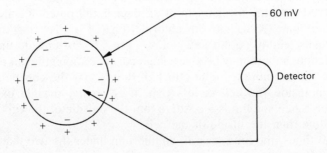

Figure 4-1   Basic cell showing measurement of transmembrane potential.

the inside of the membrane surface, the enzyme and ion split up, leaving the ion to move into the cell itself; the enzyme is free to carry in another ion from outside the cell. In mammalian nerve cells, the concentration of potassium ions is about 30 times greater inside the cell than on the outside of the cell. On the other hand, there is a tenfold differential in the concentration of sodium outside the cell compared to that inside the cell.

The relative rate of diffusion of ions through the cell membrane is called the *conductance* of the ion and is usually symbolized by the letter g. In the resting state the conductance of potassium across the cell membrane is much greater than that of sodium. Symbolically,

$$g_K \gg g_{Na}$$

As mentioned before, the selective permeability of the cell membrane is coupled by an opposing electric field gradient due to the distribution of charge caused by the concentration gradient. Although the cell membrane is almost impermeable to sodium ions, the net sodium gradient is high, causing some flow of sodium ions into the cell. Conversely, the membrane is only moderately permeable to potassium ions; the net potassium gradient is low. The net result is that the sodium and potassium currents represented by the motion of each of the ions are equal, as the net flow across the membrane boundary is zero. Thus the internal potential of the cell will not change and will remain at its -90-mV resting level. From what has been developed so far, it is seen that the three most important ions involved in the production of action potentials are Na^+, K^+, and Cl^-. Although there are other organic and inorganic molecules involved in the process of producing an action potential, Hodgkin and Katz [5] found in 1949 that the transmembrane potential in a nerve could be closely calculated using mathematical models that used only the concentrations of K^+, Na^+, and Cl^-.

MATHEMATICAL MODELS OF CELL RESTING POTENTIAL

In this ionic solution, there are the two forces of the electrical field gradient produced by the voltage V, resulting from the imbalance of charge across the cell membrane and the osmotic force as represented by the concentration gradient of the ions. The resting potential can be calculated when these two forces are in equilibrium.

The work required to move a charge, q, through a potential V is proportional to the product of q and V. Thus the energy required to move a quantity of ions of valence Z through an electrical field V is the product of the parameters:

$$ZNFV \qquad (4\text{-}1)$$

where N is the number of moles of the ion in solution and $F = 65,000$ coulombs per mole (c/mol). The transmembrane potential, V, is balanced by the forces of the concentration gradient (osmotic pressure) across the same membrane.

When this equilibrium is established, the Nernst equation (4-2) gives an approximation to the membrane potential as a function of the ion concentration. Let

C_o = concentration of ion outside the cell membrane

C_i = concentration of ion inside the cell membrane

Then the Nernst equation can be written

$$\text{potential(mV)} = \frac{RT}{ZF} \ln \frac{C_o}{C_i} \tag{4-2}$$

where R is the gas constant (0.082 liter-atmosphere per degree per mole), T the absolute temperature, $F = 1$ faraday or 96,500 c/mol, and Z is the valence of the ion. For monovalent ions and body temperature (37°C), the Nernst equation becomes

$$\text{potential(mV)} = 60.6 \log_{10} \frac{C_o}{C_i} \tag{4-3}$$

From the Nernst equation we see that the membrane potential in millivolts established by an unequal distribution of an ion is directly related to the concentration ratio of the ion on either side of the cell membrane.

Example: Using the Nernst Equation

Assume a temperature of 37°C and the cell sodium concentrations of $[\text{Na}]_o = 10$ mM/liter and $[\text{Na}]_i = 300$ mM/liter. The Nernst equilibrium potential for Na^+ at the cell membrane is

$$V(\text{mV}) = 60.6 \log_{10} \frac{10}{300} = -89.51 \text{ mV}$$

GENERATION OF AN ACTION POTENTIAL

Excitable cells can be stimulated by a variety of stimuli depending on the type of cell it is. For example, heat flow, light, and mechanical deformation are a few of types of external stimuli that can produce changes in the cell membrane. When a cell receives a stimulus of some kind, the conduction characteristics of the cell membrane at the point of stimulation will be changed, which in turn will cause a change in ionic current flow. Conductance changes in the cell membrane cause the membrane potential to change according to the following sequence of events:

1. Sodium conductance increases greatly, causing a large influx of Na^+ into the cell.

2. Potassium conductance does not change significantly during the first half of the action potential.

3. At this time in the cycle, the conductance of Na^+ is greater than that of K^+, producing an increase of ionic current due to flow sodium into

the cell which causes the internal potential to increase from -90 mV in an attempt to bring the sodium and potassium current into equilibrium.

4. When ionic current balance is achieved, the cell membrane potential has reached approximately $+20$ mV. At this point in the action potential cycle, the cell is in a depolarized state.

5. At the peak of the action potential, g_{Na}, the sodium conductance, returns to normal.

6. At this time, g_K, the potassium conductance, increases to about 30 times its normal value. This causes potassium to diffuse out of the cell very rapidly while almost no sodium diffuses out, due to the decrease in sodium conductance at this point in the cycle.

7. The rush of K^+ out of the cell is sufficient to bring the membrane potential back to its normal level and is therefore responsible for the return of the cell membrane potential back to the resting level.

8. At this time the sodium pump begins the slow process to pump out the sodium ions interior to the cell and allow the potassium back in again to bring the cell back to its resting ionic concentration.

In Figure 4-2 we can see the relative time changes of ion conductance in the cell membrane as a function of time. During the course of the action potential, the sodium pump plays almost no role in the process because it

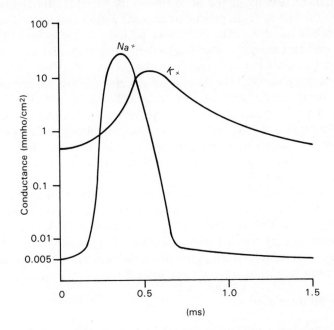

Figure 4-2 Conductance curves for sodium and potassium. (From Arthur C. Guyton, *Basic Human Physiology: Normal Function and Mechanisms of Disease,* W. B. Saunders Company, Philadelphia, 1971, by permission.)

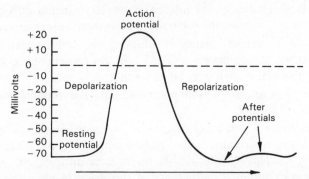

Figure 4-3 Waveform of action potential. (Time scale and amplitude characteristic vary with cell type.) (From Leslie Cromwell et al., *Biomedical Instrumentation and Measurement*, © 1973. Reprinted by permission of Prentice-Hall Inc., Englewood Cliffs, N.J.)

operates much too slowly to bring the membrane potential back to its resting value in a fraction of a millisecond. A typical action potential is given in Figure 4-3.

Refractory Periods

During the rise of the spike of voltage across the cell membrane and the initial rapid decline of the action potential, further stimulation of the cell has no further effect in producing another action potential. During this time the cell is in its absolute refractory period. At the point in the cycle when the cell membrane is returning to the resting potential level, a greater than normal stimulation can cause another response. This period is the relative refractory period. The duration of the absolute and relative refractory periods depends on the particular excitable cell and the nature of its action potential. This is apparent, for example, when one compares refractory periods of a nerve cell and a muscle cell. Since metabolic processes produce the energy associated with the action potential, it is the time for these processes to return to a prestimulus state that is, in part, responsible for the refractory period of the excitable cell.

Because of the refractory period there is a limit to the number of action potentials that an axon can carry in a given time period. The frequency of discharge is limited to an interval greater than the absolute refractory period during a continuous train of stimuli. For example, a mammalian neuron with a relative refractory period of 0.5 ms can discharge impulses at a maximum rate of about 1000 per second.

All-or-Nothing Law

If a stimulus is sufficient to generate an action potential, that potential, in terms of response time and amplitude is always the same. This is a common feature of most excitable cells. Therefore, the form of the action potential is the same regardless of the magnitude of the stimulus, just as long as it is large enough to produce an action potential.

As has been implied here, there is a minimum amplitude of stimulus required to produce an action potential. This minimum value is called the *threshold value* of the stimulus. For an action potential to be produced, a stimulus greater than the threshold value must be applied or the action potential will not be generated. This feature of excitable cells is called the *all-or-nothing law*.

Conduction of Action Potentials

The propagation of an action potential through a group of cells or along the length of a nerve cell is due to the activity of local circuit currents which are established by the inward movement of Na^+ at the region of a developing action potential (Figure 4-4). At this time, the surface of the active region of an excitable cell becomes negative with respect to the inactive regions around it. Thus positive current flows from the inactive to the active areas of the cells or local region of the nerve. This flow of current in the immediate vicinity of the action potential is large enough to cause depolarization of the adjacent cells or region of the nerve fiber. While this is taking place and the adjacent areas to the initial site are going through depolarization, the initial site does not remain active because of the transient nature of the increased Na^+ conductance, g_{Na}.

The adjacent areas are now undergoing depolarization, which in turn will produce depolarization in more cells or in areas farther down the nerve fiber. This local circuit action is responsible for the propagation of the action potential down a nerve fiber or through other masses of excitable cells. Because of the all-or-nothing response of the excitable cells, the

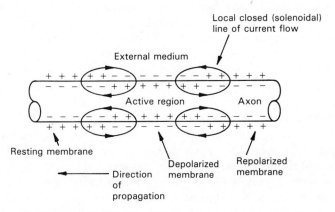

Figure 4-4 Charge distribution in the vicinity of the active region of fiber conducting an action potential. [From John Webster (ed.), *Medical Instrumentation: Application and Design*, Houghton Mifflin Co., Boston, © 1978, by permission.]

amplitude of the action potential remains unattenuated as it travels down the nerve or through excitable tissue. This means that information traveling down a nerve cannot be amplitude-modulated but must convey information in a pulse-frequency modulation form.

ELECTRICAL ACTIVITY OF THE HEART

Variations of the production of action potentials represent the manner in which information is transmitted throughout the body and produce the stimulation required for muscle cells to contract. The electrical activity of various organs is of clinical interest to medical practitioners. Some of the waveforms of interest include the electrocardiogram, which we will study in greater detail later in the chapter; the electroencephalogram; the electromyogram; the electrooculogram; and the electroretinogram.

The *electrocardiogram* (ECG) is the record produced by an electrocardiograph. It displays the electrical activity of the heart as measured at the surface of the body. The origin and measurement of the ECG is important to the clinical engineer, as the electrocardiograph is a very common piece of equipment that is found in every hospital. Thus it is important to understand the electrical activity and associated mechanical sequences performed by the heart in providing the driving force for the circulation of blood.

The function of the circulatory system is to maintain an optimum environment for cellular function. This environment is produced and maintained by the circulation of blood. The heart is the power source which provides the energy to move the blood through the body and supply cells with the nutrients, hormones, temperature, and gases that they need for cellular function and at the same time removes waste products—products of the cell's metabolism—from the cell.

The heart weighs less than a pound and is about the size of the average fist, about 6 in. long along its major axis or dimension. The heart itself is a hollow organ which is composed of four chambers with a system of one-way valves, the fluid equivalent of a diode, which provide for filling the chambers of the heart with little backflow or regurgitation of blood. The walls of the heart are made of muscle which is surrounded by a fiberlike sac called the *pericardium*. The inside of the heart is lined by a strong, thin membrane called the *endocardium*. A wall or septum divides the heart cavity into a double-pump configuration. This gives rise to the "left heart" and "right heart" concept. Each side of the heart is again divided into an upper chamber, the *atrium*, and a lower chamber, the *ventricle* (Figure 4-5). The right heart receives blood from the body after delivering nutrients and oxygen to the cells of the body and carrying away waste products of cell metabolism. This blood is pumped to the lungs, where the waste gas carbon dioxide, CO_2, is expired, and picks up oxygen, O_2, for use by the

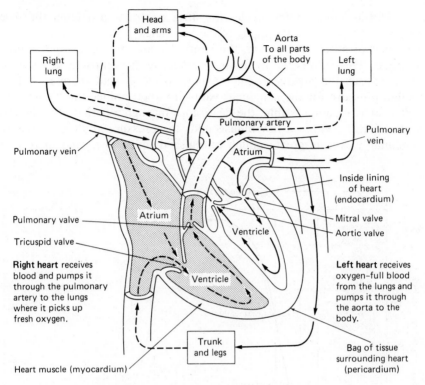

Figure 4-5 Schematic of blood flow in the heart.

cells again. The left heart receives the oxygenated blood from the lungs and pumps it out through the aorta to be distributed by smaller arteries to all parts of the body.

MECHANICAL EVENTS IN THE CARDIAC CYCLE

Consider now the following sequence of mechanical events that take place over one cardiac cycle. We begin with the heart in the resting state, known as *diastole*. The pressure in the heart is lower than the pressure in the veins leading to the right atrium, which are the *superior and inferior vena cava*. Thus there is blood flow into the right atrium and across the one-way, tricuspid valve between the right atrium and right ventricle. In the same fashion, blood flows into the left atrium and across the bicuspid valve into the left ventricle. It should be pointed out here that about 70% of the filling of the ventricles is due to the pressure differential existing across the atrioventricular valves and approximately 30% is due to contraction of the atria themselves. During this period, the semilunar valves are closed, maintaining pressure in the aorta and pulmonary artery.

At this point in the cycle, the atria contract, providing the force necessary to complete the filling of the ventricles. After about 0.15 s from the onset of atrial contraction, the ventricles are stimulated and begin to contract. This causes the hydraulic pressure in the ventricles to increase, which causes the atrioventricular valves to close—but there is still not enough pressure for the semilunar valves to open. Since the blood in the ventricles is incompressible, the heart goes through a period of "isometric contraction." Within a few milliseconds the pressure gradient across the semilunar valves is great enough to cause them to open and there is an

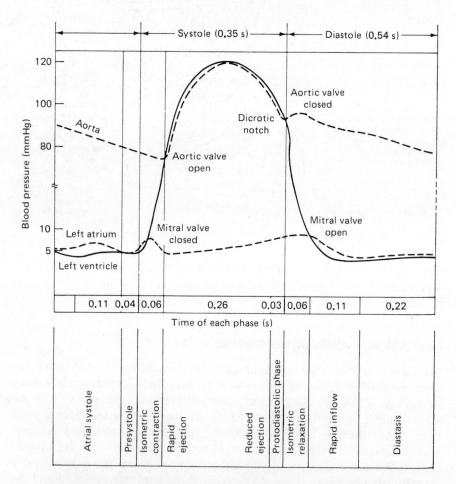

Figure 4-6 Time and amplitude relationship for aortic, left arterial, and left ventricular pressures over one cardiac cycle. (From Leslie Cromwell et al., *Biomedical Instrumentation and Measurement,* © 1973. Reprinted by permission of Prentice-Hall, Inc., Englewood Cliffs, N.J.)

outflow of blood into the system. Because of the high pressures—about 80 mmHg existing in the left ventricle and 8 mmHg in the right ventricle— the initial ejection phase of the cardiac cycle occurs quickly. About half of the total stroke volume is ejected from the ventricles within 0.1 s or less. The ventricles empty most but not all of their contents during contraction. Out of approximately 120 ml of blood in the ventricles, about 50 to 60 ml of blood remain.

After the ejection phase of the ventricles is over and the muscles of the ventricles relax, ventricular pressure begins to fall. Because the pressure in the aorta and the pulmonary artery are high at this point in the cardiac cycle the pressure difference across the aortic and pulmonary valves causes them to snap shut, which then maintains the high pressure on the blood necessary to force it through the circulatory system. Ventricular pressure is still higher than atrial pressure; thus the atrioventricular valves remain closed. Thus all four heart valves are closed at this point and the heart experiences "isometric relaxation," wherein pressure and muscular tension are changing but the dimensions of the heart are not. As the muscle tissue of the ventricles continues to relax, the elastic forces of the tissue cause the ventricles to assume their former shape before contraction. Ventricular pressure is at its lowest point, so the atrioventricular valves are forced open by the pressure gradient that appears across them and the ventricles begin to fill again. This is the beginning of a new cardiac cycle. Figure 4-6 illustrates the time and amplitude relations for aortic, left atrial, and left ventricular pressures over a cardiac cycle.

ACTION POTENTIALS OF THE HEART AND FORMATION OF THE ECG

The preceding material described the mechanical events that the heart undergoes in pumping blood through the circulatory system. There is a sequence of electrical events resulting from action potentials that are responsible for these mechanical events. To complete the discussion of the cardiac cycle, we examine next the electrical activity that is responsible for the mechanical pumping action of the heart.

To begin with, the sequence in which the myocardium contracts depends on the location of the site of the excitation and subsequent action potentials and the direction of their propagation. Events begin with a group of cardiac cells known as the *sinoatrial node* (SA node). The SA node is 25 to 30 mm in length and 2 to 5 mm thick. It is located between the epicardium and myocardium of the right atrium, in from the entrance of the superior vena cava. Fibers of the SA node radiate in all directions and integrate with common atrial fibers. The crucial property of the cells of the SA node is that the resting membrane potential is not stable; that is, it does not remain in the resting state, due to the fact that the cell membrane "leaks" sodium (NA^+).

As sodium leaks through the cell membrane due to a time-varying change of Na^+ conductance, the potential across the cell membranes of the SA node begins to increase from about -70 mV toward more positive values. As sodium flows into the cell at a predetermined rate, the associate membrane potential also increases until the threshold voltage for the production of an action potential is reached. This threshold is in the neighborhood of -40 mV. When the transmembrane cell potential reaches this threshold voltage, the membrane conductances change and an action potential is generated with a peak membrane potential reaching about $+20$ mV. The cell now goes though the sequence of rapid depolarization and subsequent repolarization shown in Figure 4-7. As the cell repolarizes, the sodium pump works in just the same way as with other excitable cells and the membrane potential drops back to about -70 mV. At this point, the membrane potential begins to increase again due to sodium leakage and the process begins again.

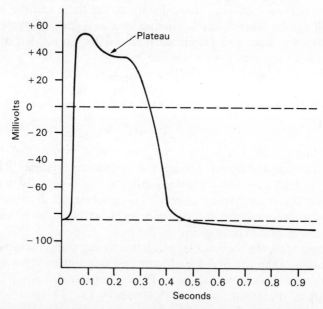

Figure 4-7 Action potential from cardiac muscle cell. (From Arthur Guyton, *Basic Human Physiology: Normal Functions and Mechanisms of Disease,* W. B. Saunders Company, Philadelphia, 1971, by permission.)

The cells of the SA node merge with the other cells of the atrial myocardium, so as the SA node cells depolarize they stimulate the adjacent atrial cells, causing them to depolarize. Thus a depolarization wave spreads over the atria in an outward-traveling wave from the point of origin. The wave of cell depolarization spreads from the upper portion of the atria downward, forcing the blood through the atrioventricular valves and into the ventricles. The wave propagates through the right and left atria at a velocity of about 1 m/s. It requires about 100 ms for excitation of the entire human atria to be completed.

Our intuition would tell us that the depolarization wave initiated in the atria would continue to propagate over the entire surface of the heart, causing the ventricles to contract. Such is not the case, as there is a fibrous barrier of nonexcitable cells that effectively prevents the propagation of the depolarization wave from continuing beyond the limits of the atria. The only excitable tissue that crosses this barrier is the *bundle of His* (pronounced hiss). The origin of the bundle of His is a mass of specialized tissue about 2 cm long and 1 cm wide called the *atrioventricular node* (AV node). The conduction velocity through the AV node is about 0.1 m/s, 10% of that of the atrial cells. The transmission delay that this represents is the key to the proper time relationships between the atria and the ventricles. As a result of reduced conduction velocity in the AV node, the atrial depolarization wave is delayed about 0.12 s as it travels through the AV node. This delay permits the atria to complete their contraction before there is any ventricular contraction.

The impulse leaves the AV node via the bundle of His, which has a higher conduction velocity than that of the fibers of the AV node. The bundle of His descends down the interventricular septum, which divides the bottom chamber of the heart into the left and right ventricles. Within a short distance, the bundle of His splits into two branches, known as the *left and right bundle branches*. Each conduction bundle goes to a corresponding ventricle. The left and right bundle branches descend down to the apex of the heart, where they curve upward and outward and begin to arborize into smaller conduction fibers that permeate the ventricular wall. These fibers are called the *Purkinje fibers*.

Conduction velocity in the Purkinje fibers is about 1.5 to 2.5 m/s. This is about six times the velocity of conduction in myocardial cells and about 40 times that of the cells of the AV node. Since the direction of the impulse propagating in the His bundles is from the apex of the heart, ventricular contraction begins at the apex and proceeds upward through the ventricular walls. This produces a quick, forceful contraction of the ventricles in a manner that produces a lifting and squeezing action which forces the blood out of the ventricles and into the arterial system. Figure

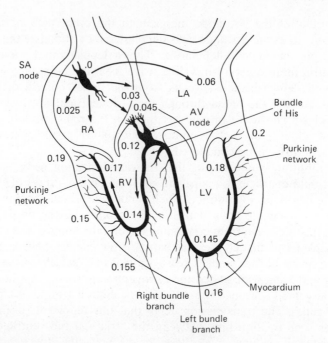

Figure 4-8 Propagation times from the SA node to different areas of the heart. (Courtesy of Hewlett-Packard, Waltham, Mass.)

4-8 illustrates the time for the action potential to propagate to various areas of the heart.

ELECTROCARDIOGRAPHY: AN ELECTRICAL ASSESSMENT OF CARDIAC PERFORMANCE

The heart can be modeled as an electrical generator enclosed in a volume conductor, the torso. Electrocardiography measures the potential differences at two points on this volume conductor and infers the clinical condition of the heart from the waveform. The record of the waveform produced by the heart, the electrocardiogram, has had more clinical interest than any other bioelectric potential recording.

The bioelectric waveform produced by the depolarization and repolarization waves as they move over the heart set up electric fields that can be monitored at the surface of the skin. This bioelectric potential monitored at the body's surface is called the ECG. A typical ECG waveform is shown in Figure 4-9. The electrocardiogram in the figure is characterized by five different "landmark" segments of the total wave shape. These are the P, Q, R, S, and T waves. Every portion of this electrocardiogram has a special interpretation. When considering the physiology of the heart, mention was made of atrial contraction, ventricular relaxation, and so on. The interpretation of the various portions of the ECG are correlated to these different events in the cardiac cycle.

P wave: a small low-voltage deflection caused by the depolarization of the atria prior to atrial contraction.

QRS complex: the largest-amplitude portion of the ECG, caused by currents generated when the ventricles depolarize prior to their contraction. It should be noted that the atrial repolarization occurs prior to ventricular depolarization, but the magnitude of the ventricular depolarization wave (QRS) overshadows it and is not seen on the ECG waveform.

T wave: marks ventricular repolarization or the relaxation portion for the ventricles during the cardiac cycle.

P-Q interval: the duration of time between the beginning of the P wave and the beginning of the QRS complex. It is the time between the beginning of the atrial contraction and the beginning of ventricular contraction. This interval is sometimes referred to as the "P-R interval" since the Q wave is not always present. It represents the delay of the electrical impulses from the atria to the ventricles.

The electrical activity of the heart may be modeled by a time-varying electrical dipole. At any instant of time the electric field vector of the dipole describes the magnitude and direction of the cardiac vector. This changes through the cardiac cycle corresponding to the activation of different regions of the myocardium. The art of electrocardiography can be simplified by considering that at any one time the cardiac dipole vector is projected

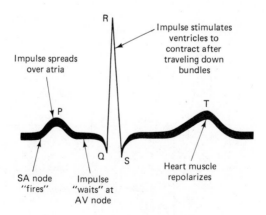

Figure 4-9   ECG waveform obtained at the body's surface.

into three orthogonal reference planes. These are the frontal plane, the transverse plane, and the sagittal plane. Figure 4-10 is a representation of this portrayal.

When the cardiac vector is projected into the frontal plane, its components in that plane are usually measured along axes that are at 60° to one another. This convention is used in electrocardiography instead of using the orthogonal set of axes used in engineering and science for standard vector representation. The triangle formed by the three 60° basis vectors is known as the *Einthoven triangle*, after Willem Einthoven, a Dutch physiologist whose pioneering contributions in electrocardiography in the first quarter of this century won him a Nobel prize.

The 60° cardiac vector projection concept allowed Einthoven to use the limbs as the contact locations for his system of electrodes. The conventions he set up are still in use today and the vector relationships for the ECG are given in *Einthoven's law*, which states that the vector sum of the projections of the cardiac vector in the frontal plane on the sides of the 60° triangle equals zero. This law applies to the cardiac vector at any instant of time, but for convenience, cardiologists have standardized on the peak of the R wave as the instant of time for determining the magnitude and direction of the frontal plane cardiac vector. The polarity convention used is such that the ventricular depolarization wave in a normal heart will cause a positive deflection of the recording by any of the three pairs of leads used to make the measurements.

Electrocardiography is still more of an art form than a hard science, even though the sophisticated engineering data acquisition methods used

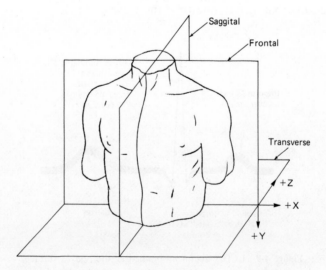

Figure 4-10   Orthogonal reference planes for the cardiac vector.

to acquire the ECG and analyze the waveform tend to obscure this fact. The clinical interpretation of the electrocardiogram is an empirical practice and the modes of analysis of the ECG can be correlated to patient condition only after an enormous number of records have been taken or examination by autopsy performed.

Frontal-Plane ECG Lead Systems

By conventions originally set up by Einthoven, the three leads used in the frontal plane are called leads I, II, and III.

Lead I: measures the potential difference between the right arm and the left arm

Lead II: measures the potential difference between the right arm and the left leg

Lead III: measures the potential difference between the left arm and the left leg

These are the three bipolar ECG lead configurations commonly used in diagnosis and monitoring of patients. Figure 4-11 illustrates the conventions used.

In addition to the three bipolar limb lead measurements, a recent addition to electrocardiography has been the use of three unipolar limb lead measurements which may be used to produce the frontal-plane cardiac vector. This system of measurement is known as the *augmented lead*

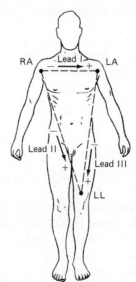

Figure 4-11   Einthoven triangle and ECG lead designation.

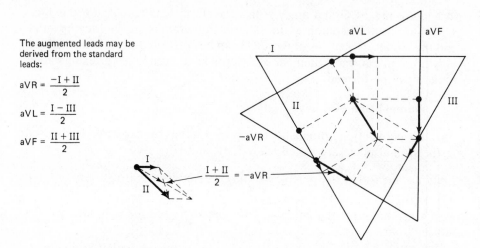

The augmented leads may be derived from the standard leads:

$$aVR = \frac{-I + II}{2}$$

$$aVL = \frac{I - III}{2}$$

$$aVF = \frac{II + III}{2}$$

Figure 4-12   Rotation of transverse plane using augmented lead system.

system. The intent behind the augmented lead convention is to provide an indifferent electrode or central terminal to be used as a reference for the remaining limb lead. What this gives the reader of the ECG is an additional look at the cardiac vector from a reference frame rotated 30° from that of the Einthoven triangle configuration (see Figure 4-12).

The designations for augmented lead system are:

1. *aVL*: augmented vector left
2. *aVR*: augmented vector right
3. *aVF*: augmented vector foot

They are formed by summing the signals obtained from the three pairs of standard bipolar leads, as shown in Figure 4-13. The vector lead relationship between the unipolar and bipolar limb leads is as follows:

$$aVR = \frac{-(I + II)}{2}$$

$$aVL = \frac{I - III}{2}$$

$$aVF = \frac{II + III}{2}$$

Figure 4-13a illustrates the concept of the augmented lead system.

The transverse plane ECG (Figure 4-13b) is obtained by recording the potential between an electrode located at one of six carefully chosen positions on the chest over the heart and an indifferent electrode formed by summing the three limb leads. These electrode locations are known as the *V leads*

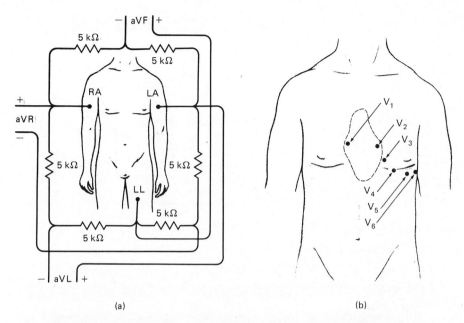

(a)   (b)

Figure 4-13 (a) Relationship between standard ECG lead system and augmented lead system showing unipolar limb leads; (b) transverse-plane ECG showing precordial lead locations. V_1, fourth intercostal space, at right sternal margin; V_2, fourth intercostal space, at left sternal margin; V_3, midway between V_2 and V_4; V_4, fifth intercostal space, at midclavicular line; V_5, same level as V_4, on anterior axillary line; V_6, same level as V_4, on midaxillary line.

and designated as V_1 through V_6 or *precordial leads* and correspond to reference locations on the chest.

Certain other configurations may be used in research hospitals or by innovative specialists. These configurations take into account the inhomogeneity of the torso to deduce an accurate rendition of the cardiac vector. The projection of the electric dipole onto the surface of the body is distorted by the lungs, which have four times the resistivity of other body tissue.

VECTOR CARDIOGRAPHY

Another method of displaying the electrical activity of the heart is the vector cardiogram (VCG). In this technique, ECGs are obtained from two orthogonal planes and the information displayed simultaneously on an oscilloscope in the X-Y mode. The first use of the term "vector cardiogram" was by Wilson in 1938. He obtained VCGs from the frontal and sagittal planes by use of a tetrahedral reference frame (Figure 4-14) based on his studies of an electrically driven dipole placed in a cadaver heart. He derived the frontal projection from the right and left arm leads vs. a central terminal

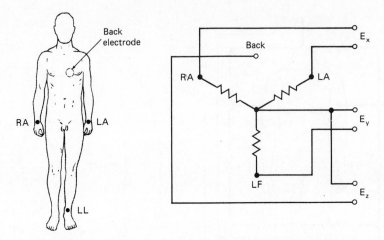

Figure 4-14   Wilson's equilateral tetrahedral reference frame.

and the sagittal projection from an electrode placed on the back vs. a central terminal.

Frank's studies on torso models indicated that the foregoing leads were not electrically orthogonal, due to the conduction inhomogeneities of the thoracic tissue. The Frank system of vector cardiography tries to achieve isotropy as well as orthogonality by using seven critically placed

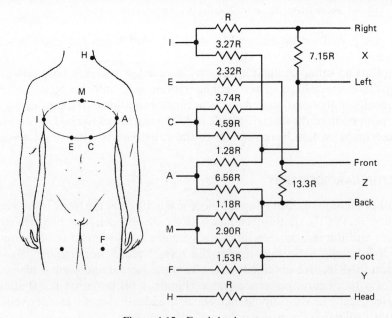

Figure 4-15   Frank lead system.

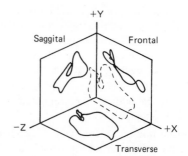

Figure 4-16   Projection of cardiac vector locus on coordinate planes.

electrodes and a resistive weighting matrix designed to compensate for the inhomogeneous resistivity of the body. In this way the Frank lead system uses seven electrodes and 13 weighting resistors to produce frontal, transverse, and left sagittal projections. Figure 4-15 illustrates the weighted resistor matrix and lead placement for the Frank lead system.

Vector cardiography is a method of displaying the instantaneous dipole vector in terms of its magnitude, direction, and polarity of the electromotive potentials generated by it. Because the spatial orientation of the cardiac vector is a periodic function of time, the time course of the tip of the vector will trace out a closed loop called a *spatial cardiac vector loop*. It is the locus of the magnitude of the cardiac vector over a cardiac cycle. The loop inscribed by atrial depolarization is called the *P loop,* ventricular depolarization produces the *QRS loop,* and the repolarization of the ventricles produces the *T loop*. Figure 4-16 illustrates the projections of the cardiac vector locus on each of the coordinate planes.

PROBLEMS

1. What are the properties of a semipermeable membrane?

2. Assume a cell environment where K^+ is the dominant ion involved in a cell's resting state. If the concentration of K^+ is 15 mM/liter on the outside of the cell wall and 150 mM/liter on the inside of the cell wall, find the magnitude and polarity of the resting membrane potential.

3. **(a)** The ease with which an ion passes through a semipermeable membrane can be stated in terms of its conductance (g). If we consider only the contributions of Na^+ and K^+ to the transmembrane potential, V_m, for any equilibrium state, derive an expression for this potential in terms of the sodium and potassium conductances and the Nernst potentials of each of the ions. (*Hint:* Apply Ohm's law to the equilibrium state and Kirchhoff's current law.)

   (b) Utilizing the expression derived in part (a), explain why $V_m = E_K$ at the resting potential and why $V_m = E_{Na}$ at the height of depolarization of the membrane.

4. Draw an action potential waveform (voltage amplitude vs. time diagram) for an excitable cell and label the amplitude and time periods as well as depolarization, repolarization, refractory period, and threshold levels portions of the diagram.

5. (a) Draw an action potential waveform for an SA node cell and label the points where conductance changes take place.
   (b) Explain the differences between an SA node (pacemaker) cell and a nerve cell in terms of their mechanism of production of an action potential. Draw a sketch of each to illustrate the differences.
6. What is the "all-or-nothing law" of action potentials? Explain how this principle is utilized by the body for unattenuated signal transmission over relatively long distances through the body.
7. (a) Describe one complete cycle of the heart's electrical activity.
   (b) Describe one cycle of the heart's mechanical pumping activity.
   (c) The action potentials of ventricular cells have a relatively long absolute refractory period as compared with other excitable cells. Explain why this is necessary.

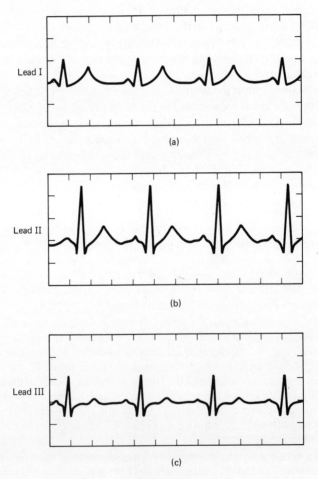

Figure 4-17

8. (a) Draw a typical ECG waveform and label the P-, Q-, R-, and T-wave portions of the diagram.

   (b) What is the electrical and mechanical significance of the recorded P, Q, R, and T waves and the QRS complex?

   (c) Develop a time diagram showing the correlation of the mechanical pumping of the heart, including the opening of the valves, with the electrical excitation events.

9. (a) For the electrocardiogram given in Figure 4-17 using the normal lead I, II, and III configuration, determine all the interval and amplitude measurements and interpret your results as to what each of these measurements means and how they related to the performance of the heart. Assume a chart speed of 25 mm/s and a voltage calibration of 1 mV/cm.

   (b) Determine the electrical axis of the heart from the electrocardiogram and interpret its meaning.

10. State Einhoven's law and draw the Einhoven triangle. Label the common ECG lead convention associated with it.

11. A lead I ECG waveform is taken from a patient and is shown in Figure 4-18. Describe what you think is the problem based on this waveform and relate this back to the mechanical and electrical characteristics of the cardiac cycle.

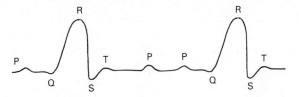

Figure 4-18

12. A cardiac vector is in the same direction as the lead vector for lead II. The scalar voltage measured in lead III is 0.7 mV. What is the magnitude and sign of the voltage measured in lead I?

REFERENCES

1. Clark, John W., "The origin of biopotentials," in *Medical Instrumentation: Application and Design,* John G. Webster (ed.) (Boston: Houghton Mifflin, 1978).

2. Friedman, H. Harold, *Diagnostic Electrocardiography and Vector Cardiography* (New York: McGraw-Hill, 1977).

3. Giese, Arthur C., *Cell Physiology* (Philadelphia: W. B. Saunders, 1962).

4. Guyton, Arthur C., *Textbook of Medical Physiology,* 4th ed. (Philadelphia: W. B. Saunders, 1971).

5. Hodgkin, A. L., and B. Katz, "The effect of sodium ions on the electrical activity of the giant axon of the squid," *J. Physiol.,* Vol. 108, 1949, pp. 37–77.

6. Kline, Jacob (ed.), *Biological Foundations of Biomedical Engineering* (Boston: Little, Brown, 1976).

7. Medtronics Currents, *An Overview of Pacing* (Minneapolis, Minn.: Medtronic, Inc., 1973).

8. Netter, Frank H., *The Ciba Collection of Medical Illustrations,* Vol. 5: *The Heart* (Summit, N.J.: Ciba Publications, 1971).

9. Ruch, Theodore C., and Harry D. Patton, *Physiology and Biophysics,* 4th ed. (Philadelphia: W. B. Saunders, 1965).

10. Selkurt, Ewald E. (ed.), *Physiology* (Boston: Little, Brown, 1963).

11. Strong, Peter, *Biophysical Measurements* (Beaverton, Oreg.: Tektronix, Inc., 1970).

5

ELECTROCARDIOGRAPHIC INSTRUMENTATION SYSTEMS

THE THEORY OF ELECTRODES

The electrocardiograph is an electronic instrumentation system used to detect, amplify, and record the electrical potentials produced by the heart that appear at the surface of the skin. As in any instrumentation system, the first part of the system is the transducer. In this case, the electrode is the transducer. The physiologic electrode is a device that provides an interface between living tissue and an electronic device, as is the case with a patient and the electrocardiograph.

The function of the electrode is to transform biochemical and physiologic phenomena into electrical currents, or conversely, to generate phenomena from electric currents. Examples of the first part of the definition are the ECG, EEG, or EMG electrodes. An example of the second part of the definition is the stimulation electrode, such as the pacemaker electrode. On the tissue side of the electrode interface, conduction takes place by the drift of ions through an electrolyte. On the device side of the interface, current flow is by electron conduction. At the interface a chemical reaction is required to convert one type of conduction into the other.

Physiologic electrodes are used to evaluate different forms of physiologic function by measuring the electrical activity produced by that function. Most physiologic activity is accompanied by electrical events that are measured more easily than the event itself. The measurement of the ECG signals at the surface of the chest wall is typically used for the assessment of cardiac performance.

ELECTRODE–ELECTROLYTE INTERFACE

Physiologic electrodes come in many sizes, shapes, and materials. ECG electrodes in particular come in disposable and reusable forms. The structure of each type is very different, but they do share some common characteristics that make them usable for electrocardiographic work. Generally, the electrode is composed of a metal, such as silver, and a salt of the metal, such as silver chloride. This construction is the silver/silver chloride (Ag–AgCl) electrode commonly used in electrocardiography and in patient monitoring. Usually, some form of electrode paste or jelly is with either type of electrode. In the case of the disposable electrode, it usually comes applied to the electrode surface.

The combination of ionic electrode gel and the metal forms a local solution of the metal in the gel at the electrode–electrolyte interface. Thus, in the case of the Ag–AgCl electrode, some of the silver dissolves into solution, producing Ag^+ ions. The chemical equation becomes

$$Ag \rightleftharpoons Ag^+ + e^-$$

When this reaction proceeds to the right it is an oxidation reaction, and when it proceeds to the left it is a reduction reaction. It is a characteristic of this simple reaction that ionic equilibrium takes place when the electric field set up by the dissolving ions is balanced by the forces of the concentration gradient. At this time there is a monomolecular layer of Ag^+ at the surface of the electrode and a corresponding layer of Cl^- adjacent to this. The combination is called the *electrode double layer* and there is a potential drop across this layer of 0.1 to 8 V (see Table 5-1). This potential drop is called the electrode *half-cell potential* or *offset potential*.

There is a marked change in the half-cell potential if the electrode system is mechanically disturbed. This change is called the *electrokinetic effect* and is the major cause of motion artifact in physiologic recordings. Motion artifact can be reduced by mechanically isolating the metal–electrolyte interface from disturbances. This is accomplished by mounting the metal

TABLE 5-1

Metal	Half-Cell Potential (V)
Al^{3+}/Al	-1.66
Fe^{2+}/Fe	-0.44
Ni^{2+}/Ni	-0.25
Pb^{2+}/Pb	-1.26
Cu^{2+}/Cu	$+0.33$
Ag^+/Ag	$+0.799$
Pt^{2+}/Pt	$+1.20$
Au^+/Au	$+1.68$
Au^{2+}/Au	$+1.50$

Figure 5-1 Cutaway view of a Ag/AgCl disposable electrode. (Courtesy of NDM Corporation, Dayton, Ohio.)

portion of the electrode into the adhesive support structure and filling the gap with a conducting gel. In this way the electrode double layer is not disturbed since the electrode moves as a unit and there is very little relative motion of the electrode relative to the skin (see Figure 5-1).

A familiar analog of the metal electrode—electrolyte interface is the $P–N$ junction in the semiconductor diode. When a region of P-type material is brought into contact with an N-type material, an electrostatic potential is formed at the interface which limits the width of the recombination region.

It is not possible to measure the electrode chemical potential directly, since any possible probe would itself introduce a potential of its own. The only way to make a measurement is to use two electrodes in contact with the electrolyte and measure the difference in potential between the two electrodes. The total voltage between a pair of electrodes is the difference in the two half-cell potentials. Because of this electrochemical fact, one of the electrodes is used as a reference electrode and the half-cell potential of the other electrodes is measured against it. The reference or standard electrode selected for making these measurements is the *hydrogen electrode*. This electrode consists of a specially prepared platinum surface in contact with a solution of hydrogen ions and dissolved molecular hydrogen. The chemical half-cell potential of most metals has been measured with respect to the hydrogen electrode. Table 5-1 is a partial list of these half-cell (oxidation) potentials for some common metals.

When electrodes are connected to a subject in order to record or display an ECG on a monitor or electrocardiograph, long-term changes in the difference in the half-cell potentials of the electrodes, sometimes called the *electrode offset potential*, appear on the recorder as baseline drift. Short-term changes appear as noise on the recording or monitor screen. The absolute value of the electrode offset potential is rarely significant unless this absolute value exceeds the maximum differential dc input of

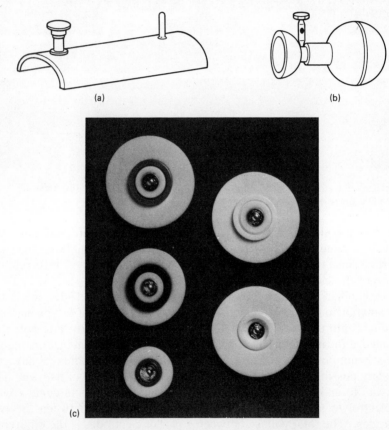

Figure 5-2 (a) ECG plate electrode, 3 × 5 cm rectangular plate made from German silver, nickel-silver, or nickel-plated steel. Must be used with electrode jelly. First used in 1910. (b) Suction cup electrode with vacuum attached for quick connection. May be used on flat or soft body areas. Disadvantage is its small area of contact, which is just the rim. (c) Disposable ECG electrode commonly used in hospitals. (Courtesy of NDM Corporation, Dayton, Ohio.)

the electrocardiograph or monitor. When this occurs, the amplifier and/or electrode system must be changed, otherwise the ECG waveform will be driven off scale and is unusable.

There are two major categories of electrodes, disposable and reusable. The disposable type is commonly used in hospitals because of its ease of use, but in order to keep price down there are limitations on their performance. This electrode is designed for single use only. The reusable electrode, on the other hand, may be used many times after having been cleaned and sterilized. This type of electrode, which may be nothing more than curved brass plates that have been silver plated, offers improved performance over disposable electrodes, but they cost more than the disposable variety. Examples of these electrodes are pictured in Figure 5-2.

CIRCUIT MODEL REPRESENTATION OF THE ECG ELECTRODE

At the electrode–electrolyte interface we have shown that there exists a half-cell potential which is produced by a double layer of charges of opposite sign (i.e., the Ag^+ and the Cl^- ions). This double layer of charges of opposite sign separated by a dielectric constitutes a form of capacitance. Since the thickness of the dielectric is molecular in dimension, the capacitance per unit area is very large. In looking at the impedance of the electrode, there is a phase angle of less than $-90°$, indicating that there is also a resistive component to the impedance. Thus a simple circuit model of the electrode–electrolyte interface can be given as a series capacitance and resistance, as shown in Figure 5-3.

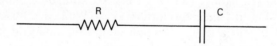

Figure 5-3   Simple *RC* model of electrode–electrolyte interface.

The next step in developing the circuit model of the ECG electrode is to identify the important components between the terminals of the electrode when two electrodes are applied to the skin for the purpose of recording an ECG. Figure 5-4 illustrates the components needed to describe a complete model of the physical system.

All of the circuit component values for resistance and capacitance are dependent on area of electrode and the materials of which the electrode is made. Thus it is difficult to assign specific values to these parameters, as they are also nonlinear functions of current and frequency. The impedance measured between the terminals of a pair of electrodes applied to the skin of a patient is high at low frequencies, decreasing with increasing frequency until a plateau is reached at higher frequencies, when reactances become

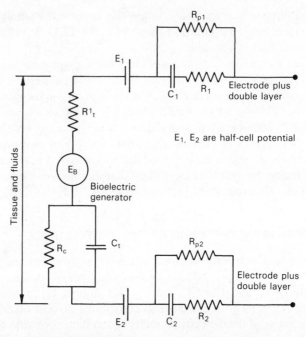

Figure 5-4 Circuit model of tissue and electrode system. (From L. A. Geddes et al., *Applied Biomedical Instrumentation,* 2nd ed., John Wiley & Sons, Inc., New York, 1975, by permission.)

small with respect to the resistances in the circuit. Figure 5-5 illustrates the relationship between impedance of the electrode–tissue system and frequency. Skin resistance can vary from a few hundred ohms for abraded or punctured skin to 1 megohm for dry skin. While deep-tissue resistance varies less, it is still a direct function of the resistivity of blood, tissue, and the path length between the electrodes.

The series *RC* model of the electrode must be modified slightly to account for the fact that the electrode impedance does not increase to infinity as the frequency approaches zero. As seen in Figure 5-4, the impedance remains finite and relatively constant over the frequency range from 0 to ∞. Because of this physical characteristic, the electrode equivalent circuit model must be modified to take this into account. This is done simply by adding a paralleled resistance R_p to the model which accounts for the electrochemical processes that occur at the electrode–electrolyte interface. As mentioned previously, the values of resistances and capacitances depend on many parameters of electrode construction, such as area and type of metal, surface condition, current density, and the type and concentration of electrolyte used.

Using this simple model of the electrode–electrolyte interface together with models of tissue resistance and capacitance, a composite circuit model

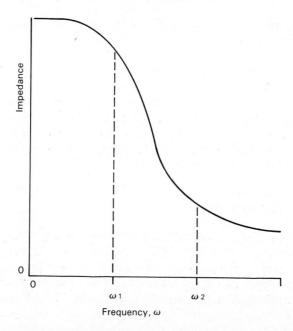

Figure 5-5 Terminal impedance as a function of frequency for electrode/skin system. (From L. A. Geddes et al., *Applied Biomedical Instrumentation,* 2nd ed., John Wiley & Sons, Inc., New York, 1975, by permission.)

can be formulated which provides some insight into the impedance system seen by any instrumentation amplifier, such as the input stage of an ECG system. A representative model of the system is given in Figure 5-6.

In summary, the important physical characteristics of electrodes for medical use are:

1. Stability in long-term offset potential changes (drift characteristics)
2. Stability in short-term offset potential changes (noise)
3. Fast recovery time to a stable state after large voltage perturbations

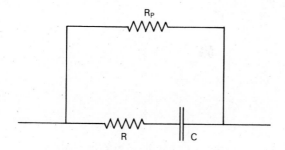

Figure 5-6 Circuit model of electrode.

produced by defibrillation or use of electrosurgery units on patients wearing the ECG electrodes

Electrodes should be chemically inert and have good mechanical characteristics, such as being easy to clean, sterilizable, easy to apply, and mechanically rugged. If silver electrodes are used, they must be 99.99% pure to prevent local battery action. This is why silver metal electrodes should not be cleaned with steel wool, as the steel slivers embed in the silver, setting up a local battery action that contributes to an unstable offset potential which shows up as noise on the ECG.

THE BIOPOTENTIAL AMPLIFIER

Physiological signals acquired either by electrodes or other transducers are typically below the 10-mV level. Thus amplification is required to increase the signal amplitude for further processing and to have significant voltage for use with recorder and display devices. The problem of amplification of the bioelectric signals as measured on the surface of the body is that it is not the only signal presented to the input of the amplifier. There is a high level of 60-Hz noise because the body is capacitively coupled to the surrounding ac power lines, lighting systems, motors, and transformers that are in the vicinity of the patient. Figure 5-7 illustrates how the patient is part of a capacitive voltage-divider network.

The patient is coupled to the power systems and other 60-Hz devices by a source coupling capacitance in the neighborhood of 0.2 pF, which is

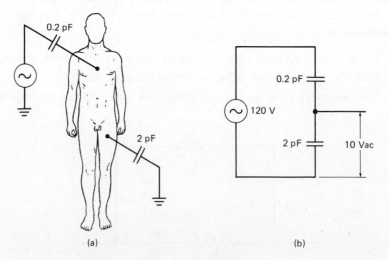

Figure 5-7 (a) Capacitance coupling of power system to patient; (b) equivalent circuit for ac coupling of power system to patient.

within an order of magnitude of the actual value. The patient is also capacitively coupled to ground. The capacitance to ground tends to be about 10 times larger than the source coupling capacitance, due to the fact that ground references are much more widespread than are the sources. So the capacitance to ground is in the neighborhood of 2 pF. From this simple circuit analysis, it is seen that the patient is at approximately 10 V ac above ground. The total voltage presented to the amplification system is the "noise" voltage due to capacitive coupling of 60-Hz voltage and the ECG voltage produced by cardiac events.

The key to extracting the desired ECG signal from the 60-Hz noise is the fact that the ECG voltage is a differential voltage, represented by the voltage difference between any two of the ECG input electrodes. On the other hand, the 60-Hz noise voltage is of the same magnitude at all points on the body and has equal magnitude at each of the electrode inputs. Since the noise voltage is common to each electrode, it is called the *common-mode voltage*.

The amplifier most commonly used to separate a differential signal from a common-mode signal is the *differential amplifier*. This amplifier responds to the difference in input voltages at its input. If the amplifier were an ideal differential amplifier, it would subtract out the common-mode signal and amplify only the differential signal. Let V_{ECG1} and V_{ECG2} be the ECG voltages that appear at inputs 1 and 2 of the differential amplifier. Let V_{cm} be the common-mode voltage that appears equally at each of the inputs of the differential amplifier. If A is the gain of the amplifier, the output voltage from this ideal differential amplifier is given by the equation

$$A(V_{ECG1} + V_{cm}) - A(V_{ECG2} + V_{cm}) = A(V_{ECG1} - V_{ECG2}) \qquad (5\text{-}1)$$

In the idealized case, the common-mode voltage is eliminated and only the differential voltage is amplified. Of course, we live in a nonideal world, so no differential amplifier is able to completely eliminate common-mode voltage, but it can significantly attenuate it. The measure of the differential amplifier's ability to reject a common-mode signal is called the *common-mode rejection* (CMR). A figure of merit for the ability of a differential amplifier to reject common-mode signals is called the *common-mode rejection ratio* (CMRR). The CMRR is defined as the differential gain of the amplifier (A_d) divided by the common-mode gain (A_{cm}).

From a laboratory measurement's point of view, this definition is hard to work with, so an alternative definition for CMRR is the ratio between the amplitude of the common-mode signal to the amplitude of a differential signal that would produce the same output voltage. Thus we have

$$V_{cm}A_{cm} = V_d A_d \qquad (5\text{-}2)$$

$$\frac{V_{cm}}{V_d} = \frac{A_d}{A_{cm}} \qquad (5\text{-}3)$$

Therefore,

$$\text{CMRR} = \frac{A_d}{A_{cm}} = \frac{V_{cm}}{V_d}$$

Example

A 1-V signal is connected to a differential amplifier in a common-mode configuration. If the output voltage is equivalent to one produced by a 10-μV differential signal, the CMRR is

$$\frac{1\text{ V}}{10\ \mu\text{V}} = 100,000:1$$

When CMRR is specified in decibels (dB), it is called *common-mode rejection*: that is,

$$\text{CMR} = 20\log_{10}\text{CMRR}$$

For the example above, the common-mode rejection is

$$\text{CMR} = 20\log_{10}(10^5) = 100\text{ dB}$$

Good common-mode rejection is only one of the important characteristics required of a biopotential amplifier. Other requirements of medical amplifier systems are the same as those of many standard voltage-amplification systems. Included in this set of useful characteristics is that the amplifier should have a high input impedance. The rationale for this is simply that we do not want the amplifier to draw current that will load down the input voltage source and cause distortion of the signal we are trying to measure. It is preferable to have the biopotential amplifier see the open-circuit voltage of the output from the voltage source, which usually is the ECG or other biopotential electrode being used. Consider the circuit shown in Figure 5-8. It is seen that with a high input impedance of the biopotential amplifier, the voltage drop across the electrode–electrolyte interface is almost zero; thus the voltage seen by the biopotential amplifier is essentially that of the bioelectric event. Modern amplifiers for medical usage have input impedances in a range from 2 to 10 megohms (MΩ).

Another characteristic of importance is the output impedance of the amplifier. Ideally, the amplifier should have a zero output impedance, $Z_o = 0$. Usually, the amplifier will be driving a load of some sort, either another stage of amplification or a display or recording device. Thus, to reduce distortion of the signal, the output impedance should be low with respect to the load impedance of the next stage and it must be able to supply the power needed by the load.

In terms of minimum distortion and maximum fidelity of the biopotential being amplified, measured, and reproduced, the bandwidth of the amplifier is of prime importance. Ideally, an infinite bandwidth would be desirable to ensure that all the frequencies in the Fourier spectrum of the measured

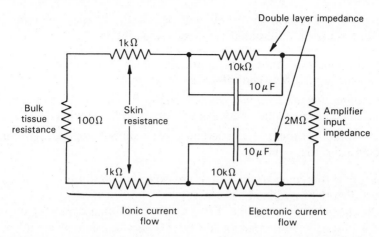

Figure 5-8   Equivalent circuit of electrode/skin system, including amplifier input impedance.

waveform are not attenuated. In the case of bioelectric potentials in general and the ECG waveform in particular, this turns out not to be the ideal characteristic. Since the ECG waveform contains frequencies only up to a few hundred hertz, it is desirable to bandlimit this amplifier to the band of frequencies contained in the ECG potential only and attenuate the higher frequencies in order to maximize the signal-to-noise ratio.

Electrocardiographic equipment usually can be operated in one of two modes: diagnostic or monitoring. The *diagnostic mode* of operation has the greater bandwidth of the two settings since the greatest fidelity of the waveform is demanded. For diagnostic purposes, a frequency range of 0.2 to 100 Hz is required with a high-frequency rolloff at 12 dB/octave above 150 Hz and below 0.1 Hz. In the *monitoring mode* a bandwidth of 0.5 to 50 Hz or even 2 to 20 Hz can be satisfactory since the purpose of the monitoring mode of operation is to determine the presence of the ECG signal, obtain a heart rate, and detect missing or grossly abnormal cardiac waveforms.

OPERATING SPECIFICATIONS FOR DIFFERENTIAL AMPLIFIERS

Dynamic Range Characteristics

When the input voltage, either common mode, differential mode, or a combination of these, is increased, a point is reached when the amplifier will be driven into a nonlinear mode of operation such as saturation. At this point, the output voltage is no longer a representation of the input voltage. If the amplifier's dynamic range specifications are not exceeded, this type of waveform distortion will not occur. Some dynamic range specifications for a differential amplifier are as follows:

Input differential dynamic range: the maximum peak differential voltage that can be applied between the differential inputs

Input common-mode dynamic range: the maximum peak common-mode voltage that can be applied between the inputs to the amplifier and ground

Output dynamic range: the maximum peak output voltage that can be expected from the amplifier

If none of the foregoing specifications are exceeded, the differential amplifier will not distort or overload.

In the process of taking an ECG or monitoring a patient during a medical procedure, one cannot always avoid a transient overload of the differential amplifier. An important characteristic of a physiologic amplifier is its ability to recover from these transient overload conditions. Dc-coupled amplifiers recover faster than do ac-coupled amplifiers. This is due to the coupling capacitors becoming charged due to the overload. The *RC* time constant of the circuits formed by the coupling capacitors determines the recovery rate of the amplifier.

Ac coupling is not used at the input of the differential amplifier due to the deleterious effect it can have on the common-mode rejection ratio of the amplifier. Ac coupling is used after several stages of dc coupling to eliminate electrode offset potential. This type of amplifier is called an *ac-stabilized amplifier.* This form of differential amplifier is used in ECG systems. It has a time constant of about 2 s and in about 4 s is almost completely recovered from a transient overload. This characteristic is important for ECG monitors used in intensive care units (ICUs), where the amplifier can be severely overloaded when the patient to whom the amplifier is attached is defibrillated. It is important that the amplifier recover quickly and continue to monitor the patient immediately after the defibrillation process is complete. In order to limit the input voltage into the amplifier during defibrillation, protection circuitry using zener diodes is used in each leg of the ECG input circuit to provide transient protection for the amplifier. In this way, the differential amplifier is not driven so "hard" into saturation, reducing the recovery time of the amplifier.

CONTROL OF FACTORS THAT REDUCE ECG PERFORMANCE

Degradation of ECG performance can come about from factors that are not due to electronic component failure. Ideally, the output from a differential amplifier should be proportional to the input without any additional frequency components being added to the output signal that were not present at the input. The areas of reduced performance of the ECG are caused by unbalanced source impedance at the input to the ECG differential amplifier, electric and magnetic induction of voltages at the input of the amplifier, and noise that originates in the electronics itself (i.e., shot noise and Johnson noise). We will consider only the first two of these areas of signal interference.

Source Impedance Unbalance

The common-mode rejection ratio is a measure of the differential amplifier's ability to reject common-mode signal. For an ECG amplifier, the CMRR is in the neighborhood of 100 dB, but still, in some cases in the hospital, 60-Hz artifact can be seen in an ECG tracing or monitor scope. A possible cause of the problem is source impedance unbalance; that is, the impedance of the electrodes as seen by the input of the ECG amplifier is not equal in all legs of the input. The problem that this creates is that some of the common-mode voltage presented at each of the inputs is seen as a differential voltage and is amplified by the differential gain of the amplifier. To illustrate this, consider the circuit shown in Figure 5-9. Let

$$V_{cm} = 1 \text{ V}$$
$$A_d = 20,000$$
$$A_{cm} = 0.2$$
$$\text{CMRR} = 100 \text{ dB}$$
$$Z_{in} = 10 \text{ M}\Omega$$

Let the differential impedance unbalance be only 150 Ω in the positive leg of the amplifier input. In this situation, the input network forms a voltage divider network in which there is a 15-μV drop across the 150-Ω resistor. Thus the input voltage in the positive leg is

$$V_+ = 1.0 - 15 \times 10^{-6} = 0.999985 \qquad (5\text{-}4)$$

The common-mode input voltage in the other leg is 1 V. The true common-mode voltage is now calculated according to the definition

$$V_{cm} = \frac{V_+ - V_-}{2} = \frac{1 + 0.999985}{2} = 0.9999925 \qquad (5\text{-}5)$$

Thus there is a 15-μV difference in voltage between V_+ and V_-. This differential voltage is amplified by the differential gain of the amplifier, A_d.

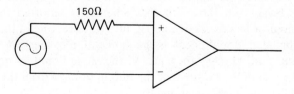

Figure 5-9 Model of source impedance unbalance.

The component of the output voltage due to differential form of the common-mode voltage is given by

$$V_o = 15 \times 10^{-6} \times 2 \times 10^4 = 300 \text{ mV}$$

We see that there is a 300-mV common-mode signal at the output due to the source impedance unbalance.

Since the differential amplifier is not perfect in its rejection of the common-mode signal, there is also a component in the output due to incomplete common-mode rejection. This voltage component is given by the equation

$$V_o = A_{cm} V_{cm} = 0.2 \times 1 \text{ V} = 200 \text{ mV} \tag{5-6}$$

The worst-case output voltage due to common-mode signal is the sum of the two components, or 500 mV.

The question now is: What is the "apparent" CMRR of the amplifier when there is this source impedance unbalance in the input of the amplifier? To calculate the apparent CMRR for this circuit, we use the basic working definition of CMRR, that is, the ratio of the common-mode voltage to the differential voltage to produce the same output-voltage level.

Since the common-mode voltage produces 500 mV at the output, the differential voltage required to produce the same magnitude voltage at the output is given by the relation

$$V_o = A_d V_d$$

Solving for V_d, we have that

$$V_d = \frac{500 \times 10^{-3}}{2} \times 10^4 = 25 \text{ } \mu\text{V}$$

Thus 25 μV of differential voltage will produce 500 mV at the output. The calculation of the apparent CMRR becomes

$$\text{CMRR}_{\text{apparent}} = \frac{V_{cm}}{V_d} = \frac{0.9999925}{25 \times 10^{-6}} = 40,000$$

$$\text{CMR} = 20 \log_{10}(40,000) = 92 \text{ dB}$$

We see that the impedance unbalance due to the electrodes on the ECG has reduced the common-mode rejection by 8 dB.

This was a worst-case analysis, but it illustrates the effect the application of dissimilar electrodes or mixed brands of electrodes can have on the ECG output. This brings out the point that when it is reported that an ECG recorder or monitor has excessive 60-Hz interference, one of the first things the clinical engineer should check is the application of the ECG electrodes, their condition, and whether they are all the same brand or not.

From the example, we see that the output voltage from the differential amplifier consists of three components:

1. Desired output due to amplification of the differential signal

2. A component of the common-mode signal due to incomplete rejection by the differential amplifier

3. A component of the common-mode signal due to source impedance unbalance

To minimize the problem of 60-Hz noise when using ECG electrodes, the input impedance of the differential amplifier must not be less than 10 MΩ in order to reduce the effect of the electrode insertion impedance.

Other Sources of Interference

Sixty-hertz interference can enter the biopotential recording and monitoring system in other ways besides the conversion of common-mode voltage to amplified differential signals, as seen in the example above. Other ways include electromagnetic induction, displacement currents in the electrode leads, and displacement currents in the body. Let us consider electromagnetic induction next.

From electromagnetic field theory, a conductive loop of wire cut by lines of magnetic flux will have a potential induced in the wire. This voltage is proportional to the magnetic flux density, the area of the loop, and the angle the loop makes with the direction of the magnetic flux lines. The equation that describes these conditions is

$$V = \frac{d\phi}{dt} = \mathbf{B} \cdot \mathbf{S} \tag{5-7}$$

where ϕ = magnetic flux in webers

$\mathbf{B}$ = magnetic field vector

$\mathbf{S}$ = area of the loop

A loop area of 0.1 m^2 can yield a potential of 10 μV in a typical hospital environment (see Figure 5-10). Since the voltage is proportional

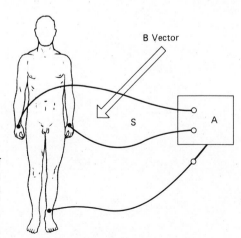

Figure 5-10 Magnetic flux coupling of ECG electrode wiring. (Adapted from J. C. Huhta and J. G. Webster, "60-Hz interference in electrocardiography," *IEEE Trans. Biomed. Eng.*, Vol. BME-20, Mar. 1973, pp. 91–101.)

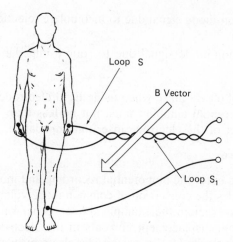

Figure 5-11   Reduction of induced voltage by twisting leads.

to the area of the loop, simply keeping the loop area as small as possible will reduce this induced voltage. This may be done by twisting the leads and running the wires close to the body. In Figure 5-11 the loop area S_1 causes only a common-mode component, since the ac potential induced in this loop is common to both inputs.

This mode of voltage interference in the hospital is more theoretical than practical because twisted leads are rarely used in the hospital. This indicates that magnetic induction of noise voltage is not a prevalent problem and represents the least of the noise problems encountered in the hospital.

Displacement current in ECG leads (Figure 5-12) is another form of noise injection in the system. Changes in electric field intensity may be capacitively coupled to the lead wires to yield a displacement current in

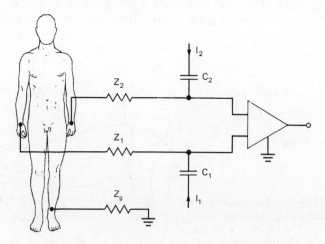

Figure 5-12   Displacement current in ECG leads.

the ECG. If the input impedance of the amplifier is very high compared to the skin electrode impedance Z_1 or Z_2, most of the displacement current will flow to ground through the path of least resistance, Z_g. The displacement voltage becomes

$$V_{dis} = I_2 Z_2 - I_1 Z_1 \qquad (5\text{-}8)$$

Now if $I_1 Z_1 = I_2 Z_2$, the interference equals zero. In test equipment, 1 m of wire picked up only 6×10^{-9} A of interference current. This current can be reduced by shielding the cable leads.

The best input technique is to have a high-input-impedance amplifier such as an emitter follower near each electrode. Long shielded leads can then be used without interference pickup. Unity-gain buffer amplifiers matched to 0.1% are available commercially.

Displacement Currents in the Body

The surface of the body can act as a capacitor and the displacement current will then flow through the body, if it is grounded. Any current through the body will set up a potential drop due to the internal impedance of the body. We can represent the potential difference by the displacement current I_d through the internal impedances R_1 to R_5 between points of electrode attachment. These displacements can be reduced by reducing I_d, moving the grounding point to another location on the body, or covering the body with a conducting, grounded blanket. The latter method is not practical and may even distort the ECG waveform; a better method is to drive the body to an equal but opposite voltage with a right-leg amplifier, as shown in Figure 5-13. The 5 MΩ resistor is necessary to limit the current through the body to 2 μA.

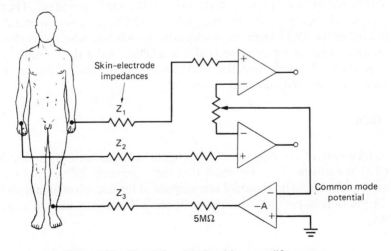

Figure 5-13 Circuit for right-leg-driven amplifier system.

PERFORMANCE CRITERIA FOR ECG SYSTEMS

Generally, electrocardiographic systems are considered to be the direct writing and monitoring devices used in hospitals. As will be discussed later in Chapter 16, performance testing of medical electronic equipment is an important and integral part of the safety program set up by the clinical engineering department. Since most electronic equipment has few, if any, moving parts, the concept of preventive maintenance does not apply and the concept of performance testing is more appropriate. In this section dealing with electrocardiographs, we will look only at performance specification and testing, deferring the topic of electrical safety of ECG equipment to Chapter 16.

Since medical devices now are governed by the Food and Drug Administration (FDA) in terms of their performance and efficacy, the FDA is working toward setting up performance standards for ECG direct-writing and monitoring equipment. They are working with various voluntary standards groups such as the Association for the Advancement of Medical Instrumentation (AAMI) and the American Heart Association. What will be cited here is based on these evolving standards of performance.

Frequency Response

The Food and Drug Administration set a frequency-response specification for ECG equipment of ±1 dB for the band of frequencies between 0.1 and 50 Hz and ±3 dB for the frequency band 50 to 100 Hz. The low-frequency cutoff value of 0.1 Hz is used to prevent significant distortion of the S-T segment of the ECG waveform. The upper-frequency cutoff of 100 Hz is required to prevent distortion of the QRS complex. There is some research being done on "high-fidelity" ECG analysis, wherein the bandwidth of the ECG amplifier is extended to 400 Hz, where it is claimed that more diagnostic information is still available. As of this writing, more work needs to be done and the previously stated bandwidths for ECG equipment remain the standard at present.

Gain

Since part of the diagnostic assessment is based on the amplitude of the ECG waveform, it is important that the equipment being used have a known gain factor. The standard recommended for use, given in mm deflection in the output per mV of input voltage, is 5 mm/mV and 10 mm/mV (±10%).

Common-Mode Rejection

The American Heart Association specifies that in the range of frequencies from 45 to 65 Hz, the common-mode rejection be at least 60 dB. At other frequencies it should be at least 40 dB.

The FDA standard sets up a test criteria for the ECG system in terms of the ability of the differential input system to reject a 28-V peak-to-peak 60-Hz signal, coupled through a capacitor that is common to all inputs and with a source impedance imbalance of approximately 2.7 kΩ (at 60 Hz) in one of the lead inputs. There should be no signal greater than 100 μV at the output of the electrocardiograph.

Paper Speed

The accuracy of the paper or chart speed in a direct-writing ECG is an important factor in the determination of the time-dependent parameters of the ECG waveform, such as the QRS duration or the P-R interval. It is essential that the ECG recording device maintain an accurate and stable rate of recording for proper diagnosis of these critical timing relationships.

The American Heart Association suggests that the standard chart speed be 25 mm/s, with the capability of using a speed of 50 mm/s with a speed accuracy of $\pm 2\%$. The FDA recommends the same chart speed with a speed accuracy of $\pm 5\%$. These values are fine as far as they go, since the majority of ECG direct-writing systems use a form of strip-chart recording. It is a question how long this will be true, as Hewlett-Packard is presently marketing a microprocessor-based ECG system that makes use of a digital plotter to record the various ECG lead inputs on an $8\frac{1}{2}$ by 11 inch piece of graph paper. Thus the time intervals on the chart are obtained from the microprocessor, and it is questionable whether these standards apply to such a new approach to ECG recording.

Linearity

Any nonlinearity in the recording device will result in distortion of the ECG waveform. Thus uniform response over the width of the recording paper is important when the ECG is used for diagnostic purposes. The proposed FDA standard specifies a maximum deviation in signal magnitude of no more than $\pm 10\%$ over any portion of the recording channel width.

These standards are designed to bring all electrocardiographic equipment to the point of uniform performance regardless of where it is used. Until some standard is finally adopted, the clinical engineer should consult the performance specifications supplied by the manufacturer to determine the performance criteria for a particular ECG system.

PERFORMANCE FUNCTION TESTING OF ELECTROCARDIOGRAPHIC MONITORS AND DIRECT-WRITING SYSTEMS

As part of a complete safety program conducted by the clinical or biomedical engineering department, performance testing of medical electronic equipment is essential. In the case of ECG equipment, one would like to determine the value of the following:

1. Frequency response
2. Bandwidth
3. Common-mode rejection ratio
4. Linearity

as well as other parameters of the ECG monitoring system, such as the following:

5. Sweep speed of monitor scope
6. QRS threshold detection
7. High/low alarm trip points
8. Accuracy of heart rate meter (cardiotachometer)

To make the appropriate tests for performance verification, a good function generator, a calibrated oscilloscope, and a shielded attenuator are required if the output voltage of the function generator exceeds the maximum input voltage of the ECG.

Determination of Frequency Response

Frequency-response measurements need not be extremely precise, but they are important and should conform to the manufacturer's specifications. A Fourier frequency decomposition of an ECG waveform shows that there are components of the waveform at very low frequencies. Consequently, the passband of the amplifier must include frequencies in the range 0.05 to 0.2 Hz at the low end of the passband and frequencies as high as 150 Hz at the high end. As frequencies go, this group is relatively low compared to the normal passband of an electronic amplifier. Thus the ECG amplifier is intentionally bandlimited in order to reduce noise in the system. The actual frequency response of the system under test will depend on the manufacturer and mode of operation (monitor/diagnostic). Present-day non-fade monitors are usually designed with a lower high-frequency response, due to digital sampling and storage considerations. There should be a flat frequency response between the low and high ends because the interpretation of cardiac waveforms often depends on small variations in the display.

To test the frequency response of an ECG, the input sinusoidal signal is varied and the results observed on the monitor screen and chart recorder to determine the 3-dB roll-off points. The low-frequency testing is performed using a square wave from the function generator instead of a sine wave, because amplitude measurements are difficult at frequencies in tenths of a hertz and are also time consuming.

Procedure. To obtain the high-frequency -3-dB point, connect the output of the function generator to the attenuator and the output from the attenuator into the lead I input of the ECG or to the inputs of a monitor. Set the lead selector on the ECG to lead I. Set the frequency of the input to approximately 4 Hz. Set the amplitude to a convenient value that will allow easy determination of 70% of the reference amplitude.

Note: Be sure that the output of the function generator is isolated. If it is not, it will ground one side of the differential amplifier and force the input of the ECG to operate "single-ended." This means that the input voltage is referenced to ground and the desirable properties of the differential amplifier will be lost.

Note this reference frequency and amplitude on the monitor and sweep the frequency upward until the amplitude of the sine wave has fallen to 70% of its original value. The frequency at which this occurs is determined from the period of the waveform on the calibrated scope. Do not use the monitor screen for this measurement. The frequency at which 70% of the reference amplitude occurs is the upper 3 dB.

To determine the lower 3-dB point, change the function generator from sine wave to square wave. Set the frequency at 1 Hz and the amplitude at 1 mV. The output display should show a 4-cm waveform amplitude on the monitor screen or on the chart recorder for a direct writer (see Figure 5-14).

Reduce the frequency of the square wave until the top of the falling trace is 2 cm lower than the top of the rising trace. This represents the point of $V_{max}/2$ (see Figure 5-14). The lower 3-dB frequency can be calculated using the formula

$$f_1 = \frac{0.11}{t} \qquad (5\text{-}9)$$

where t is the time between the two vertical traces, which is half the period P of the applied square wave. From the data obtained, one can evaluate the frequency response, linearity, and bandwidth of the ECG system to determine whether the ECG system is performing up to the manufacturer's specifications.

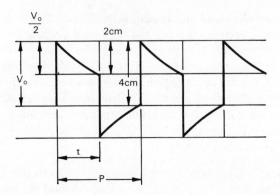

Figure 5-14   Test waveform for determination of lower 3-dB point.

Derivation of the Lower 3-dB Frequency Equation

The formula for the lower 3-dB point in frequency is derived in the following manner. The low-frequency response of the ECG amplifier system is determined by the equivalent input capacitance and the resistance of the amplifier system. The *RC* equivalent input circuit for an ECG system is shown in Figure 5-15.

A sequence of square waves can be thought of as a series of positive- and negative-going step functions. If $u(t)$ is a unit step function, the first square wave of the sequence can be characterized as

$$V_{in}(t) = V_{in}[u(t) - 2u(t - P)] \qquad (5\text{-}10)$$

where $V_o(t)$ = time-dependent output voltage

$V_{in}(t)$ = input square-wave voltage of V_{in} volts

R_{eq} = equivalent input resistance of amplifier system

C_{eq} = equivalent input capacitance of amplifier system

Writing the input loop equation results in a simple first-order differential equation whose solution is

$$V_o = V_{in}e^{-t/R_{eq}C_{eq}} \qquad (5\text{-}11)$$

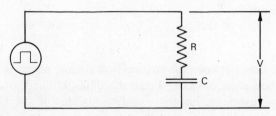

Figure 5-15   Equivalent *RC* network representing the input impedance of an ECG amplifier.

If, at a time t_1, the output voltage is half the input voltage, that is,

$$V_o = \tfrac{1}{2}V_{in} \qquad \text{at time } t = t_1$$

Then by substitution into equation (5-11), we have

$$\tfrac{1}{2}V_{in} = V_{in}e^{-t_1/R_{eq}C_{eq}} \tag{5-12}$$

Reducing the equation and taking the natural log of both sides, we have

$$\ln 0.5 = \frac{-t_1}{R_{eq}C_{eq}} \tag{5-13}$$

Solving for $R_{eq}C_{eq}$, we have that

$$R_{eq}C_{eq} = \frac{t_1}{0.69}$$

The frequency at which the -3-dB point occurs is given by the relation

$$f_{3dB} = \frac{1}{2\pi R_{eq}C_{eq}}$$

Substituting for $R_{eq}C_{eq}$ in the equation, we have

$$f_{3dB} = \frac{0.11}{t_1} \tag{5-14}$$

Equation (5-14) can be put into a form that is convenient to use in terms of sweep speed for monitors or paper speed for direct writers. This is given as

$$f_{3dB} = \frac{(0.11)(\text{speed in mm/s})}{d} \tag{5-15}$$

where d is the distance to the half-amplitude location measured in millimeters.

Example

If a square-wave input to an ECG operating at 25 mm/s has a distance to half-amplitude of 14 mm, the lower -3-dB frequency is given by

$$f_{3dB} = \frac{(0.11)(25)}{14} = 0.2 \text{ Hz}$$

Measurement of CMRR

When working with measurement of amplifier voltages in an ECG system, it is difficult to measure the output voltage directly from the differential amplifier itself. To do this would require opening up the system, which, for routine performance testing, is not feasible. Ultimately, the clinical or biomedical engineer is interested in the effect of common-mode noise voltages on the output of the system, either on a monitor screen or on the chart recorder. This points the way to looking at the effect of the input voltage at the output of the ECG equipment. Such effects are the deflection on the monitor screen or on the chart recorder. With this concept in mind, we will make small modifications to the definition of CMRR in order to produce a definition that will allow the direct measurement of an ECG common-mode rejection ratio with common electronic test equipment.

Let

$$A_d = \text{overall system differential gain}$$

$$A_{cm} = \text{overall system common-mode gain}$$

$$V_d = \text{differential input voltage}$$

$$V_{cm} = \text{common-mode input voltage}$$

The deflection at the output due to a differential voltage at the input is given by

$$V_d A_d = \text{voltage deflection at the output}$$

The deflection at the output due to a common-mode voltage at the input is given by

$$V_{cm} A_{cm} = \text{voltage deflection at the output}$$

If appropriate voltages are used that produce equal output deflection, the following equation is true:

$$V_d A_d = V_{cm} A_{cm} \tag{5-16}$$

Solving for the ratio of gains and voltages, we obtain the expression for the common-mode rejection ratio.

$$\frac{A_d}{A_{cm}} = \frac{V_{cm}}{V_d} = \text{CMRR} \tag{5-17}$$

From equation (5-17) we see that CMRR can be measured by taking the ratio of the common-mode input voltage to the differential input voltage when each produces the *same* deflection at the output of the ECG.

PERFORMANCE TEST PROCEDURES FOR CMRR

The following procedures will allow one to determine the CMRR of a monitor of electrocardiograph by first finding the differential voltage and the common-mode voltage that produce the same output deflection and taking their ratio as indicated in equation (5-17).

1. Apply a 1-mV peak-to-peak sine wave at 60 Hz between the lead inputs RA and LA (lead I).
2. Adjust the gain control for a 6-cm peak-to-peak display on the monitor or electrocardiograph. Leave the gain control in this position.
3. Remove the signal and connect the RA and LA leads together.
4. Connect the signal source between RA, LA, and RL (reference or ground) leads and apply a voltage in the range of 10 V, peak to peak,

60-Hz sine wave to these inputs. Adjust this common-mode voltage amplitude until the output again shows a 6-cm peak-to-peak display.

5. The CMRR is determined by the ratio

$$\text{CMRR} = \frac{V_{cm}}{V_d}$$

for the same voltage amplitude on the display.

PROBLEMS

1. (a) What is the function of a physiologic electrode?
   (b) Determine whether an ECG electrode can be classified as a transducer. Explain your answer fully.
   (c) What is the "half-cell potential" and the electrode "double layer" for an Ag/AgCl electrode?
   (d) How is the stability of an ECG waveform dependent on the mechanical stability of the electrode double layer?

2. Claims have been made about ECG electrodes that the lower their intrinsic impedance, the "better" the ECG waveform that can be obtained. As a clinical engineer, devise a method to test the claims of medical salespersons about the impedance of their electrodes using a minimum of test equipment.

3. An ECG electrode may be represented by an equivalent circuit consisting of a 10-kΩ resistor shunted by a 10-μF capacitor. Calculate and draw the input waveform of the ECG preamplifier with:
   (a) a 1-MΩ resistor input impedance
   (b) a 10-kΩ resistor input impedance
   (c) a 1-kΩ resistor input impedance
   when viewing a 1-s-wide square pulse through such an electrode/amplifier combination. From your results, deduce the minimum permissible amplifier impedance which must be used with these electrodes to keep distortion to 1% or less.

4. Given the precordial lead system in Figure 5-16, where the lead potential averaging circuit is shown with 5-kΩ input resistors. Redesign the input circuit of this ECG lead system so as to reduce the voltage variations at the input due to differences in electrode impedance up to a factor of 20. Show mathematically that the circuit does provide the buffer required against electrode impedance changes. Draw the equivalent circuit to accompany your equations.

5. If we consider an EEG to be a first-order medical instrumentation system that can measure cortical signals up to 100 Hz, what is the maximum allowable time constant for the instrument if an amplitude inaccuracy of less than 5% is all the instrument can have? What is the phase angle of the voltage at 50 and 100 Hz?

6. If the impedance of an ECG electrode pair is 6 kΩ each and these electrodes are used in conjunction with the inputs to a differential amplifier with an input impedance of 85 kΩ, how many decibels is the signal reduced at the input due to this configuration?

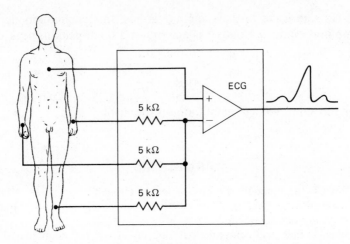

Figure 5-16

7. In the design of an ECG or patient monitor, the input impedance must be as high as possible. Why is this done, and what can happen if the input impedance is too low?

8. (a) When obtaining an ECG, there is always the problem of common-mode voltage interference. What is common-mode signal voltage as picked up on a patient, and what are some of the origins of it in the hospital environment?
   (b) Why is a differential amplifier used to measure ECG signals from the body?
   (c) Define CMRR and explain what it is a measure of.

9. Calculate the CMRR of the operational amplifier circuit given in Figure 5-17,

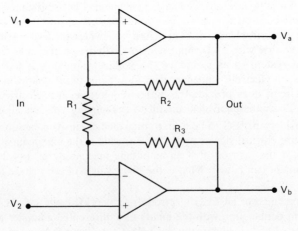

Figure 5-17

which is configured as a differential amplifier. Use as your working definition for CMRR

$$\text{CMRR} = \frac{V_{\text{out}} \text{ for } V_1 = -V_2}{V_{\text{out}} \text{ for } V_1 = V_2}$$

10. An ungrounded patient, shown in Figure 5-18, is at an ac common-mode potential of V_{cm}. Lead III of his ECG is being measured using an amplifier with a common-mode rejection ratio of C. The skin/electrode resistances are R_1 and R_2, as shown in the figure. Assuming an instantaneous lead III voltage of V_{III}, calculate the ratio of the output-signal component due to V_{III} to the output-signal component due to V_{cm}.

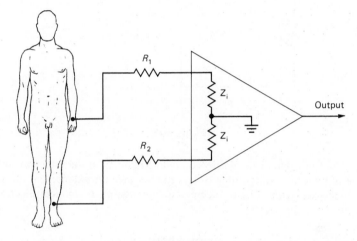

Figure 5-18

11. For the operational amplifier configuration given in Figure 5-19, derive the relationship for the output voltage of the differential amplifier:

$$V_o = A_d V_d + A_d V_{\text{cm}} \left[\frac{1}{\rho} + 1 - \left(\frac{Z_{\text{in}}}{Z_{\text{in}} + Z - Z_1} \right) \right]$$

where A_d = differential gain

ρ = common-mode rejection ratio

Assume that $Z_{\text{in}} \gg Z$ and Z_1 and $V_{\text{cm}} \gg V_d$.

12. Given the differential amplifier in Figure 5-19, where $V_{\text{cm}} = 1$ V peak to peak at 60 Hz and $V_d = 100 \ \mu\text{V}$ peak to peak at 2 Hz. The specifications on the amplifier are

$$\text{CMRR} = 100,000:1 \qquad A_d = 20,000$$

$$Z_{\text{in}} = 10,000,000 \qquad Z_d = 150$$

$$Z_1 = 0$$

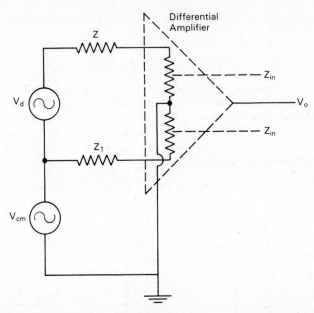

Figure 5-19

Calculate the value of the desired output component of the total waveform, the value of undesired output due to input impedance unbalance, and the value of the undesired output due to incomplete common-mode voltage rejection.

REFERENCES

1. Cromwell, Leslie, Fred J. Weibell, and Erich A. Pfeiffer, *Biomedical Instrumentation and Measurements* (Englewood Cliffs, N.J.: Prentice-Hall, 1980).

2. Dymond, A. M., "Instrumentation for bioelectric measurements," in *CRC Handbook of Engineering in Medicine and Biology,* Barry N. Feinberg and David G. Fleming (eds.) (Boca Raton, Fla.: CRC Press, 1978).

3. Strong, Peter, *Biophysical Measurements* (Beaverton, Oreg.: Tektronix, Inc., 1970).

6

MEASUREMENT
AND MONITORING
OF BLOOD PRESSURE

Arterial pressure is the driving force that causes blood to perfuse through the tissues of the entire body. Maintenance of adequate pressure is essential for the function of vital organs and the prevention of vascular collapse. Because of the importance of blood pressure to the well-being of the critically ill patient, especially those in shock or with cardiovascular problems, blood pressure is routinely monitored in critical care areas.

Blood pressure, as it is conventionally measured, is essentially the lateral pressure, or force, exerted by the blood on one unit area of wall in a blood vessel. This lateral force is assumed to be equal to the force per unit area in the longitudinal direction as a result of *Pascal's law* describing the transmission of forces in fluids. This law states that a force is transmitted equally in all directions in a contained fluid. The arterial blood pressure is constantly changing during the course of the cardiac cycle. The highest pressure or maximum amplitude of arterial pressure is called *systole*. It is the result of the ejection of blood into the aorta by the left ventricle. The lowest pressure or minimum pressure is called *diastole*. It occurs during the resting or diastolic phase of the cardiac cycle. *Mean blood pressure* represents the average pressure attained in the cardiac cycle. Mean blood pressure is not the arithmetic average of the systolic and diastolic pressure, because the arterial pressure is changing with time. The mean pressure is computed by integrating the pressure over one cardiac cycle and dividing by the time for one cycle. This is given by the equation

$$P_{\text{ave}} = \frac{1}{T} \int_0^T P(t) \, dt \qquad (6\text{-}1)$$

where T is the time for one cardiac cycle. Since systole is only about one-third of the cardiac cycle and diastole about two-thirds of the cardiac cycle, mean arterial pressure can be approximated by the equation

$$\text{mean} = \frac{\text{systolic} + (\text{diastolic} \times 2)}{3} \qquad (6\text{-}2)$$

Thus, for a normal blood pressure of 120 mmHg systolic pressure and 80 mmHg diastolic pressure, the approximated mean pressure becomes

$$\text{mean pressure} = \frac{120 + 80 \times 2}{3} = 93.3 \text{ mmHg}$$

Mean pressure is often considered the most important of the blood pressure values as it is considered to be the parameter that indicates if vital organs are adequately perfused. Mean arterial pressure should be above 40 mmHg to avoid vascular collapse. The mean pressure is not affected by the morphology of the waveform, and generally is not affected by arrhythmias causing a variation in the amplitude of the arterial waveform. Some typical ranges for normal arterial pressure are

Systolic: 100–140 mmHg

Diastolic: 60–90 mmHg

Mean: 80–100 mmHg

A number of factors, acting in concert, and dynamic equilibrium are integrated via the central nervous system, in the determination of the arterial blood pressure. Some of these are:

1. Cardiac output
2. Peripheral vascular resistance
3. Volume of blood in the arterial system
4. Blood viscosity
5. Compliance of the arterial walls

The exact contribution of each of these factors is not known, but peripheral resistance and cardiac output seem to have the greatest influence on blood pressure.

INDIRECT MEASUREMENT OF BLOOD PRESSURE

There exist both direct and indirect methods of measuring blood pressure. The method used in routine examinations in physician's offices or in a hospital is to occlude the arteries in the upper arm with an expandable wraparound pressure cuff. The cuff is wrapped around the upper arm of the patient and the pressure increased to above systolic pressure (see Figure 6-1). The pressure in the cuff is then slowly released. The theory goes that when the pressure in the cuff has fallen to a level equal to the peak, or systolic, pressure, a stethoscope can pick up the sounds of the blood squirting through the once-compressed artery. The sounds of the turbulent blood flow through the distorted lumen of the artery are called *Korotkoff sounds* after the Russian physician who described them in a paper in 1906. These sounds may be divided into five phases (Figure 6-2).

Phase one is the sudden appearance of a clear but faint tapping sound growing louder during the succeeding 10 mm fall in pressure. In phase two, the sound takes on a murmurous quality during the next 15 mmHg fall. During phase three, the sound changes little in quality, but becomes clear and louder during the next 15 mmHg fall. During phase four, the sounds take on a muffled quality which lasts through the next 5- to 6-mmHg fall. Phase five is simply the onset of silence, when all sounds disappear.

The beginning of phase one sounds usually occurs around 120 mmHg in normals and is representative of systolic pressure. The pressure at which phase five occurs is an indirect approximation to diastolic pressure. In normals, this occurs at about 80 mmHg. This method of indirect blood pressure measurement is known as *sphygmomanometry*. At its best it is accurate only to 10 mmHg. Sphygmomanometry is a fast and easy method to assess indirectly the effects of blood pressure on the measuring system. As with everything, there are drawbacks to this method. Since the operator is depending on the onset and cessation of sounds and the cuff pressure at which they occur, the method becomes heavily dependent on the frequency response of the stethoscope used in the procedure and the sensitivity and frequency response of the hearing of the operator. Thus the accuracy of this method can be dependent on the acoustical considerations noted above but is largely ignored by most practitioners.

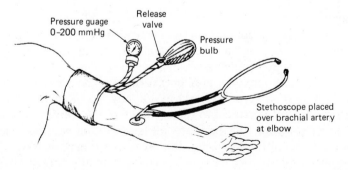

Figure 6-1 Typical sphygmomanometer using a compression cuff and stethoscope to listen for Korotkoff sounds.

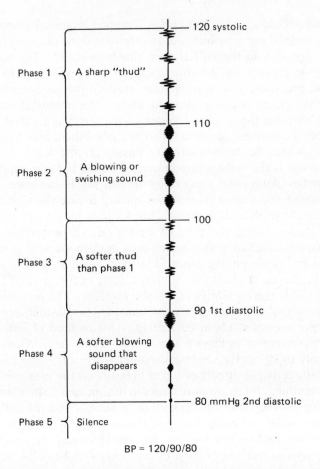

Figure 6-2   Phases of the Korotkoff sounds. (From *Primer of Clinical Measurement of Blood Pressure,* George Burch and Nicholas DePasquale, C. V. Mosby Company, St. Louis, 1962, by permission).

There are other ways of determining the systolic pressure, including watching the mercury column for the first signs of a visible pulse, and simply feeling for a pulse at the brachial artery. In some cases, as for infants or obese persons, special techniques must be used. For best results of indirect blood pressure measurement, an appropriate pressure cuff should be used. It was discovered as early as 1901 by Von Recklinghausen that the width of the cuff is critical to the accuracy of the sphygmomanometry. If the cuff is too thin, the pressure measured is on the high side, and conversely, if the cuff is too wide, the pressure estimate is low.

Automatic microprocessor-controlled equipment is now used in some hospitals to inflate a cuff at regular intervals while a microphone picks up the 400- to 500-Hz Korotkoff sounds. The pressures at the beginning and

Figure 6-3 Microprocessor-controlled automatic blood pressure monitor. (Courtesy of The Kendall Company, Boston, MA.)

end of the sounds are recorded and mean pressure calculated to provide a long-term record of a patient's blood pressure. These devices are becoming more popular for use as automated blood pressure monitors. Figure 6-3 is typical of these devices. Many of the microprocessor-controlled blood pressure devices such as the one shown in Figure 6-3 have small dot matrix printers attached to them for logging the patient blood pressure data over time. This type of indirect blood pressure monitor is beginning to be used more frequently because of the trauma and risk imposed by direct measurement of blood pressure and the lack of better indirect methods for monitoring the hemodynamic state of the patient.

DIRECT MEASUREMENT OF BLOOD PRESSURE

For the critically ill patient, invasive blood pressure measurement has several advantages over indirect methods using cuff occlusion with stethoscopic detection of Korotkoff sounds. It allows for the beat-by-beat evaluation of electromechanical function as well as indirect estimates of overall cardiac function through interpretation of the waveforms presented on the monitor screen. The shape of the tracing, the response to mechanical ventilation, and greater accuracy of the monitored arterial blood pressure have long been cogent reasons to resort to an invasive approach to patient monitoring. This approach is more accurate than the cuff method and is relatively free from motion artifact. During a hypotensive episode, cuff pressures are particularly inaccurate because of decreased cardiac output and increased vascular resistance. The cannula or catheter used for obtaining direct pressure measurement may also be used for drawing blood for blood gas analysis.

All of these advantages are not without associate risks to the patient. Any invasion of the cardiovascular system carries with it the danger of embolization, bleeding, vessel and tissue damage, and infection. Central arterial or venous catheters carry the added risk of causing arrhythmias. If the operator of the system is not well trained or fails to consider important details of the system operation, error is easily introduced and can remain undetected. In the critical care environment, these disadvantages are generally outweighed by the benefits. It is essential, however, that procedures, protocols, and materials be designed to minimize the risks involved with this form of patient monitoring.

With the increase in reliability of modern integrated-circuit electronic engineering systems and a corresponding reduction in price, and the development of plastic catheters together with the use of heparin to prevent clotting at the tip of the catheter, direct pressure monitoring has become part of the management of the critically ill patient. Accurate, continuous, and fairly safe measurements of blood pressure can be made by introducing a rubber or plastic catheter which has a blunt tip and an opening in its side into an artery or vein. The catheter is saline-filled, with the free end connected through a flush valve to a pressure transducer. The flush valve is necessary to keep a small volume of saline flowing out through the catheter opening to prevent clotting at the tip of the catheter. A typical blood pressure waveform obtained by the indwelling catheter method is given in Figure 6-4.

By using small-diameter catheters, about 1 to 2 mm, it is possible to make pressure measurements inside the heart itself. This technique can supply information on the condition of valves, the existence of shunts, and many other properties of the intracardiac environment. While the catheter is in place, it is possible to insert radio-opaque dyes or to sample blood in the chambers of the heart for oxygen content. Since the small diameter of such a catheter drastically limits the frequency response of the fluid column, cardiac catheterization is sometimes done with miniaturized transducers which will fit inside a 2-mm catheter.

When measuring blood pressure externally using a catheter, the frequency response of the fluid-filled catheter must be taken into account. This hydraulic system has distributed inertia, compliance, and resistance

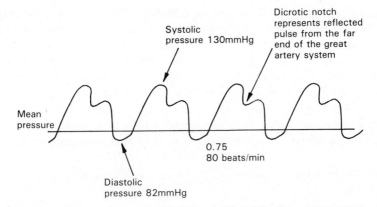

Figure 6-4 Blood pressure waveform that can be obtained from an indwelling arterial catheter.

in the same manner that an electrical transmission line has distributed resistance, inductance, and capacitance. For clinical applications, the distributed-parameter model is not used because a simplified lumped-parameter model represented by ordinary differential equations is sufficiently accurate to describe the response of the system to different pressure inputs.

Thus the hydraulic transmission system can be thought of as a lumped second-order system where the system's resonant frequency and transient response must be considered when configuring the catheter system. The catheter must be wide enough so that its resonant frequency is above the maximum frequency to be measured. This resonant frequency is determined by the following system parameters:

1. The diameter of the fluid-filled column
2. The length of the column
3. The compliance of the tubing and catheter

The equation for the resonant frequency, as derived later in this chapter, is given by equation (6-3) and is repeated later as equation (6-10).

$$f_r = \frac{d\sqrt{3}}{8\sqrt{\pi}} \sqrt{\frac{\Delta P}{\Delta V} \Big/ L\rho} \qquad (6\text{-}3)$$

where d = diameter of the fluid column
$\quad L$ = length of the fluid column
$\quad \Delta P$ = pressure change in the fluid column
$\quad \Delta V$ = volume change produced by the pressure change ΔP
$\quad \rho$ = density of the medium

The ratio $\Delta V/\Delta P$ is the compliance of the fluid column.

Damping is obtained in the catheter-tubing fluid column system by reducing the caliber of the tubing or inserting a section of smaller caliber in the line. The problems of transient response, phase response, and zero drift are all important if accurate results are to be obtained. The pressure inside the heart can change as rapidly as 2000 to 6000 mmHg per second. The phase response affects the shape of the pressure waveform, and the zero-level errors will result if the transducer is not on the same level hydraulically as the catheter tip. If there is a height difference between the transducer and the catheter tip, there will be a pressure bias or offset from zero which is equal to the pressure head produced (positive or negative) by the height of the saline above or below the catheter tip. This pressure is equal to the height of the column times the density of the fluid.

BLOOD PRESSURE MONITORING SYSTEMS:
DIRECT MEASUREMENT

A blood pressure monitoring system is an instrumentation system designed to record and display the pressure variations with time of arterial pressure. As pointed out in Chapter 2, the components of any such instrumentation system are a transducer, an amplifier, and a display. The display can consist of a cathode ray tube, digital display, and/or a chart recorder. Figure 6-5 is a typical blood pressure monitoring system.

The components of a blood pressure monitoring system are:

Catheter: a semirigid tube, usually filled with saline, inserted into the location where pressure is to be measured

Pressure transducer: an electromechanical device that converts mechanical energy into electrical energy

Monitor: the amplifiers and signal-processing electronics combined with some type of display system

The catheter is a hydraulic coupling device between the artery or vein where the pressure is to be measured or monitored and the pressure transducer. The key component of the pressure transducer is a thin, flexible diaphragm which terminates the "plumbing" from the catheter. When there is a pressure gradient across the diaphragm, it will deflect in the direction of the lower pressure. Figure 6-6 is a cross-sectional view of the diaphragm.

The deflection of the diaphragm can be converted into an electrical signal by two common means: the strain gauge and the linear variable differential transformer (LVDT). The characteristics of strain gauges and the associate bridge circuits were taken up in Chapter 2. There are two basic types of strain gauges, bonded and unbonded. The *unbonded strain gauge* has one end of the strain element fixed while the other end is attached to the diaphragm. Thus motion of the diaphragm in response to pressure changes applied to it cause corresponding changes in the resistance of the

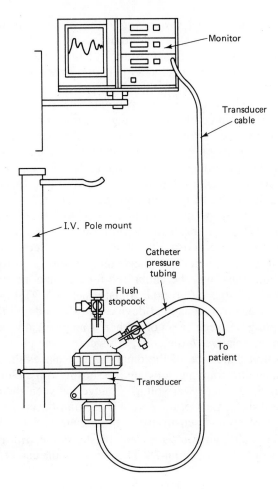

Figure 6-5 Blood pressure monitoring system.

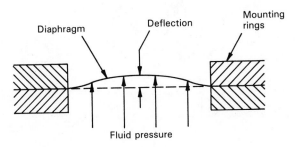

Figure 6-6 Schematic of diaphragm used in a blood pressure transducer.

strain gauge. In the *bonded strain gauge,* the strain element is bonded to the diaphragm. Any elongation or compression of the member due to diaphragm deflection causes the resistance of the element to change. The strain element is usually connected as part of a Wheatstone bridge, where changes in resistance of one of the elements unbalances the bridge and produces a voltage difference that can be amplified.

Some silicon semiconductors change their resistance in response to being stressed, as does wire, but differs in that they require very small displacement of the diaphragm to produce large changes in voltage with low bridge excitation voltages. The gauge factor for semiconductor strain gauges can vary from 50 to 200 with a typical value for silicon of 120. Gauge factors for wire are typically on the order of 2 to 4. Some solid-state strain gauges can be made small enough that they can be placed in the catheter tip. There are problems with these types of transducer in terms of fragility, temperature sensitivity, and adverse effects of sterilization. Semiconductor strain gauges are more temperature sensitive than other strain gauges and thus must be calibrated for baseline and true zero.

The linear variable differential transformer (LVDT), as discussed in Chapter 2, is an electromechanical device that depends on the changes of magnetic coupling between two coils when a magnetically permeable core material is moved relative to the coils. This core is coupled to the pressure diaphragm so that deflection of the diaphragm displaces the core and thus changes the magnetic coupling between the sensing coils. Since the sensing coils are ac excited, there is a voltage change between the coils that is a function of the displacement of the core. These transducers are used because of their high stability and sensitivity at the expense of a larger mass.

On top of the diaphragm is a clear plastic dome with two connector fittings or ports on it (see Figure 6-7). This dome on the end of the transducer

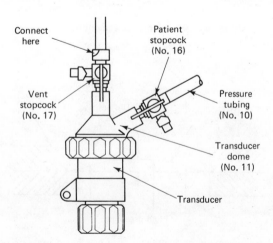

Figure 6-7 Blood pressure transducer with plastic dome.

is the reservoir for the saline-filled system. Its transparent construction helps in the detection and removal of air bubbles, which can degrade the frequency response of the monitoring system. One port on the dome is coupled through the connecting tubing to the catheter while the other is used for venting air from the dome. Figure 6-7 shows the characteristics of a typical blood pressure dome.

DYNAMIC RESPONSE OF BLOOD PRESSURE MONITORING SYSTEMS

As with any measurement system, the frequency response or bandwidth of the system will determine the fidelity of the reproduced blood pressure wave. The obvious goal of the monitoring system, which includes all the hydraulic interfacing between the transducer and the patient, is to introduce negligible or no distortion to the blood pressure waveform. The electronic portion of the monitoring system can achieve almost any design bandwidth that is desired. Generally, the electronic amplification, signal processing, and display portion of any blood pressure monitoring system is intentionally bandlimited to increase the signal-to-noise ratio. Thus the response and performance characteristics of the blood pressure monitoring system are dominated by the response of the hydraulic interface system between the patient and the transducer. It is therefore of value to look at the components of this system and model them so as to see how each of these components influences the response of the blood pressure system. The following derivation is based on a development by Geddes [4].

For a second-order system, the resonant frequency of the system gives us information about the frequency characteristics of the system. This information, together with the damping ratio or damping, provides information on the dynamic performance of the system. There are two important pieces of information provided by the resonant or natural frequency: (1) it provides information as to whether any of the Fourier frequency components of the pressure signal will cause resonant distortion of the displayed waveform; and (2) when one considers the Bode plot (log magnitude vs. frequency) for a second-order system, the resonant frequency is the corner frequency of the Bode plot and the amplitude rolls off at 12 db/octave after the corner frequency. From the system damping, we obtain information about the dynamic response to transient changes at the input of the system. From a Laplace transform point of view, we can have some feeling for the proximity of the roots of the characteristic equation to the $j\omega$ axis. Thus natural frequency and damping provide insight into the ability of the blood pressure system to track the periodic changes in the pressure that are applied to the system.

Figure 6-8 is a representation of a fluid-filled catheter and transducer. An assumption made here is that for small deflections of the diaphragm,

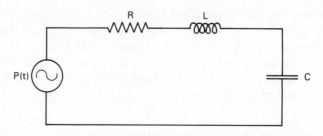

Figure 6-8 Equivalent circuit for a fluid-filled catheter. R, flow resistance; L, fluid inertia; C, total compliance of the catheter and pressure transducer system.

its motion can be considered uniform across the transducer cross section, as if it were a piston in its action.

Let M be the mass of the fluid in the transducer, A the cross-sectional area of the piston model of the diaphragm, K the spring constant of the flexing diaphragm, R the internal frictional loss of the fluid filling the transducer, L the length of the fluid connection, and a the cross-sectional area of the fluid connection to the transducer.

The spring constant, K, is a measure of the pressure required for a unit volume displacement. The ratio $\Delta P/\Delta V$ is called the volume modulus of elasticity and is expressed in units of dynes/cm^2/cm^3.

If a pressure, ΔP, is applied to the system, the force, F, on the diaphragm is equal to the pressure times the area of the diaphragm or equivalent piston. Thus

$$F = \Delta PA \tag{6-4}$$

This force causes the piston to be displaced by an amount X. Writing this in terms of the spring constant K, we have

$$F = K\,\Delta X = \Delta PA \tag{6-5}$$

This represents a volume displacement equal to $\Delta V = \Delta XA$. By eliminating the variable ΔX, we can write the spring constant in terms of the volume modulus of elasticity:

$$K = \frac{A^2 \Delta P}{\Delta V} \tag{6-6}$$

Consider now the calculation of the equivalent mass for the system. Let M_L be the mass of the fluid in the connection to the transducer and let M be the mass of the fluid in the transducer. It can be shown through the use of kinetic-energy equations for the system that the effective mass of the system can be given by

$$M_{eq} = \tfrac{4}{3} \left(\frac{A}{a} \right)^2 M_L + M \tag{6-7}$$

The distributed parameter system can now be modeled by a single second-order ordinary differential equation which relates the time response of the transducer and the fluid-filled catheter–transducer system to a pressure input $P(t)$.

$$M_{eq} \frac{d^2x}{dt^2} + R \frac{dx}{dt} + A^2 \frac{\Delta P}{\Delta V} x = P(t) \qquad (6\text{-}8)$$

The natural frequency for this system is obtained from the coefficients of the differential equation and is given as

$$f_r = \frac{1}{2\pi} \sqrt{\frac{A^2 \Delta P}{\Delta V} M_{eq}} \qquad (6\text{-}9)$$

Substituting equation (6-7) for the equivalent mass, M_{eq}, into equation (6-9) gives the natural frequency of the hydraulic system in terms of the system parameters.

$$f_r = \frac{1}{2\pi} \sqrt{\frac{A^2 \Delta P}{\Delta V} \Big/ \left[M + \tfrac{4}{3}\left(\frac{A}{a}\right)^2 M_L \right]}$$

This formulation can be simplified by noting that if the mass of the fluid-filled catheter (M_L) is much greater than the mass of fluid in the transducer dome (M), which tends to be the case, we may drop the term under the radical containing M and obtain the approximate formula:

$$f_r = \frac{a\sqrt{3}}{4\pi} \sqrt{\frac{\Delta P}{\Delta V} M_L}$$

$$= \frac{d\sqrt{3}}{8\sqrt{\pi}} \sqrt{\frac{\Delta P}{\Delta V} \Big/ L\rho} = 0.122 d \sqrt{\frac{E}{L\rho}} \qquad (6\text{-}10)$$

where d = internal diameter of the tubing, cm
L = length of the tubing, cm
ρ = density, g/cm^3
$E = \Delta P/\Delta V$, volume modulus of elasticity, dynes/cm^5

Equation (6-10) shows the relationship between the various physical variables in the system and natural frequency of the system. The importance of this is that we prefer that the dominant frequencies in the blood pressure (BP) waveform, represented by the first few terms of the Fourier expansion of the wave, do not correspond to the natural frequency of the system. If they do, this will cause large oscillations to be superimposed on the BP waveform (see Figure 6-9). Second, we know that the natural frequency of the system corresponds to the corner frequency of the Bode plot. This means that the amplitude response of the system will roll off at 12 dB/octave, which will cause distortion in the waveform. Thus, to keep waveform distortion to a minimum, it is necessary to have the resonant frequency of the system as high as practicable.

Selecting the physical parameters of the hydraulic interface between the patient and transducer so that the natural frequency of the system is

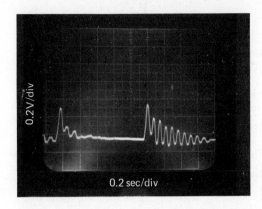

Figure 6-9 Second-order oscillations of a fluid-coupled blood pressure measurement system.

high enough is not the total criterion for system performance. Together with the natural frequency, an appropriate degree of damping is required (Figure 6-10). If the damping is too large, the system will not respond quickly enough to changes in pressure and the waveform will be distorted. On the other hand, if there is too little damping, the system response tends to overshoot, producing distorted peaks in the pressure waveform. It is important to identify the system parameters that determine the damping in the system. Damping is the result of viscous drag of the fluid in the catheter and connecting lines of the system and other forms of energy dissipation that may be present.

It can be shown, making use of the laws of fluid flow [4], that the resistance coefficient of the first-order derivative in equation (6-8) can be given by

$$R = \frac{8nLA^2}{\pi r^4} \tag{6-11}$$

where n is the viscosity of the fluid in the catheter.

If we make the assumption, as before, that $M_{eq} \gg M$, it can also be shown from the solution of the second-order differential equation that models the blood pressure system that the damping, D, can be obtained as

$$D = \frac{R}{\sqrt{4M_{eq}A^2E}} \tag{6-12}$$

Substituting from equation (6-7) for M_{eq} and remembering that $M_L = LA$, $A = \pi r^2$, and $r = d/2$, we have the expression for the damping as

$$D = \frac{16n}{d^3} \sqrt{\frac{3L}{\pi \rho E}} \tag{6-13}$$

We see from this expression that system damping is very dependent on the diameter, d, of the tubing as the damping varies with the inverse of the cube of d and varies less as a function of length, L, or the elasticity, E, of the tubing. The optimal degree of damping is 0.7, critical damping. This

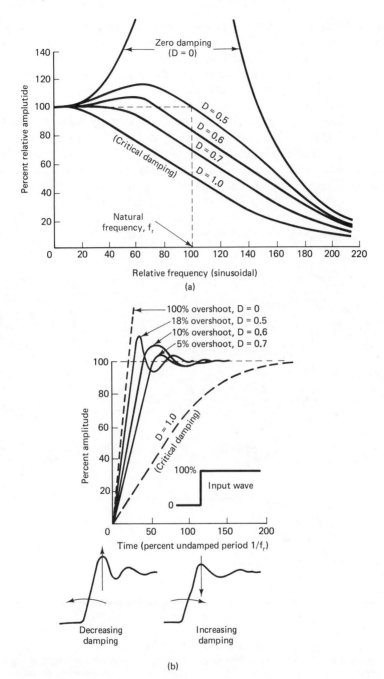

Figure 6-10   Sinusoidal (a) and step-wave (b) responses in a resonant system with various degrees of damping. (Reproduced with permission from L. A. Geddes, *The Direct and Indirect Measurements of Blood Pressure*. Copyright © 1970 by Year Book Medical Publishers, Inc., Chicago.)

produces the fastest system response to pressure variations without undue oscillation. Because the frequency response of the catheter–transducer system is important in determining whether a system is suitable for use, many manufacturers specify the frequency response of their transducers with common-size catheters attached.

What this analysis has done is provide a mode of understanding for a distributed-parameter hydraulic transmission system by imposing a lumped-parameter model on the system. This model resulted in a second-order differential equation model that can be visualized as an *RLC* circuit (Figure 6-11a), as those with an electrical engineering background tend to do, or as a mass, spring, and damper system (Figure 6-11b), as those with a

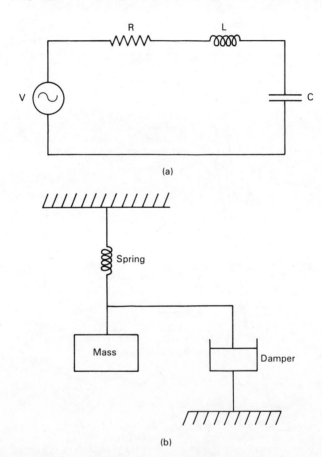

(a)

(b)

Figure 6-11 (a) *RLC* equivalent circuit for the distributed-parameter hydraulic pressure transmission system for an indwelling blood pressure system; (b) mass–spring–damper mechanical equivalent system representing the distributed-parameter hydraulic pressure transmission system for an indwelling catheter blood pressure system.

mechanical engineering background tend to do. Regardless of how you tend to think of this blood pressure monitoring system, it will respond to inputs closely resembling the classical response of a second-order system, as discussed in Chapter 2.

We have seen how the diameter, length, and volume modulus of elasticity determine the resonant frequency and the damping of the blood pressure monitoring system. This is important from the standpoint of understanding how these physical parameters determine the system response because if the system is not responding properly, the clinical engineer will know what changes need to be made in the system in order to make the system respond in an acceptable manner. For example, if the resonant frequency is too low, it is possible that the connecting tubing is too long or too compliant. If the damping of the system is too low, possibly the diameter of the tubing is too large.

One problem that can occur with the blood pressure monitoring system is to have a trapped air bubble in the lines connecting the patient to the transducer or in the transducer dome itself. Since air is compressible, the change in volume with respect to pressure, or $\Delta V/\Delta P$, is larger, so the compliance of the air adds to that of the connecting tubing. This represents a decrease in the volume modulus of elasticity. Using the RLC model for the system, the air bubble can be represented as a capacitor in parallel with the system capacitance. Since the two are in parallel, the total capacitance is the sum of the two. Since $\Delta V/\Delta P$ increases, the reciprocal value, $\Delta P/\Delta V$, decreases. Thus the volume modulus of elasticity, E, is smaller. From equations (6-10) and (6-13) for resonant frequency and damping, we see that f_r is decreased but the damping, D, is increased. Figure 6-12 is a schematic representation of the effect of an air bubble in the BP monitoring system.

Figure 6-13 is a Bode plot, log magnitude vs. frequency, representation of the effect of an air bubble on the frequency response of the monitoring system. It shows a decrease in the natural frequency of the system with an increase in the system damping. The introduction of any air into the

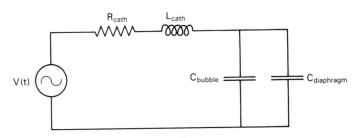

Figure 6-12 Equivalent circuit showing the added capacitance produced by a bubble in the fluid-filled line of a blood pressure monitor.

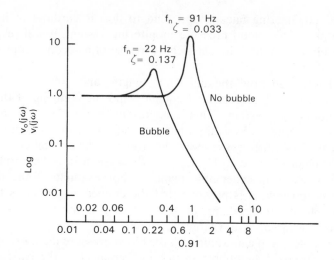

Figure 6-13 Frequency-response curves for catheter–transducer system with and without bubbles. Natural frequency decreases from 91 Hz to 22 Hz and damping ratio increases from 0.033 to 0.137 with the bubble present. (From John G. Webster (ed.), *Medical Instrumentation: Application and Design.* Copyright © 1978, Houghton Mifflin Company, Boston. Used with permission.)

system results in significant waveform distortion and could result in incorrect systolic and diastolic pressure readings.

The reason an air bubble distorts the pressure waveform is a matter of energy transfer. The kinetic energy of the pulse wave goes into compressing the air bubble. By compressing the bubble, kinetic energy is partially stored as potential energy in the compressed gas, and some of the energy is lost due to its conversion into heat. When the potential energy of the compressed gas is released, there is a net energy loss. This accounts for the increased damping of the system response. The higher-frequency and larger-amplitude components are mainly responsible for altering the energy content of the pressure wave since the higher frequencies compress the gas more often. The more energy that is lost from any Fourier frequency component of the pressure wave, the more waveform distortion is produced. The extent of the waveform distortion will be related to the amount of gas trapped in the lines or in the dome.

To reproduce the blood pressure waveform with minimum distortion, the tubing that is the interface between the arterial system of the patient and the transducer must have a high resonant frequency. Tubing and catheters with as large an internal diameter, as short a length, and as small a compliance as possible should be used. Air bubbles must be eliminated by flushing the system and by ensuring tight connection to prevent air infiltration. Also, unnecessary stopcocks should be eliminated as air bubbles can lodge here and go undetected visually.

PERFORMANCE TESTING OF BLOOD PRESSURE MONITORING SYSTEMS

In the case of blood pressure monitoring instruments, the inspection and performance testing represents a detailed examination of the electromechanical performance and electrical safety of the system. This is done to ensure that the device functions reliably and safely in order to provide the maximum benefit to the patient. Performance testing for a blood pressure monitoring system is composed of two parts, static and dynamic testing.

Static Testing

Static testing of a blood pressure monitoring system is a check of the linearity of the pressure transducer and associated electronics. It is also

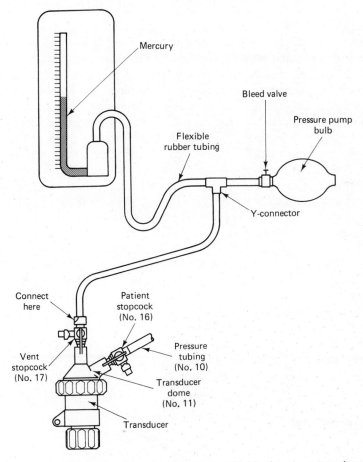

Figure 6-14   Test configuration for static testing of a blood pressure transducer.

Figure 6-15 Test equipment for dynamic testing and calibration of a blood pressure system. (Courtesy of Bio-Tek Instruments, Inc., Burlington, Vt.)

a test for the accuracy of the system in a static mode of the correspondence between a calibrated applied pressure and indicated pressure by the blood pressure system. Figure 6-14 is a picture of the test equipment and arrangement required for static testing of the blood pressure equipment.

Static testing of the blood pressure system is based on the introduction of a known, calibrated pressure which acts on the transducer and causes an appropriate deflection of the signal on the monitor or recording device. For this test, a manometer is required. A mercury type is preferred, as it is self-calibrated. An aneroid manometer can be used if it is calibrated. A sphygmomanometer can be used if the cuff is disconnected and, in its place, the pressure transducer is connected in the manner shown in Figure 6-14. At least three test pressures should be used to determine the linearity of the transducer system. The error should be less than or equal to that specified by the manufacturer or if none apply here, an error of plus or minus 5% or less is acceptable. From this simple test, the linearity and static accuracy of the transducer system can be determined.

Static pressure testing is fast and simple to perform, but a blood pressure monitor is subject to time-varying pressures. Thus a more realistic performance test is to introduce into the system a time-varying test pressure. This can be done with any one of the blood pressure waveform simulators that are on the market at present. Examples are the Fogg Systems ECG/Blood Pressure Simulator and the Bio-Tek Instruments Blood Pressure Systems Analyzer. Figure 6-15 is an example of one of these devices.

Dynamic Testing

With a simulator like the Bio-Tek, the entire blood pressure system is tested by introducing hydraulic pressure signals resembling a choice of clinical blood pressure waveforms that are stored in the simulator's memory. Using sine-wave input methods, the system bandwidth, resonant frequency, and system damping can be determined. In this way the dynamic response of the fluid-filled system can be observed. Since the test instruments will produce simulated blood pressure waveforms of various kinds, the system performance can be observed under simulated dynamic conditions. Calibrated input waveforms can also be used to check systolic, diastolic, and mean pressure determination made by the blood pressure monitoring system under test.

The main value of this form of testing over the static testing methods is that the entire system performance can be assessed under conditions close to those that the system will see in use on patients. The static testing method using a manometer allows all the transients to die out and thus the effect of system parameters on the time-varying response to pressure inputs is lost. Therefore, performance testing under dynamic conditions would have much greater value to the clinical engineer in assessing a blood pressure monitoring system in terms of the fundamental characteristics of accuracy, linearity, transient, and frequency response.

RELATIONSHIP BETWEEN DIRECT AND INDIRECT BLOOD PRESSURE MEASUREMENT

In the selection and purchase of blood pressure equipment that uses the indirect compression cuff method of measurement, comparisons are sometimes made as to the closeness systolic, diastolic, and mean pressures come to the same parameters determined with an arterial line system. Let us look to see whether this is even a valid comparison to make, as we may be talking about "apples and oranges." Both methods claim to measure systolic and diastolic pressure, yet in comparison tests performed by Van Bergen et al. [9] there is a definite difference in the values of the pressure parameters obtained using these methods.

In terms of fluid mechanics, one can consider the heart as a pulsatile energy source that imparts most of the kinetic energy of ventricular contraction to potential energy stored in the elasticity of the aorta. Only a small percentage of the energy of contraction goes directly into the kinetic energy of bulk transport of the blood. The reason for this is that the inertia of the mass of blood in the arterial system cannot be accelerated instantaneously. Thus the energy in the ejected blood is stored in the aorta and is manifested as increased arterial pressure. When the aortic valve closes due to a reverse pressure gradient across it, the potential energy stored in the elasticity of the arterial walls is converted to kinetic energy in two ways: (1) as bulk fluid flow with a velocity of 0.5 m/s, and (2) as a traveling pressure wave down a distributed parameter hydraulic transmission line. The wave velocity, dx/dt, of the pressure wave is approximately 10 m/s. The initial slope of the pressure pulse is related to the blood acceleration out of the ventricle. The character of the traveling pressure wave, as it moves through the arterial tree, changes in wave shape and amplitude. This is due to changes in the ratio of elastic fiber and muscle in the arterial wall and the decreasing diameter of the branching vessels. These parameters affect the characteristic impedance of the transmission system, which in turn affects the waveform of the traveling wave.

At a point of bifurcation or at any point where the characteristic impedance of the transmission line changes, wave reflection will occur in the reverse direction from the incident wave. The amount of energy reflected is a function of the characteristic impedance mismatch at the junction. Zero energy is reflected at a location where there is a perfect impedance match. Total reflection of the wave takes place at a point where the vessel is occluded, which represents infinite character impedance. Thus it is seen that the nature of the traveling-wave pulse is a function of many variables and is not as simple to characterize nor the straightforward technical task that some would have us believe. All of these topics are being mentioned so that we can look at the nature of two methods of blood pressure measurement in terms of what they are measuring.

A fundamental principle of any measurement system is that it should not disturb the system in which the measurement is made. Certainly, the compression cuff method violates this principle, as it totally occludes the brachial artery. When this occurs, a number of things happen which will directly affect the opening pressure of that artery.

First is the fact that blood no longer flows, which is a major alteration of the system. Second, there is total reflection of the arterial pulse wave at the occlusion site, which produces higher pressure due to a reinforcement between the incident and reflected waves at the site of the occlusion. Van Bergen et al. showed an increase of between 20 and 30 mmHg of directly measured systolic pressure when the brachial artery was compressed immediately distal to an indwelling needle transducer. If we combine these ideas with the fact that systolic pressure increases as the pressure pulse proceeds from major vessels to more peripheral vessels, we see that the location site of measurement plays a part in the values obtained in the measurement of blood pressure. The fact that the cuff and catheter methods do not measure blood pressure at the same location is a factor in the differences seen in the determination of systolic and diastolic pressure. There is also the problem of pulse-wave reflections at the catheter due to the change in arterial characteristic impedance produced by the insertion of the catheter into the artery.

From what has been developed, it is questionable whether each system is really measuring the same thing. The complexity of the pressure wave and the many factors that affect its time course should at least point out that each measurement method should be considered separately as an indication of the hemodynamic state of the patient, and not try to force a congruency of pressure values between the two methods. The reason for this is that the manner in which each affects and interacts with the arterial system and the propagating arterial wave affects the measurement made. It is unreasonable to expect that the two methods will produce the same systolic and diastolic pressure values when measured on the same patient.

PROBLEMS

1. What are some of the physiological parameters that play a part in the determination of blood pressure? Why do you think that blood pressure is routinely measured during surgery and in intensive care?

2. Describe the function of a sphygmomanometer and how it is used to estimate diastolic and systolic pressure.
   (a) What are the Korotkoff sounds, and how are they used to determine blood pressure?
   (b) How does the frequency response of the stethoscope used with the sphygmomanometer affect the determination of systolic and diastolic pressure using this method?
   (c) How would you expect that the attenuation of high frequencies would affect the determination of diastolic and systolic pressures? How would the low-frequency attenuation affect this same measurement?
   (d) How does the hearing of the person making the blood pressure measurement affect the accuracy of the measurement? How does the combination of frequency attenuation in the operator and the stethoscope affect the accuracy of the measurement of diastolic and systolic pressure? Explain your answers.

3. Draw a schematic illustration of an invasive blood pressure system which illustrates the function of the system.
   (a) How does the height of the pressure transducer relative to the heart affect the determination of diastolic and systolic pressure?
   (b) Derive an equation that expresses the error in the measuring of blood pressure as a function of the height above or below the level of the heart.

4. If an invasive blood pressure system can be considered as a simple *RLC* second-order system, develop a test for the frequency response and the natural response for the blood pressure system as well as a determination of the damping in the system.

5. If, in a test of the time response of an invasive blood pressure system, the system is determined to have an underdamped waveform, what might be the cause or causes of this condition, and how will it affect the blood pressure waveform? Draw a sketch of the affected waveform.

6. What are the requirements, in terms of bandwidth and phase relationship, for a blood pressure transducer/amplifier system? How does the tubing connecting the arterial catheter to the pressure transducer affect the total system response as a function of its length, diameter, and compliance?

REFERENCES

1. Brunner, John M. R., *Handbook of Blood Pressure Monitoring* (Littleton, Mass.: PSG Publishing Co., 1978).

2. Brunner, John M. R., et al., "Comparison of direct and indirect methods of measuring arterial blood pressure, Part I," *Med. Instrum.,* Vol. 15, No. 1, Jan.– Feb. 1981.

3. Fox, Forest, "Blood pressure instruments: direct systolic/diastolic measurements," in *Hospital Instrumentation Care and Servicing for Critical Care Units,* Robert B. Spooner (ed.) (Pittsburgh, Pa.: Instrument Society of America, 1977).

4. Geddes, L. A., *The Direct and Indirect Measurement of Blood Pressure* (Chicago: Year Book Medical Publishers, 1970).

5. Greatorex, C. A., "Indirect methods of blood pressure measurement," in *IEE Medical Electronics Monograph 1-6,* B. W. Watson (ed.) (London: Peter Peregrinus, 1971).

6. Nichols, W. W., "Instantaneous aortic blood flow derived from pressure gradient," in *Advances in Cardiovascular Physics—Cardiovascular Engineering,* Part II: *Monitoring,* D. N. Ghista et al. (eds.) (Basel: S. Karger, 1983).

7. "Pressure measurement," Application Note AN710 (Waltham, Mass.: Hewlett-Packard, 1970).

8. Rushmer, R. F., *Cardiovascular Dynamics,* 3rd ed. (Philadelphia: W. B. Saunders, 1970).

9. Van Bergen, F. H., et al., "Comparison of indirect and direct methods of measuring arterial blood pressure," *Circulation,* Vol. 10, 1954, pp. 481–490.

7

ELECTRONIC PACEMAKER SYSTEMS

Under certain circumstances, the heart may lose the ability to conduct the depolarization wave initiated by the SA node through its normal conduction pathways. When this happens, there is a need to stimulate the ventricles to contract artificially in a synchronized fashion in order to produce an adequate pumping action that will maintain a person. The pacemaker is an electronic muscle stimulator used to produce the required ventricular stimulation when the heart's own pacing system is malfunctioning. This malfunction is manifested as some form of conduction abnormality in the His bundle. This can happen when this tissue becomes damaged, as can happen during heart surgery or as the result of a heart attack. Some forms of heart disease can also impair the conduction process in the bundle of His. These problems may cause a total block of the action potential initiated by the SA node.

When the action potential is not conducted well by the AV node or the bundle of His, the self-pacing of the ventricular cells becomes dominant and will produce ventricular contraction, but at a slower rate than would be produced normally. The result is a much lower heart rate, as the ventricles beat on their own. Since the filling of the ventricles is mostly a function of the pressure differential across the valves separating the atrium from the ventricles and is dependent on atrial contraction only in a minor way, the ventricles are filled without exact synchrony with the atria. The result of the reduced number of beats per second is that the rate of blood flow through the body is reduced. Resultant lack of oxygen and nutrients can cause physical symptoms such as fatigue, shortness of breath, dizziness, fainting, and blacking out.

HEART BLOCK

The defects of the conduction system can be classified according to their temporal character into permanent and temporary defects. There are three main types of *atrioventricular* (A-V) *blocks* that are used to describe the conduction abnormality in the heart (Figure 7-1). These are the first-degree, second-degree, and third-degree A-V blocks. In a first-degree A-V block there is a prolongation of the conduction time for the action potential wave to pass through the A-V node; thus each impulse is delayed in arriving at the ventricles. The term "block" is somewhat of a misnomer since all of the atrial impulses reach the ventricles. Arrhythmia is more to the point, as we are talking about a transmission delay in the conduction system rather than a block. This type of arrhythmia usually results from some abnormality in the tissues of the A-V node or the bundle of His. This type of conduction abnormality can be seen on the ECG as an increase in the P-R interval longer than 0.2 s.

Second-degree A-V block represents the conduction of some action potentials to the ventricles. This means that the SA node will "fire" and trigger an atrial wave represented by the P wave on the ECG, but there is no ensuing QRS complex for some P waves. This may decrease the pumping efficiency of the heart. Clinical experience has shown that this condition may progress to complete heart block at any time.

A third-degree A-V block is the most severe, as there is no conduction of the SA action potential to the ventricles. On an ECG this would appear as a sequence of P waves with no QRS complex and then a QRS complex, indicating ventricular contraction, but with no associated P wave. This is the most common cause of pacemaker implantation and accounts for about 95% of pacemaker use.

Another form of conduction abnormality is called a *bundle branch block*. In this case the action potential is not conducted normally to the respective ventricle. The spread of the contraction in the ventricles occurs abnormally over a longer period. Bundle branch block may be either complete or incomplete. Generally, this is defined by determining how wide the QRS complex is on the ECG. This is a measure of the conduction delay in the ventricles.

PACEMAKER CONSTRUCTION

The cardiac pacemaker is nothing more or less than a physiologic stimulator. The device consists of three basic parts:

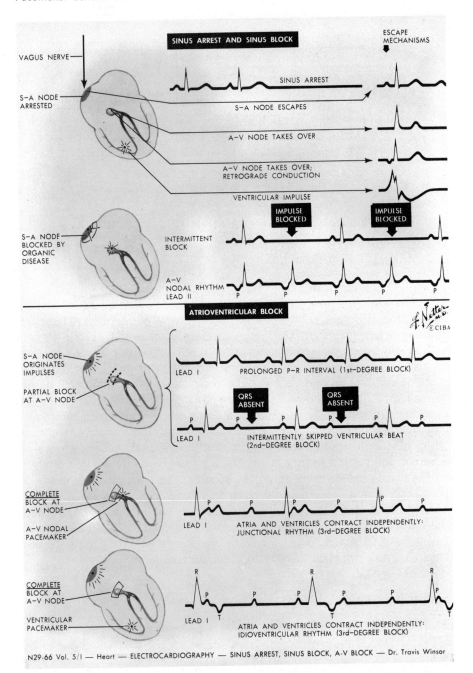

Figure 7-1 Atrioventricular block. (© Copyright 1969 CIBA Pharmaceutical Company, Division of CIBA-GEIGY Corporation. Reprinted with permission from *The CIBA Collection of Medical Illustrations*, illustrated by Frank H. Netter, M.D. All rights reserved.)

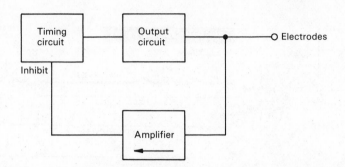

Figure 7-2 Block diagram of pacemaker.

1. An oscillator or other form of pulse-producing electronics
2. A power amplifier
3. An isolation or pulse transformer

Figure 7-2 is a block diagram of a typical pacemaker.

The modern, implantable pacemaker was made possible by the invention of the transistor but is now made up of integrated circuits. The exact circuitry used by pacemaker companies is not made public, but we will look at some representative circuits that have been used in electronic pacemakers. The history of pacing began with the work of Galvani in 1780, when he did experiments dealing with electrical stimulation of living tissue. Vassali in Italy worked with decapitated bodies of criminals. Using an early form of battery, he reported producing strong heart contractions with electrical stimulation. In 1932, Hyman successfully used a mechanical magnetogenerator to resuscitate a patient. Zoll, in 1952, was able to keep alive a patient recovering from a heart attack by applying 75 to 150 V in 2-ms pulses to the chest. This external pacemaker was operated by manually pressing a button every time the heart was to beat. Weirich et al., in 1957, applied electrodes directly to the myocardium, with leads exiting through the chest to pace a patient. The first fully implantable pacemakers were the work of Chardack, Zoll, and Kantrowitz, in 1960. They used stainless steel electrodes applied directly to the myocardium.

PACING MODES OF OPERATION

There are three basic modes in which a pacemaker may be designed to operate. One is the *asynchronous mode,* where the pacemaker pulses the heart at a fixed rate. This was the first type of pacing used. It was soon discovered that in some cases, atrioventricular conduction recurs spontaneously, causing dangerous competition between the pacemaker and the heart's own pacing system. This could be dangerous since the pacemaker could issue a pulse to the ventricles during the vulnerable period and cause ventricular fibrillation. To prevent this problem, a noncompetitive or *demand mode* of operation was developed. A demand pacemaker issues a pulse only after a preset time has elapsed and it has not sensed a ventricular contraction. If the ventricles beat, the electrical activity of their contraction is sensed by the pacemaker and its timing circuits are reset to wait for another pulse.

Another useful mode of pacing the heart is the *atrial synchronous mode*. Here the depolarization of the atrium is sensed and, after an appropriate delay, the ventricle is stimulated. This mode has the advantage of retaining the use of the body's autonomic and sympathetic control of heart rate.

PACING LEADS AND ELECTRODES

The type of material and the location of pacemaker electrodes are both important, as they determine the magnitude and stability of the depolarization threshold. The safety of the installation procedure is also involved, since transvenous implantation may be done under local anesthesia, but myocardial electrodes must be sutured onto the exposed heart.

Transvenous electrodes may be bipolar (anode and cathode in the heart) or unipolar (cathode in the heart, anode elsewhere). If the anode is placed in the heart, polarization, anodal erosion, and eventual failure may result. Some locations for transvenous electrodes are shown in Figure 7-3.

Physical requirements for intracardiac electrodes are electrochemical inertness, resistance to flexation fatigue, and low electrical resistance. Some currently used materials are platinum + 10% iridium, silver + stainless steel, gold-plated stainless steel, and stainless steel cable wound on a fabric

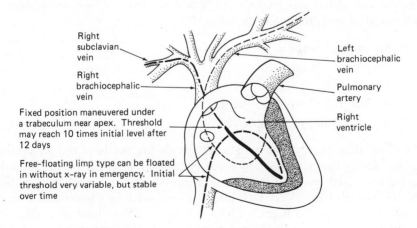

Figure 7-3 Placement of transvenous pacemaker electrode.

core welded to a platinum electrode tip. Unipolar configurations often employ large stainless steel anodes to achieve low current density with correspondingly low corrosion. The unipolar geometry is some 10 times more sensitive than the bipolar configuration in sensing a P wave, which indicates atrial contraction. For this reason, bipolar electrodes are usually used for temporary pacing. They can be installed with less difficulty, since only one catheter must be inserted.

Pacing lead failure is the most common cause of pacemaker failure. Consider that within one year's time, with the human heart beating about 70 beats per minute (BPM), a heart will beat 36.8 million times. Add to this just over 11 million respiratory movements. The mechanical characteristics of the pacing leads must be able to withstand this physical abuse in addition to a hostile environment of high humidity and numerous ions affecting them electrochemically.

In the design of catheter electrodes, there is a trade-off between flexibility and resistance to myocardial perforation and its durability. Those electrodes that possess superior flexibility and fatigue resistance are usually so flexible that they require a stiffening stylette for initial placement in the myocardium, which is removed after placement. The catheter electrode is usually a helical form, allowing the guide-wire stylette to slide through to the tip for placement. The wire may be bent to facilitate moving around corners and rotated for a change of direction. Another design of a pacing lead is a multistrand conductor, each strand consisting of a polyester yarn wrapped with a foil-thin conductor. This catheter design has shown excellent fatigue-failure resistance.

PACEMAKER SYSTEM

The general functional anatomy of this electronic engineering system is composed of essentially five parts:

1. An energy source
2. An output circuit
3. A timing circuit
4. An electrode delivery system
5. A sensing and processing circuit

The ideal energy sources should supply the needs of the pacemaker for the life of the patient. Unfortunately, in our physical world, we have no batteries that can hold that much energy in the very small space within the chest of the user. Very little power is consumed by a pacemaker. Each pulse delivered to the heart is about 4 V into a 500-Ω load for a duration of about 1.7 ms. At a stimulated heart rate of 70.6 beats per minute, this comes to only 64 μW of power average drain. At 50% efficiency, there are only 175 mA-hours to battery drain per year. In practice, however, cells rated at 1000 mA-hour last only three years, due to internal battery drain.

For a long time, the energy source used in pacemakers was the mercury-zinc cell. A new cell starts with an electromotive force (emf) of 1.4 V, which drops during the first 10% of its life to 1.3 V. As mentioned before, the mercury cell has a life expectancy of about 30 months. Within recent years, the lithium-iodide cell has become the power source of choice since it has a much higher energy density than the mercury cell. Thus it can last up to 10 years as compared with only $2\frac{1}{2}$ years for the mercury battery. This is very significant because the pacemaker is a sealed unit, with batteries and electronics encased together. When the batteries of the pacemaker are exhausted, the entire unit must be replaced. This is a hazard for the patient and expensive because the patient must undergo surgery again to replace the pacemaker, with all of the dangers associated with any form of surgery. There is lost time for the patient and the whole replacement operation is expensive, which adds to medical costs. We see that the advent of the lithium-iodide battery with its extended life represents a significant improvement for the patient.

The requirements of pacemaker circuit design are very stringent, as it can affect the life of the user. The system should work even if one or two cells should fail or develop high internal resistance. There should be some means of detecting battery failure. This is often done by causing the rate to decrease from 71 to 70 beats per minute as the battery voltage drops 5%.

ENGINEERING DESIGN CONSIDERATIONS

The Output Circuit

Even though cardiac pacemakers are now made up from integrated circuits, we must not lose sight of the fact that the circuits on the substrates are still transistor circuits. The output circuit delivers the stimulating pulse to the myocardium via the catheter electrode; it is the interface between the electrode and the power source. The parameters of this circuit determine the final waveform and energy delivered to the electrodes. Figure 7-4 is a general form of a transistor output circuit. The transistor acts as a switch which is closed for the duration of the pulse. The on/off characteristics of the output stage are controlled by the timing circuit. The output circuit should be ac-coupled to the electrode to prevent electrolytic action due to a constant dc current. Also, there is the need to isolate the electrode system from the timing circuit so that resistance changes in the electrodes

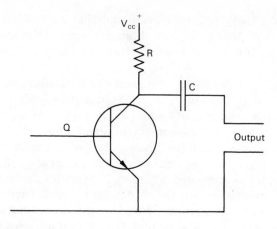

Figure 7-4   General form of a transistor output circuit.

do not affect the *RC* time constants of the timing circuit or interfere with a crystal oscillator if used in the timing circuits.

As an example of what can happen when a direct-coupled circuit is used in the output of a pacemaker, consider the circuit of Figure 7-5. Here the load resistance, R_1, is the average value for tissue electrical resistance, 500 Ω, and let R_s represent the source impedance. The load voltage and current are V_1 and I_1, respectively. The values for V_1 and I_1 are given by equations (7-1) and (7-2). The source voltage is given by *E*.

$$I_1 = \frac{E}{R_s + R_1} \tag{7-1}$$

$$V_1 = \frac{R_1 E}{R_s + R_1} \tag{7-2}$$

If the source impedance is low, $R_s = 0$ and the load voltage $V_1 = E$.

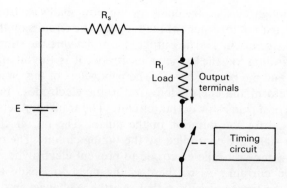

Figure 7-5   Direct-coupled output circuit for pacemaker.

When the timing circuit makes the transistor conduct, current flows through the transistor and the load. If the resistance of the transistor and load is low, the pacemaker acts as a constant-voltage-source pulse generator, with pulse voltage approximately equal to the battery voltage.

Circuit Drawbacks

Following are some common circuit disadvantages:

1. Current flow is unidirectional, which can lead to possible electrolytic damage to the electrode.
2. Transistor leakage current would flow through the heart and add to the polarization problem.
3. In the event of output transistor failure, usually a collector-to-emitter short, we would have the full battery voltage continuously across the heart, which might cause fibrillation.

Consider the circuit of Figure 7-6. The battery here is not part of the discharge circuit; thus changes in the battery internal resistance do not affect the output of the circuit. When the transistor is on, the capacitor discharges to ground reference through the heart. In this circuit, transistor failure does not cause a fibrillation problem to the patient, as it is isolated from the pacemaker leads by the capacitor. If the capacitor is "leaky," it will cause excessive battery drain and cause the pacemaker to fail prematurely.

Timing Circuits

Pacemaker timing circuits belong to a class of circuits known as *relaxation oscillators*. These oscillators operate at a frequency determined by an *RC* or *LR* circuit rather than the oscillation *LC* tank circuit used in communications circuits. The two major circuit types that are used are:

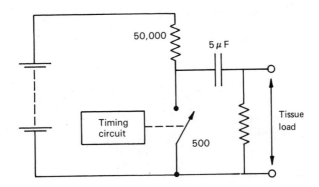

Figure 7-6 Capacitor discharge output circuit for a pacemaker.

1. The astable multivibrator
2. The blocking oscillator

Astable multivibrator. Most pacemaker timing circuits now use this type of circuit configuration. The advantages include very low power drain and the ability to be subminiaturized on thick- or thin-film integrated circuits. Figure 7-7 is a typical circuit configuration for an astable multivibrator. In this circuit, a *PNP-NPN* complementary pair are used for the oscillator. This allows for efficient use of the batteries, as there is power consumption only during the pacing pulse duration. The energy stored in the timing capacitors can be made small by using large values of R and small values of C for the RC time constant of the circuit. Another advantage of this circuit is that it can be made relatively invariant to battery voltage or it can be designed to act in a predictable manner, avoiding pacemaker runaway or sudden battery failure.

An operational description of the astable multivibrator is as follows. First, we assume that Q_1, Q_2, and Q_3 are biased off. Capacitor C_2 charges through resistor R_5. The rate of charging is determined by the RC time constant of that circuit. The rise in voltage at point A in Figure 7-7 forward-biases transistor Q_1, causing it to conduct, making point B low and C low relative to point A. This in turn drives transistor Q_2 on, making point D go positive. When D goes positive it biases Q_3 on. During this time, C_1

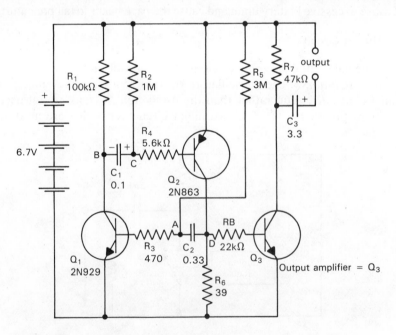

Figure 7-7 Multivibrator timing circuit. (Courtesy of Cordis Corp., Miami, Fla.)

is charging through R_4 until the voltage at point C has biased transistor Q_2 off. When Q_2 is off, Q_3 becomes reversed biased and is also turned off, which in turn causes Q_1 to stop conducting. This gets us back to the starting condition and the sequence begins all over again. Transistors Q_1 and Q_2 and their associated resistor and capacitors comprise the astable multivibrator circuit and transistor Q_3 is an output amplifier transistor which can supply power to the pacing electrodes.

Blocking Oscillator

Figure 7-8 illustrates a typical configuration for a blocking oscillator circuit. When power is initially applied to this circuit, the capacitor C is charged through the resistor R_1. Transistor Q_1 is turned on when the voltage on the capacitor forward-biases the transistor. The resulting emitter current flowing through resistor R_2 produces a base current for Q_2, making it conduct and amplifying the current from the base. Q_2 performs the same output function as Q_3 did for the astable multivibrator. The collector current in Q_1 also flows through the primary of the transformer. This induces a voltage in the secondary with a polarity that causes the capacitor C to discharge.

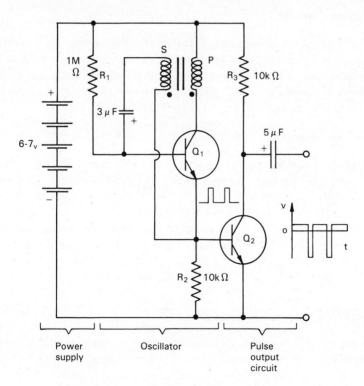

Figure 7-8 Diagram of a blocking oscillator.

When this happens, Q_1 becomes reversed biased and is turned off. At this point the cycle is repeated for the production of the pacing pulses.

The advantage of the blocking oscillator is the reduced number of components required in the circuit and the fact that no power-consuming voltage dividers are used. The disadvantages of the circuit are that it requires a relatively large transformer, which is heavy and expensive as components go, and the capacitor required in the timing circuit is large. Also, the frequency of oscillation is sensitive to the battery voltage in this circuit, and a runaway condition can be produced by a low battery condition.

Pulse Width

Using the minimum effective pulse width to accomplish ventricular stimulation lowers the current drain on the batteries, which increases the effective life of the implanted unit. A pulse width that is too narrow results in a significant increase of the stimulating threshold. The pulse duration of the pacemaker simulates the QRS interval and is usually set for a pulse width of 0.8 to 2 ms. Its duration and waveform, together with the current level, determine the level of cardiac stimulation provided to the patient.

Output Current

The amount of current to be applied as an artificial stimulus is controlled by a calibrated dial on an external pacemaker and is set or programmed into internal pacemakers. Current levels are usually set empirically, where the current level of the output pulse is increased gradually until the heart responds to the pacing pulses. This is termed *capture*.

Demand Threshold

Thresholds are commonly expressed in milliamperes assuming constant voltage and impedance. The stimulating threshold is defined as a current value of 20-ms duration at which ventricular capture is intermittently lost as the strength of the impulse is being reduced from the level of complete capture. The circuitry within a demand pacemaker usually senses the presence of an R wave and inhibits the output pulse and produces a pulse in the absence of an R wave.

PACEMAKER REFRACTORY PERIOD

Following the QRS complex, the ventricles are, for a short period, insensitive to another stimulus. This short period of insensitivity is called the *refractory period* and is the same type of phenomenon as that seen in excitable cells when they go through a depolarization. Following this portion of the cardiac cycle is the T wave and its associated vulnerable period. For this reason, a minimum refractory period is usually built into the pacemaker so that a pulse is not issued by the pacemaker during the vulnerable period of the T wave. This will also prevent the pacemaker from interpreting a large-amplitude T wave as a QRS complex and thus cause the timing circuit to be reset prematurely.

ELECTROMAGNETIC INTERFERENCE

Electromagnetic interference (EMI) is a concern for cardiac pacemaker wearers because, in large enough quantities, electromagnetic radiation may adversely affect the function of their pacemaker. Pacemakers operating in the demand mode are particularly susceptible. Asynchronous or fixed-rate pacemakers are generally not affected by EMI. Generally, it is safe to say that most appliances found in the hospital and at home will not interfere with demand pacemaker operation. Pacemakers are designed with input filters to reduce outside interference. When outside interference becomes too great, the pacemaker reverts to continuous pacing. Electrosurgical units, radio-frequency diathermy, and microwave ovens are likely candidates for concern for cardiac pacemaker wearers in the hospital.

The two dominant factors in EMI with pacemakers are the frequency of the radiated energy, which dictates antenna length for efficient capture of energy; and the shielding capability of the human body at the given frequency. The frequency at which most commercial microwave ovens operate is 2450 MHz. At this frequency, the antenna length necessary for efficient capture of available radiated energy is small, perhaps a few centimeters. The leads of a transistor or an integrated circuit in a pacemaker may make a good antenna. At distances beyond 1 ft, the electric field intensity from a microwave oven is usually not great enough to pose a problem for pacemaker users unless the microwave oven is defective.

PROGRAMMABLE PACEMAKERS

Like everything else in electronics today, medical electronics also makes use of microprocessors. In this case it is used in a line of programmable pacemakers and pacemaker programming devices. Because of the constant engineering advances in microcircuitry, pacemakers can be made and programmed to respond to any changing patient parameters both at the time of implantation and later as changes take place in the patient. The basic function of the pacemaker is as described previously except that instead of having different pacemakers to perform the varied functions desired by the physician, a more universal programmable unit designed to function in any of the previous modes is now possible within the same pacemaker unit.

An example of this new breed of multipurpose pacing system is shown in Figure 7-9. This type of pacemaker is a programmable, two-chamber pacemaker that has been developed for the treatment of sinus bradycardia, A-V block, or both. The pacemaker automatically shifts to the atrial-synchronized, ventricular-inhibited pacing mode when the atrial rate accelerates above the minimum pacing rate. The pacemaker can also be programmed to function in the ventricular-inhibited mode regardless of the existing atrial rhythm, or it can be externally overdriven for the temporary treatment of a cardiac arrhythmia.

One form of design provides for separate sensing and stimulating circuits for the atria and ventricles. Utilizing the corresponding programming device, the following pacemaker parameters can be adjusted while the pacemaker is inside the patient.

1. *Minimum rate* (ppm):   the set levels of 50, 60, 70, and 80 beats per minute, below which the pacer will not allow the heart rate to drop.

2. *Ventricular sensitivity* (mV):   input sensitivity, ventricular channel; the minimum amplitude of a rectangular voltage step test signal which the pacemaker will sense. Sensitivities ranging from 0.8 to 5.5 mV can be programmed into the pacemaker. The actual pacemaker sensitivity may vary with the slew rate of the sensed QRS complex.

3. *A-V delay* (ms):   electronic delay between sensed P wave or atrial output pulse and the pacer ventricular output pulse. Delays of 80, 120, 165, and 250 ms can be programmed into the pacemaker.

4. *Fallback rate* (ppm):   protection provided by the pacemaker design against atrial-synchronized ventricular pacing at undesirable rapid rates during an episode of atrial tachycardia. When a tachycardia is detected, the pacemaker can switch to the ventricular-inhibited mode and pace at a gradually decreasing rate. The rate of ventricular pacing can be programmed at rates of 55, 65, 75, and 85 ppm.

5. *Maximum rate* (ppm):   the atrial rate which will represent to the pacemaker that a tachycardia is present, causing the pacemaker to revert to the ventricular-inhibited mode. The rates that can be programmed into the pacemaker are 100, 130, 160, and 180 pulses per minute.

6. *Ventricular output duration* (ms):   the pulse width of the 5.5-mA ventricular output pulse. The programmable values are 0.5 to 2.0 ms in 0.5-ms increments.

7. *Atrial sensitivity* (mV):   minimum input sensitivity on the atrial channel for a rectangular voltage step test signal which the pacemaker will sense. This can be programmed for a value of "off" and 0.8, 1.5, and 7.0 mV. The pacemaker sensitivity may vary with the slew rate of the P wave.

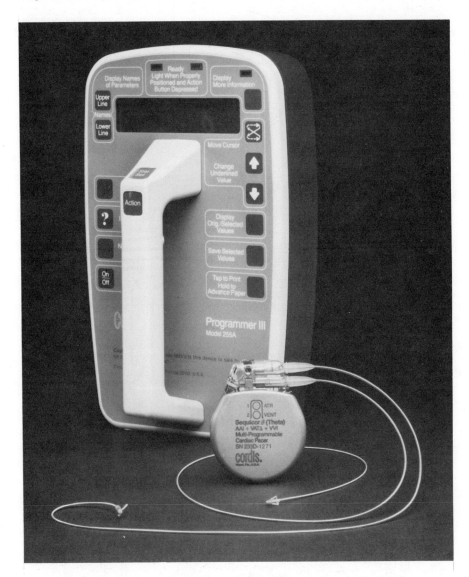

Figure 7-9 Modern programmable pacemaker system. (Courtesy of Cordis Corp., Miami, Fla.)

8. *Atrial output* (ms):   the pulse width of a 5.5-mA atrial output pulse. The programmable values are "off" and 0.5, 1.0, and 1.5 ms.

Figure 7-10 shows one form of pacemaker programming device. The programming device is electromagnetically coupled to the pacemaker, allowing the pacemaker to be programmed through the skin. The programmer emits

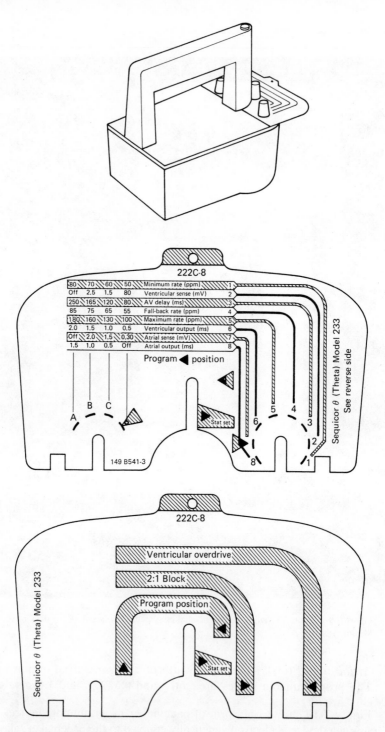

Figure 7-10 Two sides of the control panel used with Cordis Omnicor Programmer to program the parameters and mode of operation of Cordis Sequicor pacemakers. (Courtesy of Cordis Corp., Miami, Fla.)

a train of electromagnetic pulses which the pacemaker detects. The coded information received represents the program selection made by the user. In the Cordis model of pacemaker programmer the pacemaker's operating parameters can be programmed one step at a time, or a standard set of parameters can be selected and transmitted in one single transmission.

Programmable pacemakers are developed around a low-power drain microprocessor with associated input/output circuits, single-chip amplifiers, and CMOS logic. Asynchronous data may be put into the CPU from the sensing amplifiers, programming circuit, and low-battery indicator. Figure 7-11 is a simplified functional block diagram of a Cordis Sequicor programmable pacemaker circuit. The programmable microprocessor-controlled pacemaker represents the current status of pacemaker electronic engineering. Basically, all the pacemaker functions previously mentioned are embodied in the current line of programmable pacemakers from manufacturers. As more is learned about cardiac pacing, electronics, implantable materials,

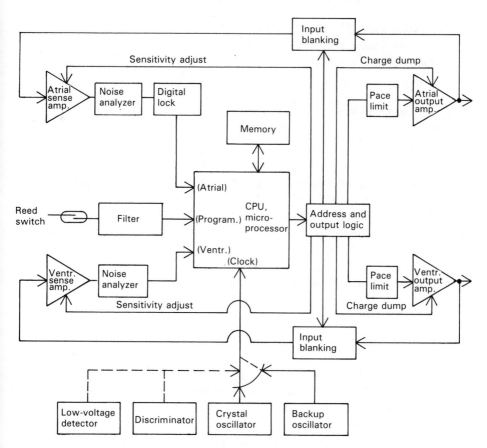

Figure 7-11 Simplified functional block diagram of Sequicor pacemaker circuitry.

batteries, and chemical sensors, we can expect to see further advances in this area of cardiac pacemaker systems.

INSPECTION AND PERFORMANCE TESTING OF CARDIAC PACEMAKERS

A number of commercial pacemaker testers are now available. Typically, these testers check the pacemaker output in milliamperes, pacemaker rate in beats per minute, ability to sense and synchronize with simulated QRS complexes, electrode integrity, and battery operation. These pacemaker testers are intended to test both external and implantable pacemakers prior to implantation.

If a specialized pacemaker test instrument is not available, complete testing of the pacemaker system can be specified by the clinical engineer using other forms of electronic test equipment. Following is a list of equipment necessary to test the performance of an external pacemaker:

1. Oscilloscope, storage type if available
2. Function generator
3. Ohmmeter
4. Pulse generator

Physical Condition

The physical condition of the external pacemaker should be checked in the following manner:

1. Note the overall cleanliness and physical condition of the pacemaker. Check for cracked case, missing screws, unreadable markings, terminal insulation integrity, and so on.
2. Note the condition of controls and indicators. Check for smooth operation and physical integrity. Check the pacing indicator, lamp, or meter. The pacemaker may need a minimum output amplitude to indicate; or if indication is with current, a 500-Ω resistor load should be connected to the output terminals of the pacemaker.
3. Check the battery, battery terminals, and battery test feature, if available.

Rate Accuracy

Connect a 500-Ω or 510-Ω off-the-shelf resistor to the output of the pacemaker. Set the pacemaker output to about 5mA in the continuous mode. Connect an oscilloscope to the 500-Ω load resistor and measure the actual delivered pacing rates with the pacemaker set at 60 and 120 beats per minute (BPM). The measured pacing rate should be within 5% of the dial indication.

Pulse Shape

Using the same setup, set the pacemaker rate at 60 BPM. Adjust the oscilloscope to fill most of the viewing area with the output pulse waveform. Record the waveform and the pulse width. Compare the waveform recorded with that given by the manufacturer of the pacemaker you are testing.

Amplitude Accuracy

Still using the same setup, set the pacemaker rate to about 60 BPM and the output amplitude to an accurate 1 mA with a 500-Ω load. (This would be 0.5 V with 500 Ω ± 1% load). Calculate the average amplitude (voltage or current). If the pacemaker produces an exponential waveform, use the same technique as if it were a trapezoidal waveform. Of course, this is an approximation, but it will aid in quick computation. Figure 7-12 illustrates the types of waveform outputs.

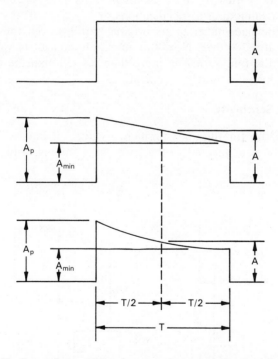

Figure 7-12 Pacemaker output pulses may be rectangular (top), trapezoidal (middle), or exponential (bottom) in shape. A, average amplitude; A_p, peak amplitude; A_{min}, minimum amplitude; T, pulse width. The average amplitude is defined as

$$A = \frac{A_p + A_{min}}{2}$$

This relationship is only approximate for the exponential decay.

The foregoing procedure should be repeated with the pacemaker amplitude set at 5 mA (2.5 V). Without changing any of the pacemaker controls, record the average amplitude with 100- and 1000-Ω resistors. If the pacemaker is designed to be a constant-current output device, the average readings should be 0.5 V and 5.0 V, respectively. If the pacemaker is a constant-voltage type, the voltage should remain at 2.5 V.

Interaction between Rate and Amplitude Adjustments

Interaction between rate and amplitude adjustments is a test to determine if the rate control has any effect on the output amplitude of the pacemaker and if the amplitude control has any effect on rate control. This is accomplished in the following way.

Put a 500-Ω load on the pacemaker and set the amplitude control at approximately 5 mA (2.5 V) at 60 BPM. Then vary the rate control over its full range while observing the output of the pacemaker on an oscilloscope. If the amplitude varies by more than 5% (0.13 V), the unit may not be meeting the manufacturer's specifications.

Return the pacemaker to its original settings. Set the oscilloscope sweep speed at 0.2 s/div. Now the amplitude control is varied over its full range. If the rate varies by more than 5% (50 ms), the unit may not be within the manufacturer's specifications.

Demand Sensitivity

A test setup for this measurement is given in Figure 7-13. Here a pulse generator is connected to the pacemaker through a high impedance

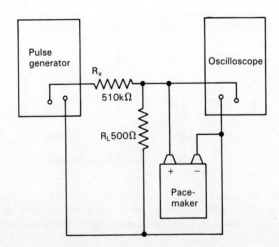

Figure 7-13 Demand sensitivity setup. Pulse output: 1.25 Hz = >40-ms pulse on Tektronix.

such as 510 kΩ resistance. The load resistance on the pacemaker should be 500 Ω. Adjust the pulse generator for a 45-ms rectangular pulse with a period of 800 ms. The 45-ms rectangular pulse is being used here to simulate a QRS complex. Although this is not an accurate simulation it is convenient to generate and serves the function of being a reproducible waveform.

Slowly increase the pulse generator output from zero until the pacemaker is completely inhibited. Recording the pulse generator amplitude now provides a record of the sensitivity of the unit to the simulated cardiac activity.

60-Hz Interference

Since 60-Hz coupling into the pacemaker is likely either in the hospital or at home, it is important to determine the pacemaker's ability to reject this form of interference. The basic test configuration as illustrated in Figure 7-13 is modified by the addition of a sine-wave generator coupled into the test circuit through a resistance network (see Figure 7-14). With the sine-wave generator connected across the 500-Ω load, set the pacemaker for demand or standby mode at approximately 5 mA and 60 BPM. The pulse generator is adjusted to produce a 45-ms pulse width at a rate slightly higher than the pacemaker and at an amplitude which causes the pacemaker to inhibit. The output of the sine-wave generator is now increased from zero to 250 mV peak to peak in order to see if the 60-Hz signal would cause the pacemaker to revert to asynchronous pacing as the 60-Hz signal amplitude is increased.

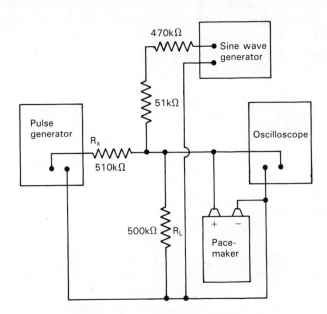

Figure 7-14   Interference test setup, ECG simulator.

The same test is then performed with the pulse generator turned off to determine whether the pacemaker will be inhibited or caused to operate in the asynchronous mode by the 60-Hz test signal.

Rate Hysteresis

Rate hysteresis in a pacemaker is the difference between the set rate of the pacemaker and the heart rate below which the pacemaker will begin to pace when in the ventricular inhibit mode. The test setup is the same as that in Figure 7-12. Using an oscilloscope, set the pacemaker rate at exactly 60 BPM. The output of the pulse generator is increased to the point where the pacemaker is inhibited. Using the oscilloscope to observe, disconnect the pulse generator at the beginning of a generator pulse from the 500-Ω load. Note the time from the beginning of the last generator pulse to the beginning of the first pacemaker pulse. This quantity is known as the *escape interval* and it is used to obtain the rate hysteresis from the chart presented in Figure 7-15.

Refractory Period Determination

The pacemaker refractory period is the short period of time following a pacing pulse or a sensed beat from the heart during which the pacemaker will not sense any input signal. Using the same test configuration as before (Figure 7-14), set the pacemaker at an accurate 60 BPM. Using a 45-ms pulse width and a period of 2 s as the output of the pulse generator, measure the pulse generator amplitude until the pacemaker is experiencing inhibition of some pulses. If the period of the input pulses is decreased, the number of pacing pulses issued by the pacemaker will decrease to the point where the pacemaker will become erratic.

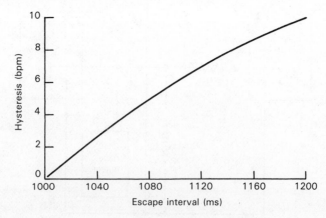

Figure 7-15 Rate hysteresis can be determined from the measurement of escape interval.

When this point is reached, the pacemaker is turned off and the period of the pulse generator can be measured on the oscilloscope. The refractory period is then calculated by subtracting the escape interval obtained previously from the period of the pulse generator pulses. Thus

refractory period = pulse generator period − escape interval

PROBLEMS

1. Define the following terms:
   (a) Bradycardia
   (b) Tachycardia
   (c) First-degree heart block
   (d) Third-degree heart block

2. How is a pacemaker used as part of the method of treatment for the medical conditions in Problem 1?

3. What is meant by the terms *capture, threshold,* and *sensitivity* as applied to a cardiac pacemaker?

4. Explain the difference between the bipolar and unipolar electrode systems and the pros and cons of each type.

5. Draw a block diagram and describe the functional operation of the following pacemaker types:
   (a) Asynchronous pacemaker
   (b) Demand pacemaker
   (c) External pacemaker
   (d) Atrial programmed pacemaker

6. Explain why the wearers of implanted pacemakers should avoid obvious sources of electromagnetic interference (EMI).

7. Describe, and use any circuit equations to demonstrate your reasoning, what, if any, changes in battery voltage and associated internal battery resistance of a pacemaker power source will have on both the blocking oscillator and multivibrator types of pacemakers when the batteries are near exhaustion. Explain how these changes might be used to indicate impending pacemaker failure so that the user or physician might be alerted to this condition.

8. Fibrotic growth around the cardiac leads of a pacemaker can represent a change in the load to the output stage of the system. How does that occurrence affect the output characteristics of an astable multivibrator pacemaker?

9. The leads of a pacemaker are attached to the heart. The output of the pacemaker is 10 V into a tissue resistance of 250 Ω. If the pacemaker is set to beat at 70 BPM, calculate the energy delivered in one year to the cardiac muscle. Assume that the resistive load presented to the pacemaker remains constant and that the pulse duration is 3 ms.

10. A lead I ECG is shown in Figure 7-16. Describe what you think is the problem based on this ECG waveform and relate this back to the mechanical and electrical portions of the cardiac cycle.

11. In the testing of the sensing function of a pacemaker by clinical engineering,

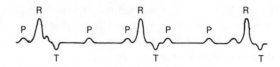

P - P interval = 0.0132 sec;   R - R interval = 0.033 sec;

Figure 7-16   P-P interval = 0.0132 s; R-R interval = 0.033 s.

a 45-ms rectangular pulse can be used to simulate an intracardiac QRS complex and cause the pacemaker to sense. This signal is not equivalent to the action potential appearing at the catheter tip. The pacemaker response could conceivably be different from that obtained in actual use. Why is such a pulse used in testing procedures? What could be used if improved circuit designs avoid sensing pulses other than those closely resembling intracardiac action potentials?

12. An engineer with an atrial synchronous pacemaker enters a high-energy physics laboratory, where he is subjected to a *B* field of strength 4.0*T*. If the pacemaker wires form a loop of area 5 cm^2, what voltage is induced in the wires assuming that the *B* field is oriented so as to do the most harm? Would this be considered a hazardous situation for this engineer? Why or why not? Explain your answer.

REFERENCES

1. Furman, S., and D. J. Escher, *Principles of Cardiac Pacing* (New York: Harper & Row, 1970).

2. Gill, Carl, William Jakobi, Thomas Morton, and Andrew Wechsler, *The Cardiovascular System as It Relates to Heart Pacing* (Minneapolis, Minn.: Medtronic, Inc., 1975).

3. Roy, O. Z., "The current status of cardiac pacing," *Crit. Rev. Bioeng.,* 1974.

4. Samet, P., *Cardiac Pacing* (New York: Grune & Stratton, 1973).

5. Schaldach, M., and S. Furman (eds.), *Advances in Pacemaker Technology* (New York: Springer-Verlag, 1975).

8

THE RESPIRATORY SYSTEM AND THE MEASUREMENT OF PULMONARY MECHANICS

THE RESPIRATORY SYSTEM

The exchange of gases in any biological system is termed *respiration*. To sustain life the human body must take in oxygen, which is utilized on the cellular level with other essential nutrients in the metabolic oxidation process. As part of this process, CO_2 is given off as a by-product of cellular metabolism. The hemoglobin in the blood is the dominant transport mechanism by which O_2 is brought to the cells, and the CO_2 produced by the cells is dissolved in the blood plasma and carried off for disposal to the atmosphere via the lungs. Thus the lungs perform two basic functions, that of moving bulk gases into and out of the lung tissue and as a location where gas exchange can easily take place between the air and blood at this tissue interface.

The primary function of the lungs is to oxygenate blood and to eliminate carbon dioxide in a controlled manner. During inspiration, fresh air enters the respiratory tract, becomes humidified and heated to body temperature, and is mixed with the gases already present in the trachea and bronchi. This mixture is then further mixed with the gases in the alveoli, the tissue in which pulmonary gas exchange takes place. Oxygen diffuses from the alveoli into the pulmonary blood supply while carbon dioxide diffuses in the reverse direction. Both CO_2 and O_2 are driven by diffusion gradients caused by concentration differences across the alveolar membrane. Oxygenated blood is then returned to the heart, which circulates it to the remainder of the body. The processes of bulk gas transport into and out of the lungs and the diffusion of the gases across the alveolar membrane is known as *pulmonary function*. Tests performed to determine parameters of system efficiency are called pulmonary function tests. We will first look at the anatomy and physiology of the respiratory system and then discuss some of these measurements.

ELEMENTS OF RESPIRATORY PHYSIOLOGY

Air enters the respiratory system via the mouth, nose, or both openings. From there it passes through the pharynx and larynx, then into the trachea. The trachea is a cartilagenous tube about 1.5 to 2.5 cm in diameter and approximately 11 cm long, extending from the larynx to the upper boundary of the thoracic cavity. The air then passes into the left and right main stem bronchi. The bronchi arborize for many generations until they are very small passages called *bronchioles*. The bronchioles then enter small "air sacs" called *alveoli*, where gas exchange takes place (see Figure 8-1). It is estimated that there are about 300 million alveoli in the lungs, giving rise to a total gas exchange surface area of about 80 square meters. The lungs are covered by a covering called the *visceral pleura*. There is a similar smooth, slippery lining on the inner surface of the chest wall called the *parietal pleura*. The lungs adhere to the parietal pleura due to liquid interface between them, in the same manner that two wet plane surfaces stick together due to atmospheric pressure. The liquid interface provides a mechanism of adherence of the lungs to the inside of the thoracic cavity and at the same time provides lubrication for one to slide over the other when there is a difference in relative velocity between the two surfaces during the respiratory cycle.

Inspiration and expiration are accomplished by musculature that literally changes the volume of the thoracic cavity, and in so doing creates negative and positive pressures which move air into and out of the lungs. The lungs themselves play a passive role in respiratory movements. The muscles involved in performing this function are the diaphragm and the internal and external intercostal muscles. On inspiration the length and diameter of the thorax is increased. This increase in diameter is caused by elevation of the ribs. The resultant increase in volume creates a negative pressure in the thorax. Since the thorax is a closed chamber and the only opening to the outside is via the conducting airways in the lungs, the negative pressure produced causes air from the outside at atmospheric pressure to respond to the negative pressure and flow into the lungs.

The length of the thoracic cavity is determined by the diaphragm. At rest the diaphragm is dome-shaped, convex upward, but with inspiration the diaphragm contracts and flattens out by moving downward. This has the effect of increasing thoracic length. Expiration is essentially passive

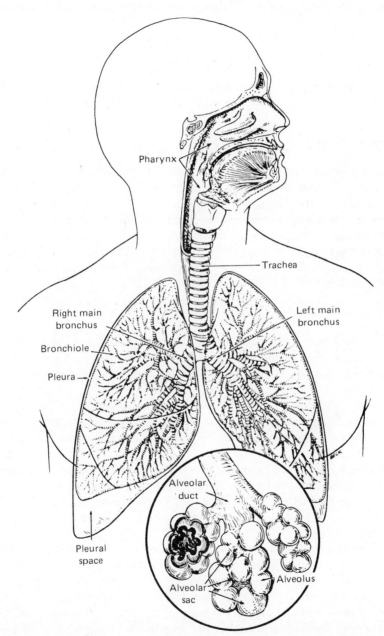

Figure 8-1 Cutaway view of the respiratory system showing anatomical location of the conducting airway and pleural surfaces. (Reproduced by permission from Anthony, Catherine Parker, and Thibodeau, Gary A.: *Textbook of Anatomy & Physiology*, ed. 11, St. Louis, 1983, The C. V. Mosby Co.)

during quiet breathing and is due to the elastic properties of the lungs. During inspiration the elastic tissue in the lungs is stretched. When inspiration ceases, the energy stored in the elastic fibers of the lungs is used to retract the system to its initial length. This act creates a positive pressure in the lungs and creates a pressure gradient which forces gas out of the lungs. During normal inspiration the pressure inside the lungs, intra-alveolar pressure, is about -3 mmHg, whereas it is about $+3$ mmHg during expiration.

RESISTANCE AND COMPLIANCE OF THE LUNGS

The lungs inside the thoracic cavity can be thought of as an elastic sac where lung volume depends on the pressure difference between the inside and outside of the lungs. If the lungs are inflated in stages, there is a relationship between internal volume change and the driving pressure required to produce the volume change in the lungs, called *lung tissue compliance.* Mathematically, if the change in pressure is ΔP and the corresponding change in lung volume is ΔV, the compliance of the lungs, C, is given by the expression

$$C = \frac{\Delta V}{\Delta P} \qquad (8\text{-}1)$$

Thus compliance is the ratio of the change in volume of the lungs to the change in pressure producing the volume change. This defines static compliance and can be thought of as a gross measure of distensibility of the lung tissue.

Since both the lungs and the thorax are acted upon by opposing elastic forces, each has a "natural volume" at which the structure is neither stretched nor compressed. This is the *relaxation volume* of the structure. The relaxation volume of the thorax is larger than the relaxation volume of the lungs. Since the lungs and thorax are coupled together at the pleural surface, forces are set up due to the difference in the relaxation volumes of the lungs and thorax. The forces are in opposition to one another, and the equilibrium position obtained determines the resting or natural size of the chest cavity.

During respiration the thorax moves up and down in a pump-handle fashion. This motion pulls on the surface of the lungs, causing it to move with the ribs, with its associated change in volume. This is the coupling mechanism by which respiratory muscle forces are transmitted to the lungs in the form of pressure changes at the pleural surface. The increase in lung volume with inspiration stretches the lung tissue elastic fibers, causing the lungs to attempt to pull away from the thoracic wall. This decreases the pleural pressure, producing an increase in lung volume.

The decreased pressure in the alveolar region causes air to rush into the lungs. The rate at which air enters the conducting airways, to a first approximation, depends on the pressure drop from the mouth to the alveoli (trans-airway pressure) and on the resistance to flow in the passageways. Expiration begins when the inspiratory muscles relax, allowing the elastic fibers in the lungs to shorten. The lungs continue to shrink until their own elastic forces are counterbalanced by those of the thorax. Airflow is greatest in the early phase of expiration, since elastic tension in the lungs is greatest and the compression of the thorax is least at this time. The relaxation volume of the lungs alone is approximately 1 liter and the relaxation volume of the thorax is approximately 5 liters. When the lungs are in the thorax, the equilibrium volume that results from the balanced forces between these two is 2.5 liters for normals. At this point, the elastic forces of the lungs and the thorax are equal and opposite each other.

The compliance of each of these structures, lungs and thoracic cage, act in series with each other. Thus the compliance of the total system can be computed on the same basis as capacitors in series. A simple mathematical relationship for total pulmonary compliance is given by the equation

$$\frac{1}{C_{total}} = \frac{1}{C_{lungs}} + \frac{1}{C_{thoracic\ cage}} \tag{8-2}$$

So if the compliance of the lungs is 0.2 liter/cmH$_2$O and the compliance of the thoracic cage is 0.2 liter/cmH$_2$O, from equation (8-2) the total compliance of the system is 0.1 liter/cmH$_2$O.

Besides the mechanical parameter of compliance, there are other parameters that influence the bulk movement of air into and out of the airways of the lungs. The volume flow of air ($\dot{V}$) in inspiration and expiration depends on the resistance to airflow and the pressure difference between the atmospheric pressure at the mouth and the alveolar pressure. If we consider atmospheric pressure as zero, the alveolar pressure swings positive and negative about this point and alveolar pressure then becomes the driving pressure for the system. Resistance represents the loss of energy in the system of conducting airways. Thus the higher the airway resistance to the bulk flow of air, the more energy it takes to move that air into and out of the lungs.

As a first approximation in characterizing airflow, we assume that flow is laminar in nature. This provides us with a simple "Ohm's law" type of relation between pressure and flow as given by

$$\text{resistance} = \frac{\text{driving pressure}}{\text{airflow}} \tag{8-3}$$

This relationship is based on laminar flow through a tube, which also depends on a number of factors, such as the viscosity, equivalent radius, and length of the conducting airways. Equation (8-3) holds only for laminar or streamline flow. In the bronchial tree, some turbulent flow occurs particularly at points of bifurcation. For our purposes here, we do not need to improve over the simple model of equation (8-3), as it is adequate for present medical purposes. Therefore, at the present time we neglect the effect of turbulence and impedance mismatch in the conducting airways.

PULMONARY FUNCTION TESTS AND MEASUREMENTS

The assessment of pulmonary function can be broken down into two major classes. The first is the evaluation of the purely mechanical aspects of pulmonary function, which affects the bulk gas transport into and out of the lungs. The second is the evaluation of gas exchange or diffusion at the alveoli. We will take up each of these areas in turn.

The ability of the pulmonary system to move air and exchange oxygen and carbon dioxide is affected by the various components of the air passages, the diaphragm, and the rib cage and its associated muscles, and by the characteristics of the lung tissue itself. Tests have been devised to measure each of these parameters separately but no single test has yet been devised to measure the total performance of the pulmonary system. In this section several tests which are in current use to assess pulmonary function will be described.

Among the basic tests performed are those to determine the volumes and capacities of the respiratory system (see Figure 8-2). The components of lung volume are calculated in the following manner. We first determine a reference level, chosen to be at the end of a passive expiration. Here the respiratory muscle activity is at a minimum. The volume of the lungs at this time is determined primarily by the elastic forces in the lung tissue and the thorax. This lung volume, the *resting expiratory level,* is a convenient reference point from which changes in lung volume can be measured. From this level it is possible to inspire or expire at various depths. There are four lung volumes, which are given below together with their "normal" value and percent of the total lung volume. These volumes are defined as follows:

A. Lung Volumes
1. *Tidal volume* (TV): the volume of air inspired and expired with each normal breath (600 ml − 10%)
2. *Inspiratory reserve volume* (IRV): the maximal volume of gas that can be inspired beyond the end of a normal tidal volume (3000 ml − 50%)
3. *Expiratory reserve volume* (ERV): the volume of gas expired by forceful expiration after the end of a normal tidal volume (1200 ml − 20%)
4. *Residual volume* (RV): the volume of air remaining in the lungs after a maximal expiration (1200 ml − 20%)

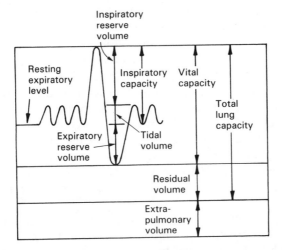

Figure 8-2 Relationship between lung volume and capacities.

There are four capacities, each of which includes two or more of the primary volumes. These lung capacities are given as follows:

B. Capacities
   1. *Vital capacity* (VC): maximum volume of air that can be expelled from the lungs by forceful effort after inspiration (4800 ml − 80%)
$$VC = IRV + ERV + TV$$
   2. *Total lung capacity* (TLC): the amount of air contained in the lungs after maximal inspiration (6000 ml − 100%)
$$TLC = VC + RV$$
   3. *Inspiratory capacity* (IC): the maximum amount of air that can be inspired after reaching the end expiratory level (3600 ml − 60%)
$$IC = TV + IRV$$
   4. *Functional residual capacity* (FRC): the volume of air remaining after a normal expiration (2400 ml − 40%)
$$FRC = ERV + RV$$

All pulmonary volumes and capacities are about 20 to 25% less in females than in males.

In addition to the static volumes and capacities already given, several dynamic measures are used to assess pulmonary mechanics. These measures are important because breathing is a dynamic process and the rate at which gases can be exchanged within the lungs is a direct function of the rate at which air can be transported.

A measure of the overall output of the respiratory system is the *respiratory minute volume* (RMV). This is the measure of the amount of air inspired during 1 minute at rest. It is calculated by taking the product of the tidal volume and the respiratory frequency.

A number of forced breathing tests are also used to assess the muscle power associated with breathing and the resistance of the airway. Among them is the *forced vital capacity* (FVC), which is nothing more than a vital capacity maneuver performed as quickly as possible. By definition, the FVC is the total amount of air that can be forcibly expired as quickly as possible after taking the deepest possible breath. If the measurement is made with respect to the time required for the maneuver, it is called a *timed vital capacity*. A measure of the maximum amount of gas that can be expelled in a given number of seconds is called the *forced expiratory volume*.

A resting person inhales about $\frac{1}{2}$ liter of air with each breath and takes between 12 and 18 breaths per minute. With vigorous exercise, this volume may increase 8 to 10 times and the rate may reach 40 breaths per minute. A respiratory disease may be suspected if these volumes, rates, or capacities are not "normal."

Respiratory *dead space* is the space or volume of the respiratory system in which no gas exchange occurs. This would include the nasal passages, pharynx, trachea, and bronchi. Physiologically, the dead space would include the volume of any inactive alveoli. In a healthy young male the dead space equals about 150 ml. During illness, the physiological dead space may be as high as 1 to 2 liters, due to decreased air permeability or decreased blood circulation.

MEASUREMENT OF LUNG CAPACITIES AND VOLUMES

The instrument used to measure lung capacity and volumes is called a *spirometer*. The record obtained from this device is called a *spirogram*. Figure 8-3 is an illustration of a water filled spirometer. It is a simple mechanical device consisting of a counterweighted bell canister inverted over a tank of water with a pipe for carrying expired air inside of and to the surface of the water in the spirometer. The spirometer is a mechanical integrator, since the input is airflow and the output is volume displacement. An electrical signal proportional to volume displacement can be obtained by using a linear potentiometer connected to the pulley portion of the spirometer (see Figure 8-3). The spirometer is a heavily damped device with appreciable hysteresis, so that small changes in inspired and expired air volumes are not recorded well with this device. A variation of this device is the Collins spirometer. It records the motion of the floating bell directly by the use of a pen attached to the counterweight which records its motion on a rotating drum with calibrated graph paper taped to it.

The spirometer is used to obtain measures of the mechanical properties

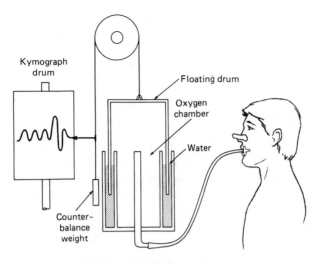

Figure 8-3   Water-filled spirometer.

of the lungs in conjunction with the thorax, since they are a coupled system. Tests made using the spirometer are not completely objective because they require the cooperation of the patient and the coaching efforts of a good respiratory technician. Also, these tests are not analytical because they measure a number of the properties of the lungs and thorax simultaneously.

Spirometric Tests

The assessment of the lungs' ability to act as a mechanical "pump" for air and the ability of the air to flow with minimum impedance through the conducting airways is grossly estimated using spirometric testing. These tests are classified into two major groups: single-breath tests and multiple-breath tests. Under the category of single-breath tests are three types: (1) tests that measure expired volume only, (2) tests that measure expired volume in a unit time, and (3) tests that measure expired volume/time.

Forced vital capacity (FVC):   a maneuver that requires that the patient inspire maximally and then breathe out through the spirometer. The patient expires air as forcefully and as rapidly as possible until he or she cannot move any more air. The spirometer records the total volume of expired gas from the patient.

Forced expiratory volume (FEV_t):   the volume of forced expiratory gas that is measured over a particular time or time interval while performing a vital-capacity maneuver. A normal person can expire 83% of his or her vital capacity in 1 s, 94% in 2 s, and 97% in 3 s. Thus, from an engineering point of view, this is a measure of the patient's pulmonary time constant.

Forced expiratory flow ($FEF_{200-1200}$):   the average rate of flow for a specific portion of the forced expiratory volume curve that is between 200 and 1200 ml of volume displaced. The slope of a straight line between these two points on the FEV spirogram represents the average expiratory flow in liters per second.

Maximal midexpiratory flow [(MMF) − ($FEF_{25-75\%}$)]:   the average rate of flow during the middle half of the FEV maneuver.

In the class of multiple-breath test maneuvers is *maximal voluntary ventilation* (MVV). Here the patient breathes in and out for 15 s as hard and as fast as he or she can. The spirometer records the total volume of gas moved by the lungs. The value is multiplied by 4 to produce the maximum volume that the patient breathed per minute by voluntary effort.

These tests are always performed in conjunction with a spirometer. Although the water-filled spirometer is very common in most pulmonary function laboratories, electronic spirometers are being used in conjunction with electronic integration of the flow. Such a system uses a pneumotachograph consisting of a fine mesh screen in a 1-in.-diameter tube with the airflow directed so that a small pressure differential develops across the screen. This pressure difference is detected by a differential pressure transducer and is used as a measure of airflow through it (see Figure 8-4). The resulting record of the flow rate is useful in itself, or it may be used as a spirogram. The circuit shown in Figure 8-5 can be used to process the transducer signal to give flow and volume measurements. A microprocessor may be used to calculate the parameters from the data provided by the pneumotachograph and associated electronics.

The previous tests cover all the capacities and volumes which are measured except *residual volume*. This is the volume of gas that remains in the lungs after a maximal expiration. To determine residual volume, a simple technique called the *nitrogen washout method* is used. The technique is based on the premise that the lung volume is unknown, but since the patient is breathing air, the gas in the lungs, whatever its volume, must be 78% nitrogen, N_2. From this fact we deduce that if we knew how much N_2 were in the lungs, the total volume of alveolar gas could be calculated. We determine the amount of N_2 in the lungs by washing it out by having

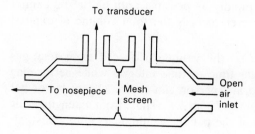

To transducer

To nosepiece

Mesh screen

Open air inlet

Figure 8-4 Schematic of pneumotachograph.

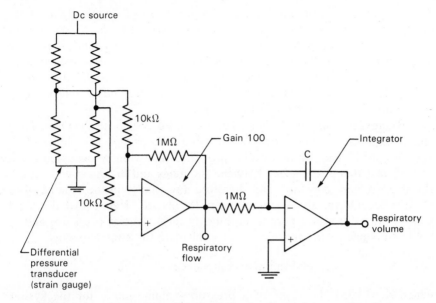

Figure 8-5 Operational amplifier circuit for the electronic processing of pneumotachograph signals for flow and volume.

the patient breathe nitrogen-free oxygen and collect expired gas in a spirometer that has been previously purged of nitrogen. The collected gas in the spirometer allows us to measure the volume of expired gas and the concentration of N_2. At the beginning of the test, the spirometer is void of N_2, and any N_2 in the system is in the lungs. At the end of the test the situation is just the opposite.

Consider that 35,000 ml of expired gas is obtained that has a measured N_2 concentration of 6%. Thus the spirometer contains

$$0.06 \times 35,000 = 2100 \text{ ml } N_2$$

all of which came from the lungs. Since 2100 ml of N_2 represents 78% of the volume of gas within the lungs, the total alveolar gas volume in the lungs at the time when the N_2 washout began is the residual volume and is given by the relation

$$2100 \times \frac{100}{78} = 2692 \text{ ml}$$

A correction is sometimes made for the effect of nitrogen diffusing from the blood and tissues due to a lower partial pressure of N_2.

There are also other parameters which are useful in diagnosing pulmonary dysfunction, such as lung compliance and airway resistance. The calculation of these parameters requires the measurement of thoracic pressure. This measurement must be made indirectly, as it is not obtainable directly.

The lungs and the thorax can be represented schematically as shown in Figure 8-6. If we let P_0 equal atmospheric pressure, the pressure inside the lungs can be computed as $P_0 - \Delta P_t$, where ΔP_t is the change in pressure due to a change in the volume of the thoracic cavity. It is seen that the transthoracic pressure, ΔP_t, is the pressure across the lungs plus the pressure across the chest wall. This is the alveolar pressure and its indirect measurement is discussed later in this chapter.

The expansibility of the lung/thorax-coupled system is called *compliance* and is expressed, as defined previously, as the change in volume of the system per unit change in alveolar pressure. The volume occupied by the *interpleural space*, the space between the lungs and the chest wall, can be shown to remain constant, and therefore any change in the volume of the chest is equal to the change in volume of the alveoli. Hence if we determine the pressure difference across the lung system, alveolar pressure, and the volume change of the thorax, we can calculate the compliance as

$$\text{total thoracic compliance} = \frac{dV_{\text{chest}}}{dP_{\text{alv}}}$$

which represents the slope of a pressure–volume curve for the system. How this curve is obtained and how alveolar pressure is indirectly measured is the next topic of discussion.

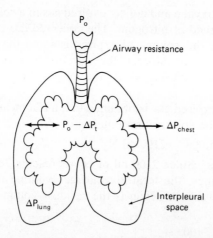

Figure 8-6 Schematic representation of lungs and thorax.

MEASUREMENT OF VOLUME OF THORACIC GAS

Because alveolar pressure cannot be measured directly, an indirect method was developed by Dubois [2] using a device known as a *total body plethysmograph*. This device consists of an airtight chamber of known volume, which is large enough to accommodate an adult patient. Since a person seated inside the plethysmograph is in a closed system, one would expect that inspiration of a quantity of air from the "box" into the patient's lungs would produce no pressure fluctuation in the box, neglecting any thermal and humidification effects. To the contrary, the pressure in the plethysmograph rises during inspiration. The reason for this is that gas flows when a pressure gradient exists. At the beginning of inspiration, muscular action has enlarged the thorax and lowered alveolar pressure below atmospheric pressure. The alveolar gas, which was previously at atmospheric pressure, is now at a subatmospheric pressure. The inspired gas expands to fill the new volume of the thorax, thus adding an increment of gas volume to the closed plethysmograph. The pressure change within the plethysmograph is registered by a pressure-sensitive transducer. The reverse process takes place during expiration, when alveolar gas is compressed. From the measurement of pressure changes within the plethysmograph, alveolar pressure can be approximated for any moment in the respiratory cycle.

In order to compute alveolar pressure using the plethysmograph pressure response to thoracic volume changes, the volume of thoracic gas, VTG, must be known. Consider the following derivation for VTG. A patient sits inside the plethysmograph (see Figure 8-7) with a pneumotachograph flow meter in his or her mouth. Both mouth and "box" pressures are displayed on the x and y axes of an oscilloscope for one breath cycle against a total mechanical obstruction to airflow. This is accomplished using a solenoid-operated shutter across the pneumotachograph. The theory here is that when a patient breathes against a closed shutter, the airflow is zero, and thus the pressure measured at the mouth is the same as the alveolar pressure. Let

ΔP = change in pressure of gas in the lungs

ΔV = change in volume of gas in the lungs

From Boyle's law we have, for constant temperature,

$$P_1 V_1 = P_2 V_2 \tag{8-4}$$

$$P_2 = P_1 + \Delta P$$

$$V_2 = V_1 + \Delta V$$

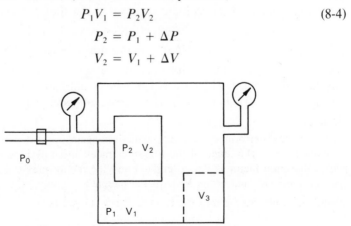

Figure 8-7   Schematic of total body plethysmograph.

Substituting into equation (8-4), we have

$$P_1 V_1 = (P_1 + \Delta P)(V_1 + \Delta V)$$

$$P_1 V_1 = P_1 V_1 + P_1 \Delta V + V_1 \Delta P + \Delta P \Delta V$$

Rewriting this equation, we have

$$P_1 \Delta V + V_1 \Delta P + \Delta P \Delta V = 0$$

or

$$V_1 = -\frac{\Delta V}{\Delta P}(P_1 + \Delta P)$$

Now, if ΔP is very small compared to P_1,

$$P_1 + \Delta P \cong P_1$$

Thus we have that

$$V_1 = -\frac{\Delta V}{\Delta P} P_1 \qquad (8\text{-}5)$$

where V_1 = volume of thoracic gas (VTG)

P_1 = atmospheric pressure at sea level of dry air (970 cmH_2O)

We can compute VTG from the output from the plethysmograph as follows:

$$\text{VTG} = -970 \frac{\Delta V}{\Delta P} \text{ or } 970 \frac{\Delta P}{\Delta V} \qquad \text{(using the negative reciprocal)}$$

Rewriting this equation and inserting the pressure transducer calibration factors for the mouth and box pressure, we have

$$\text{VTG} = \frac{970}{\Delta P \times \text{mouth pressure calibration}/\Delta P^* \times \text{box pressure calibration}}$$

where ΔP^* is the pressure at the mouth with the mechanical shutter closed and obstructing airflow. Pressure changes in the plethysmograph ("box") are proportional to the volume changes in the thorax. Therefore,

$$\text{VTG} = \frac{970(\text{box pressure calibration})}{(\Delta P/\Delta P^*)(\text{mouth pressure calibration})}$$

The term $\Delta P/\Delta P^*$ in the formula above is the slope of the oscilloscope display with mouth pressure (P) on the Y axis and "box" pressure (P^*) on the X axis. To illustrate this process, consider the following example.

Example

A patient has performed a panting maneuver against a closed shutter on a pneumotachograph. The angle of the pressure traces on an oscilloscope is 30°. If the box calibration factor is 9.6 cc/cmH_2O and the mouth pressure calibration factor is 4 cc/cmH_2O, what is the VTG of the patient?

Solution.   Substituting into the formula for VTG, we have

$$\text{VTG} = \frac{970 \times 9.6}{\tan 30° \times 4} = 4032.21 \text{ cc} = 4.032 \text{ liters}$$

MEASUREMENT OF AIRWAY RESISTANCE

The conducting airway system of the lungs is an arborizing labyrinth of tubes whose diameters and lengths change with each generation of arborization. Flow characteristics of a gas flowing through such a system have been only partially modeled yet still require a complex mathematical model using the partial differential equations of fluid flow to describe them. To be of value to clinicians for use in patient assessment, a simpler model is required. Thus a simple mathematical model is used instead of a more complex one to characterize a pressure–flow relationship for the lungs. What is done, as a first approximation, is to attribute all forms of energy dissipation within the system of conducting airways to frictional loss and assume that the gas flow is Poiseuillean. Thus energy loss is due to laminar flow resistance R. A simple "Ohm's law" model can then be used to relate pressure to flow in the measurement of airway resistance as given by

$$R = \frac{P}{\dot{V}} = \frac{P_{atm} - P_{alv}}{\dot{V}} \tag{8-6}$$

Airway resistance is measured under dynamic conditions, that is, during a panting maneuver performed by the patient while inside a body plethysmograph. At this time box pressure and airflow at the mouth are measured and displayed on an oscilloscope with flow on the Y axis and box pressure on the X axis. If atmospheric pressure is considered to be a reference "zero," the ratio of alveolar pressure (P_{alv}) to airflow at the mouth ($\dot{V}_{ao}$) can be used to characterize the airway resistance for a particular patient. Instantaneous airflow is easily measured using a pneumotachograph placed in the patient's mouth, and pressure proportional to alveolar pressure is measured during flow with the total body plethysmograph.

In the earlier section on VTG, a relationship between box pressure and mouth pressure (assumed to be equal to alveolar pressure under a no-airflow condition) was given such that we can express alveolar pressure in terms of box pressure. We can make use of this in the development of a formula for calculating airway resistance. In computing VTG we have the information that

$$\tan \phi = \tan \phi_1 = \frac{\Delta P}{\Delta P^*} = \frac{\Delta P_{alv}}{\Delta P_{box}} \tag{8-7}$$

but

$$R = \frac{\dot{V}}{P_{alv}} = \frac{\Delta P_{alv}/\Delta P_{box}}{\Delta V/\Delta P_{box}} \tag{8-8}$$

The relationship between box pressure and airflow at the mouth is nonlinear, necessitating standardization of the method of finding the slope of the pressure–flow curve. The measurement of the slope of the pressure–flow trace from an oscilloscope has been standardized by drawing a line at ± 0.5 liter/s of flow. The slope of this line ($\tan \phi_2$) is used to compute airway resistance (see Figure 8-8). Since

$$\tan \phi_2 = \frac{\Delta \dot{V}}{\Delta P_{\text{box}}}$$

then airway resistance is given by

$$R = \frac{\tan \phi_1}{\tan \phi_2} \tag{8-9}$$

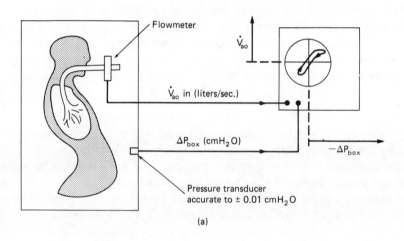

(a)

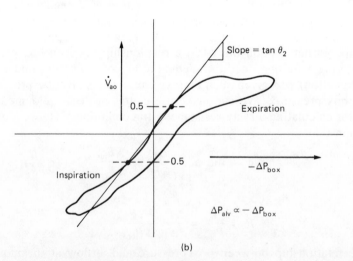

(b)

Figure 8-8 Total body plethysmograph (a) and pressure-flow loop (b) used to calculate airway resistance.

Since this will give the total resistance of the flow system, the resistance of the apparatus must be subtracted to provide the airway resistance, R_{aw}. Thus the clinical formula arrived at is given by

$$R_{aw} = \frac{\tan \phi_1}{\tan \phi_2} - R_{app} \qquad (8\text{-}10)$$

where R_{app} is the resistance of the flowmeter used in the measurement system. At present this is the standard clinical method for computing airway resistance in hospitals that use a total body plethysmograph. In other hospitals, pulmonary function testing is reduced mainly to the use of spirometry.

At this point, most of the important pulmonary function measurements have been discussed. These measurements are used to help diagnose obstructive pulmonary disease as well as other respiratory problems. We now move on to other forms of respiratory equipment used to aid in patient treatment.

We consider now the topic of artificial respirators. During anesthesia, when the patient's respiratory muscles are temporarily not functioning due to the use of muscle relaxants, a device is required that will carry out the breathing function for the patient. The characteristics of a respirator that performs this function are that the changeover from inspiration to expiration must be smooth and regular. The incoming air must be humidified, a function that is normally performed by the mucous membranes of the pulmonary passages. There should be a nebulizer for administering drugs or additional humidity.

There must be provisions for periodic "sighing." This is due to the elastic nature of the lungs; they tend to shrink away from the thoracic walls. This process is called *atalectasis*. In normal persons this is compensated for by taking an extra deep breath every few minutes; hence an artificial respirator must also perform this function. There must be a bacteria filter in the respirator. This function is usually performed by the cilia located in the conducting airways and by the macrophage dust cells in the alveoli. It may be desirable to have a spirometer on the system to measure the volume of expired air. A pressure-relief valve is necessary to avoid high pressures, which might literally explode the patient's lungs. Adjustable control over the rate and duration of the inspiratory and expiratory functions to meet the demands of various individual characteristics is required.

Two basic modes are used for controlling the action of respirators. In the *pressure-controlled mode* the tidal volume is independent of lung compliance because a constant volume of gas is delivered to the patient. *Volume-controlled systems* have the advantage of compensating for changes in lung compliance but are more susceptible to leaks than is the pressure-controlled mode.

BRIEF HISTORY OF RESPIRATORS

The first landmark in respirator design was in 1928 when the "iron lung" demonstrated that patients who had lost the muscular power to breathe could be sustained for long periods and often brought to various levels of recovery. The iron lung was a chamber in which the patient was sealed with only his or her head protruding from the structure. A bellows at the end of the chamber would increase and decrease the transthoracic pressure, causing air to move in and out of the patient's lungs, ventilating the alveoli.

It was in the early 1940s that the first positive-pressure ventilators came into widespread use. It was found that patients could be maintained for long periods by positive-pressure ventilation through tracheal intubation. Coupled with the development of effective antibiotics, this became a major step in the field of respiration design.

Intermittent positive-pressure breathing (IPPB) machines were developed after World War II that were based on systems developed for high-altitude aircraft crew breathing systems. The IPPB system was used as a means of providing therapy for persons with chronic lung disease. These were essentially ventilators triggered by a negative pressure generated by the patient that would force air or oxygen into the patient's lungs to provide alveolar ventilation. Anesthesia machines, driven mechanically, were developed to provide total respiratory support to the nonbreathing anesthetized patient. Both have emerged to form the basis of the respiratory industry.

The third major advance in ventilator use was the introduction in the 1960s of rapid blood gas analysis systems. Prior to that time respiratory management was by clinical signs, such as cyanosis (bluish coloration of the skin due to lack of oxygen). By the time the signs became evident it was many times too late to do anything for the patient. Blood gas analysis gave a clear indication of acid/base relationships in the blood and could be used to determine the level of oxygen saturation in blood which in turn, could indicate the presence of diffusion abnormalities in the lungs and whether the use of respirators would be useful. In the future, look for the use of microcomputers to control the rate, duration, and O_2 content of the inspired air.

PERFORMANCE AND SAFETY TESTING OF SPIROMETERS AND PLETHYSMOGRAPHS

Spirometers are normally used in intensive and coronary care units, respiratory therapy units, surgical units, and heart stations for the measurement of lung volumes and capacities, as described previously. These units are used to measure and display respiratory volumes by mechanical, electronic, or photographic means. We have discussed the water-filled spirometer, consisting of a moving bell inverted over a chamber of water and counterbalanced to reduce the force required to move it. Currently, electronic spirometers, sometimes called *waterless spirometers,* which work on a variety of principles, are becoming more prevalent in hospitals. These instruments are normally based on an electrical signal derived from a breath-operated transducer such as a hot-wire anemometer.

Other types of spirometers make use of a pair of light sources and detectors. The light beams are chopped by a fan, thus generating a pulse signal that is proportional to airflow. The signal is passed into a fiber-optic bundle transmission line. The output of the fiber-optic bundle is used to drive an electromechanical, photoelectric display which exposes film for a permanent record of the respiratory maneuver. Microprocessors are used to compute the various respiratory volumes and their rates.

Potential Hazards Associated with Spirometers

As with any electrical or electronic device used in conjunction with patient care, there is always a risk of injury due to electric shock to medical personnel or the patient. The mechanism of this hazard is discussed in greater detail in Chapter 16. A brief discussion will be given here for the spirometer, but many aspects of this discussion apply equally well to many electromedical devices.

An electrically defective device becomes potentially hazardous when its chassis is isolated from ground. Faulty line cord attachments are the major cause of this form of failure. The ground can fail at the chassis, however, resulting in the identical hazard. When the ground is lifted (not connected to the chassis), there is no path for leakage currents to flow which will raise the chassis voltage to a level above ground level. Also, if the device is ungrounded, a reversal of line cord polarity may raise the chassis voltage even higher above ground. In this case, if either the patient or operator touches the chassis while in contact with ground, they will probably receive an electrical shock, as they have become part of an electrical circuit.

When a spirometer is operated in conjunction with other electrical devices, such as a ventilator, IPPB, or a monitor, special care must be taken to ensure that the device is properly grounded, or possible device interaction can take place, putting the patient or operator in jeopardy.

Many electronic spirometers are relatively lightweight and portable. If they are not mounted securely or are handled carelessly by medical personnel, they can be dropped or in some other way receive a sharp blow. If this happens and it is reported to the clinical engineer, the device must be removed from service and immediately subjected to performance and electrical safety testing. The rationale for this is that the spirometer may appear to be operating correctly after being subjected to maltreatment and yet have sustained sufficient damage to create a serious electrical hazard for the patient or operator.

In the case of spirometers, safety and performance testing comprise a detailed examination of the device's mechanical and electrical condition to ensure that the device is functioning reliably and safely.

Performance and Safety Testing of Spirometers

The performance test for a spirometer combines tests for functional operation and safety, which is really no different philosophically from the assessment of any other item of electronic medical equipment. That is, the device should perform the function it was designed to perform within the specifications set up by the manufacturer, and the device should be safe to use.

Spirometers can be classified into two categories: water-filled and waterless. Both types measure airflow from the mouth and integrate the flow to produce volume measurements. How this is accomplished is a function of the type of spirometer under test. Thus performance testing of a spirometer involves simply injecting a calibrated volume of air into the system and determining if this volume is measured correctly. This applies more to waterless spirometers, which use electronic integration, than to water-filled spirometers, which operate on volume displacement. One can assume that if the correct "bell" is on a water-filled spirometer, it will always remain in calibration. The only part that needs to be tested for performance is the rate of revolution of the kymograph, to ensure that the timing is correct for calculating flows.

In the case of electronic spirometers, calibration can be done in conjunction with a water-filled spirometer. The flow sensor can be put in series with the input hose of the water-filled spirometer. Known volumes of air can be injected into the system, or a person can perform a vital-capacity maneuver into the system of two spirometers, and the results compared. Also, manufacturers of electronic spirometers may have their own calibration procedure for the clinical engineer to use in testing their system. Using the two spirometers in series will also provide information on the accuracy of the computational algorithms being used in the electron spirometer.

Safety testing of spirometers is similar to safety testing of most medical devices. The rationale for what will be taken up here will be explained in more detail in Chapter 16. In the meantime, the safety testing of spirometers should include the following items:

1. Indicators, such as pilot lights, meter movements, and recorders, should operate correctly as specified by the manufacturer's specifications in the operating manual.

2. The ground impedance from the chassis to the power connector should not exceed 0.1Ω.

3. Leakage current (to be elaborated on later) should be measured in the following modes:
   a. Power off
   b. Power on
   Each mode (as stated above) will be tested with the line cord attachment
   c. Properly grounded
   d. Ungrounded, normal polarity
   e. Ungrounded, reversed polarity

Environmental Inspection

Where applicable, the environment where the spirometer is to be operated should be inspected visually and tested electrically to ensure operator and patient safety. The following inspection and test factors should be included.

1. The electrical distribution should be inspected and tested to ensure that:
   a. The voltage level is adequate.
   b. The receptacle polarity is correct.
   c. A receptacle ground is present.
   d. The receptacle has mechanical integrity.

2. The structural integrity of mounting brackets and other equipment used in conjunction with the spirometer should be checked to assure that they are in good mechanical condition.

3. The mechanical clearance between the spirometer and other instruments, apparatus, and devices should be assured to be adequate.

The total body plethysmograph, as used in the pulmonary function laboratory, consists of the box itself, the pneumotachometer (airflow meter), a mouth pressure transducer, and a box pressure transducer, together with associated amplification and display electronics. Since the system contains electronics, the same principles of electrical safety testing apply here as were developed for the spirometer. Basically, this means testing for leakage currents, ground wire resistance, and chassis voltage. These items of electrical safety were detailed in the section on spirometers.

The calibration of the body plethysmograph becomes a calibration of the component transducer system involved with this device. The pneumotachometer can be calibrated against any type of flow standard in series with the pneumotachometer. At least three flow points should be used to determine calibration and linearity.

The pressure transducers used in conjunction with measuring pressure changes in the plethysmograph can be calibrated using a manometer, either mercury or water type, to check the static pressure response of the transducers.

PROBLEMS

1. Describe the anatomy and physiology of the cardiopulmonary system, discussing the function of the various parts, including gas dynamics across the membranes accounting for gas transfer into and out of the blood in the lungs.

2. Consider a simple model of the lungs to be as shown in Figure 8-9. Using the concepts embodied in equations (8-2) and (8-3), draw an equivalent circuit to represent pulmonary airflow, Q, produced by differential pressure across the airways. This pressure is considered to be alveolar pressure, P_{alv}, if atmospheric pressure is considered to be the zero reference pressure.

   (a) If the compliance of the chest wall is 0.15 liter/cmH$_2$O and that of tissue is 0.15 liter/cmH$_2$O with an airway resistance of 1.5 cmH$_2$O/liter per second as measured at the mouth, write the differential equation for this circuit model and determine the time constant for this simple model of the pulmonary system.

   (b) If the airway resistance were to increase to 4 cmH$_2$O/liter per second due to the onset of an obstructive pulmonary disease, what happens to the time constant of the system? How does the change in mechanical time constant affect the blood oxygen saturation provided that the diffusion constants remain normal in the disease state? Explain your answer.

3. Consider the lungs modeled as a two-compartment model, as shown in Figure 8-10.

   (a) Draw a new equivalent circuit for this more complex model of pulmonary mechanics.

   (b) In what way is the airflow to the alveoli affected if the airway resistance of just one side is twice the normal value of the second conducting airway?

   (c) What will be the effect of the phase difference on airflow in the conducting airways produced by the difference in time constants between the left and right sides of the airways for the conditions expressed in part (b)? Explain

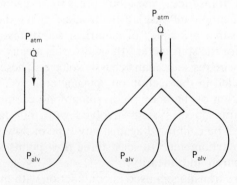

Figure 8-9   Figure 8-10

your answer in terms of airflow in each branch of the system, the central airway, and the pressures in each of the chambers as a function of time over one breath cycle.

4. The spirometer is a simple and useful instrument for the measurement of capacities and volumes. What capacities and volumes does it measure, and which ones does it not measure? Which of the capacities and volumes are static measurements, and which are dynamic measurements?

5. Given the spirometer tracing in Figure 8-11, find the following quantities:
   (a) Tidal volume
   (b) Inspiratory reserve volume
   (c) Vital capacity
   (d) Expiratory reserve volume
   (e) Inspiratory capacity

6. A patient has performed a panting maneuver against a closed shutter on a pneumotachograph inside a total body plethysmograph. The plethysmograph calibration factor is 8.2 cc/cmH₂O and the mouth pressure calibration factor is 3.5 cc/cmH₂O. The resistance of the flowmeter used in the system is 0.5 cmH₂O/liter per second. The oscilloscope trace is given in Figure 8-12.

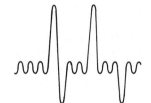

Figure 8-11

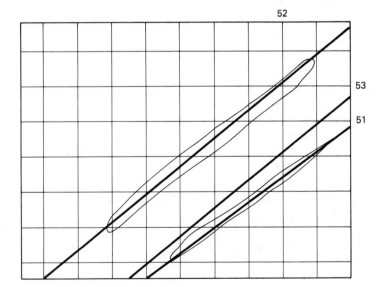

Figure 8-12

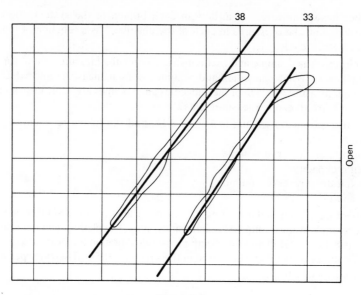

Figure 8-13

(a) Compute the patient's volume of thoracic gas (VTG).
(b) Compute the airway resistance of the same patient in the plethysmograph if the pressure-flow loop for the patient is as given in Figure 8-13.
7. Since the volume of the total body plethysmograph is constant but the volume displaced by each patient in the plethysmograph varies from patient to patient, develop a simple way to calibrate the box pressure for a given volume displacement for each patient who enters the plethysmograph.

REFERENCES

1. Comroe, Julius, *The Lung: Clinical Physiology and Pulmonary Function Tests,* 2nd ed. (Chicago: Year Book Medical Publishers, 1962).
2. Dubois, A. B., et al., "A rapid plethysmographic method for measuring thoracic gas volume," *J. Clin. Invest.,* Vol. 35, 1955, p. 322.
3. Feinberg, Barry N., et al., "Parameter estimation: a diagnostic aid for lung disease," *Instrum. Technol.,* Vol. 17, 1970, pp. 40–46.
4. Primiano, Frank P., "Measurements of the respiratory system," in *Medical Instrumentation: Application and Design,* John Webster (ed.) (Boston: Houghton Mifflin, 1978).
5. Sackner, Marvin A., "The respiratory system," in *Biomedical Foundations of Biomedical Engineering,* Jacob Kline (ed.) (Boston: Little, Brown, 1976).
6. Van de Woestijne, K. P., "The human respiratory system," in *CRC Handbook of Engineering in Medicine and Biology,* Sec. B, Vol. I, Barry N. Feinberg and David G. Fleming (eds.) (Boca Raton, Fla.: CRC Press, 1978).

9

MEDICAL ULTRASOUND SYSTEMS

Sound is simply the mechanical vibration of matter. Apart from the audible frequencies that the human ear can detect, one can divide the sound spectrum into two major categories. Those sounds below the range of human hearing are called *infrasound,* and those above the range of human hearing are called *ultrasound.* When the ultrasound frequencies are high enough where the wavelength of the sound wave in a medium of propagation becomes small compared to most physical objects, it is found that the behavior of wave patterns follows the laws of optics. Ultrasound is transmitted through a medium by means of particle vibration. There is no net movement of the medium, since each activated particle vibrates around a resting position. Medical ultrasound systems make use of frequencies in the band 1 to 20 MHz.

The propagation of ultrasound waves in a fluid is as a longitudinal wave represented as a series of rarefactions and compressions or condensations. In solids, propagation may occur as a transverse wave or shear mode wave. For our purposes we will assume that only bulk waves are propagated in biological material. The interactions of biological materials and ultrasonic energy are very complex and depend on many parameters. Classical wave mechanics, which is used to characterize the motion of ultrasound in a medium, tends to be based on such assumptions as linearity, continuity, homogeneity, and isotropy in various combinations. Few biological materials exhibit suitable combinations of these parameters.

The velocity of the longitudinal sound wave in a homogeneous medium is a function of the elasticity and density of the medium and is given by

$$v = \sqrt{\frac{E}{\rho}} \qquad (9\text{-}1)$$

where E is the modulus of elasticity and ρ is the density of the medium. This is the theoretical equation for the velocity of the acoustic wave in any material. Biological tissues are complicated structures, and their elastic moduli are not easily measured. Therefore, the most reliable estimates of velocity in these materials are those made by direct measurement. In some nonbiological media, the velocity of propagation is partly dependent on the frequency of the wave. This is known as *velocity dispersion*. The velocity of ultrasound in biological materials exhibits only a small dispersion. It is independent of frequency for most practical purposes. Table 9-1 shows the velocity of sound through biological and nonbiological material.

TABLE 9-1 Approximate Velocity of Sound in Different Media

Medium	Velocity (m/s)
Fat	1450
Human tissue	1540
Blood	1570
Eye lens	1670
Skull bone	4080
Water	1480
Air	331
Brass	4500
Aluminum	6400
Plexiglass	2680

INTENSITY OF ULTRASOUND WAVES

The *intensity* is defined as the power per unit area. *Power* is the rate of energy transfer and is in units of watts or joules per second. The unit of intensity is watts per square meter. In medical diagnostic ultrasound, the watt or milliwatt per square centimeter is used for the unit of intensity. Energy travels through a medium with a wave velocity c; thus the energy which passes through a unit area in a unit time is equal to the total energy contained in a column of unit area and length L, equal to the wave or propagation velocity divided by the unit time. This is expressed as

$$L = \frac{c}{\text{unit time}} \qquad (9\text{-}2)$$

If E is the energy density defined as the total energy of all particles in a medium in a unit volume, then the intensity of the acoustic wave is given by

$$I = cE \tag{9-3}$$

ULTRASOUND ATTENUATION

When an ultrasound wave travels through a medium, its intensity is reduced as a function of distance. Since intensity is a measure of the energy moving through a unit area, anything that reduces the energy per unit area of the ultrasound beam attenuates the beam. There are three main components of attenuation: scattering, beam divergence, and absorption. Both scattering and divergence from a parallel beam reduce the intensity of the ultrasound beam as a function of distance.

As the ultrasound plane wave travels through a tissue medium, energy is imparted to the tissue as inelastic deformation and represents an energy loss due to the transformation of energy to random molecular motion (heat). Unlike scattering and beam divergence, absorption is a function of both distance and frequency.

If we assume that the ultrasound wave is adequately characterized by a one-dimensional wave equation for a plane wave propagating in the x direction, the solution for the amplitude of the wave as a function of distance is given by

$$A_x = A_0 e^{-\mu x} \tag{9-4}$$

where A_0 is the amplitude of the wave at $x = 0$, A_x is the amplitude at distance x, and μ is the amplitude attenuation coefficient. The intensity at any point x can be described by

$$I_x = I_0 e^{-2\mu x} \tag{9-5}$$

where I_0 is the initial beam intensity and I_x is the intensity at some tissue depth x. The amplitude attenuation coefficient varies proportionally with frequency. Thus the higher the frequency, the greater the attenuation of the ultrasound beam in a given distance. For this reason, the upper limit on the medically useful frequency is in the neighborhood of 20 MHz.

CHARACTERISTIC IMPEDANCE

In the solution of the partial differential equation that describes the propagation of the ultrasound acoustic wave, there is a quantity that appears between pressure (p) and particle velocity (v) that relates these two variables in the same way that voltage and current are related by characteristic impedance in transmission lines. This relationship is given by

$$p = \rho c v \tag{9-6}$$

where ρ is density (kg/m^3) and c is the velocity of propagation of the wave in the medium. Thus the ratio of pressure to velocity defines the characteristic impedance of the medium and is given by

$$Z = \frac{p}{v} = \rho c \qquad (9\text{-}7)$$

The characteristic impedance, Z, is a property of the medium and is independent of the thickness of the medium. The impedance is a real quantity in the case of nonspreading waves in a lossless medium. For most practical purposes, the imaginary component of the complex characteristic impedance is also negligible in biological materials.

WAVE BEHAVIOR AT BOUNDARY BETWEEN TWO MEDIA

Let us now consider what happens when an acoustic wave traveling in a homogeneous medium of a specific characteristic impedance (ρc) encounters a boundary where the characteristic impedance of the second medium is different from that of the first. In general, part of the wave is reflected and part is transmitted, with the ratio of these two being governed by the media impedances and the angle of incidence of the acoustic wave at the boundary between the two mediums. This is still very analogous to the electrical transmission-line model, where there is no reflection at the termination of the transmission line if the impedance of the termination load equals the characteristic impedance of the line. When there is an impedance mismatch, there is a reflection and partial transmission of energy into the load. A similar phenomenon takes place in the arterial system relative to pressure. As mentioned in Chapter 5, there is a traveling pressure wave down the arterial system, and the application of a compression cuff used in sphygmomanometry causes an impedance mismatch at the point of compression. This causes some of the pressure wave to be reflected back down the artery because of the impedance mismatch at the point of compression.

For the sake of this development, let us assume that the wavelength of the ultrasound wave is small compared with the dimensions of the interface, that the interface can be approximated by a plane, and that it is perpendicular to the propagation plane. Because of the law of conservation of energy, we have continuity at the boundary between the two media in that there are no sudden discontinuities in particle velocity, pressure, or displacement. Thus the pressure, velocity, and displacement of particles on one side of the boundary are equal to those on the other side (refer to Figure 9-1).

Let the subscripts i, r, and t stand for incident, reflected, and transmitted waves. Since the ultrasound waves follow some of the laws of optics, we have, as in optics, that the angle of the incident wave is equal to the angle

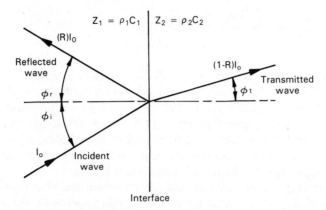

Figure 9-1 Basic wave behavior at the boundary between two materials of different characteristic impedance.

of the reflected wave, i.e.

$$\phi_i = \phi_r \tag{9-8}$$

From the boundary conditions we have

$$p_i + p_r = p_t \tag{9-9}$$

and

$$v_i \cos \phi_i - v_r \cos \phi_r = v_t \cos \phi_t \tag{9-10}$$

The negative sign in equation (9-10) is due to the fact that the reflected wave is traveling in the opposite direction from the incident wave.

As given in equation (9-6), we have that

$$p = \rho c v = Z v$$

Substituting into equation (9-10), we obtain

$$\frac{p_i}{Z_1} \cos \phi_i - \frac{p_r}{Z_1} \cos \phi_i = \frac{p_t}{Z_2} \cos \phi_t \tag{9-11}$$

From equations (9-6), (9-9), and (9-10), we have the relations

$$\frac{p_r}{p_i} = \frac{Z_2 \cos \phi_i - Z_1 \cos \phi_t}{Z_2 \cos \phi_i + Z_1 \cos \phi_t} \tag{9-12}$$

and

$$\frac{p_t}{p_i} = \frac{2Z_2 \cos \phi_i}{Z_2 \cos \phi_i + Z_1 \cos \phi_t} \tag{9-13}$$

If the acoustic wave is normal to the boundary, $\phi_i = \phi_r = \phi_t = 0$. Thus equations (9-12) and (9-13) reduce to

$$\frac{p_r}{p_i} = \frac{Z_2 - Z_1}{Z_2 + Z_1} \tag{9-14}$$

and

$$\frac{p_t}{p_r} = \frac{2Z_2}{Z_2 + Z_1} \tag{9-15}$$

From these equations we see that if $Z_1 = Z_2$, there is no reflected wave. If $Z_2 > Z_1$, the reflected pressure wave is in phase with the incident wave. If $Z_2 < Z_1$, the reflected wave is 180° or π radians out of phase with the incident wave. Once again the similarity between ultrasound transmission in a medium and that of electrical transmission lines holds, as we see that the reflection and transmission coefficients derived above are the same as those developed for an electrical transmission line.

In a similar manner we can find the reflection and transmission coefficients pertaining to the intensity of the acoustic source by making a few substitutions in the formulas just derived. From equation (9-6) we have that

$$p = \rho c v \tag{9-6}$$

For simple harmonic motion, the energy of a particle alternates between kinetic and potential energy states. We know from physics that the kinetic energy of a mass in motion is given by

$$\text{K.E.} = \frac{mv^2}{2} \tag{9-16}$$

which also holds in this case of the motion of the particles making up the transmission medium, with the exception that the mass (m) is replaced by the density of the medium (ρ). This gives us

$$E = \frac{\rho v^2}{2} \tag{9-17}$$

To find the intensity of the wave in the medium we have, by multiplying equation (9-17) by c, that

$$I = Ec = \frac{\rho c v^2}{2} = \frac{p^2}{2\rho c} = \frac{p^2}{2Z} \tag{9-18}$$

Substituting into equations (9-12) and (9-13), we have

$$\frac{I_r}{I_i} = \left(\frac{Z_2 \cos \phi_i - Z_1 \cos \phi_t}{Z_2 \cos \phi_i + Z_1 \cos \phi_t}\right)^2 \tag{9-19}$$

and

$$\frac{I_t}{I_i} = \frac{4Z_2 Z_1 \cos \phi_i \cos \phi_t}{(Z_2 \cos \phi_i + Z_1 \cos \phi_t)^2} \tag{9-20}$$

Equations (9-19) and (9-20) are the intensity reflection and transmission coefficients for an incident ultrasound beam at a boundary.

If the ultrasound beam is normal to the boundary, then $\phi_i = \phi_t = 0$

and equations (9-19) and (9-20) become

$$\frac{I_r}{I_i} = \frac{Z_2 - Z_1}{Z_2 + Z_1} \tag{9-21}$$

and

$$\frac{I_t}{I_i} = \frac{4Z_2Z_1}{(Z_2 + Z_1)^2} \tag{9-22}$$

Equation (9-21) is the amplitude reflection coefficient. When there is perfect impedance matching (i.e., when $Z_1 = Z_2$), the reflection coefficient is zero, indicating that all of the acoustic energy is transmitted into the second medium. Listed in Table 9-2 are the characteristic impedances for some commonly encountered materials. It is precisely the discontinuities of the characteristic impedance that occur in the human body which enable ultrasound to be a useful diagnostic tool.

TABLE 9-2   **Characteristic Impedances of Common Materials**

Medium	Z (g/cm^2-s) $\times$ 10^5
Air	0.0004
Water	1.48
Fat	1.38
Vitreous humor	1.52
Aqueous humor	1.50
Eye-lens material	1.84
Blood	1.61
Muscle	1.70
Human tissue	1.63
Skull bone	7.80

Consider the following example of the calculation of the reflection coefficient between tissue and air, as might be the case when the ultrasound wave impinges on the lungs. From Table 9-2 we have that $Z_{air} = 0.0004$ and $Z_{tissue} = 1.63$. The amplitude reflection coefficient, R, for this combination, assuming that the angle of incidence is $0°$, is written as

$$R = \frac{Z_2 - Z_1}{Z_2 + Z_1}$$

Substituting for Z_1 and Z_2, we have

$$R = \frac{0.0004 - 1.63}{0.0004 + 1.63} = -1$$

In this case we see that the energy in the ultrasound wave is reflected π radians with respect to the incident wave.

The reflection coefficient for bone and soft tissue is approximately 0.3. If most of the energy from the acoustic wave passes through the interface between these two media, bone becomes a partial obstruction to the structure behind it. This is as opposed to x-ray, where tissue structures are shadowed by bone due to their high absorption of x-rays.

ULTRASOUND TRANSDUCERS

Ultrasound for diagnostic purposes is invariably generated and detected by piezoelectric transducers. The well-known *piezoelectric effect,* which is bilateral, is illustrated in Figure 9-2. Briefly, a piezoelectric crystal undergoes mechanical deformation when subjected to an electrical field. Conversely, if the piezoelectric material is mechanically deformed, it will generate a voltage related to the mechanical deformation of the crystal. The amplitude and frequency of mechanical response are directly proportional to the applied voltage, for deformation within the elastic limits of the material. Narrow-beam ultrasound radiation is almost always used in ultrasonic diagnosis either as a single transducer or as an array of narrow-beam transducers. Such a beam is best generated by a disk of piezoelectric material excited by means of two electrodes, one on each side of the parallel surfaces. Suitable piezoelectric materials include quartz, lithium, and sulfate crystals, which are single-crystal materials. Some ferroelectric materials are also used, such as lead zirconium titanates (PZT). This material has a high permittivity, a high efficiency as a transmitter, and a high sensitivity as a receiver.

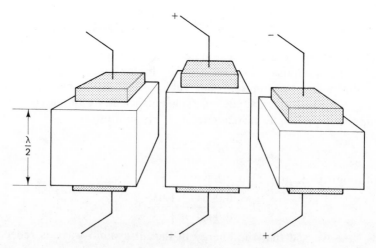

Figure 9-2 The piezoelectric effect enables a crystal to change its dimensions in response to an electrical stimulus, or to produce an emf in response to strain.

Ultrasound Transducer Construction

Piezoelectric material is the heart of the transducer. By design it transmits the ultrasonic wave and receives its reflections. A transducer can also be used solely as a receiver or a transmitter, depending on the manner in which it is connected to the associated electronic equipment. Figure 9-3 is a cross-sectional view of a basic single-crystal transducer. The components that make up the transducer are:

1. Transducer crystal
2. Backing layer
3. Faceplate
4. Case of housing
5. Front and back electrodes
6. Tuning or impedance-matching network

When a voltage is applied to the piezoelectric material, both sides of the crystal will produce an acoustic wave. It is desirable to have only the

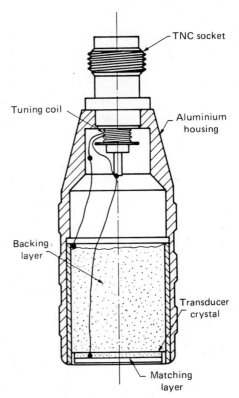

Figure 9-3 Cross section of a single-crystal unfocused transducer.

front face of the crystal produce the ultrasound waves, as waves from the back side of the crystal will produce interference with the front face wave since they are out of phase with one another. Thus a backing material is used to absorb the acoustic energy coming from the back side of the crystal. To do this, the characteristic impedance of the material must be close to that of the crystal or the waves at the interface between the crystal and the backing material will be reflected. This ensures that the ultrasound waves coming from the transducer come from the front face of the crystal.

The backing material also affects the mechanical performance of the piezoelectric crystal by damping the mechanical vibrations of the crystal. This reduces the time constant of the mechanical oscillations produced by electrical voltages applied to the crystal. It also reduces spurious echoes from inside the transducer, which is particularly important when the transducer is used in a pulsed ultrasound mode of operation. The backing material affects the Q of the transducer. The quality factor, Q, of a transducer is a measure of its frequency characteristics, and internal mechanical losses. This is analogous to the Q of a tuned electrical circuit. Transducers with a high Q factor have an output which is more critically dependent on frequency than do those with low Q. Generally, the Q of the crystal, which is determined by its cut, can be modified by an appropriate choice of backing material. A more viscous backing increases bandwidth and damping and reduces the sensitivity and time constant of the mechanical oscillations. Figure 9-4 shows the difference in response of a high-Q and a low-Q system.

The piezoelectric material in the transducer is coated with an electrical conductor on two parallel surfaces forming the front and back electrode pair. If a voltage is applied to the conductors, the piezoelectric crystal will change its thickness by an amount dependent on the applied voltage. If the crystal is subjected to a mechanical stress, as is the case with a returning ultrasound echo, such that it is deformed, a voltage will appear between the two electrodes that is dependent on the strain.

An electrical impedance-matching circuit is used to match the transducer to the electronics at the desired nominal frequency. This provides for maximum power transfer to the transducer from the electronics and avoids electrical wave reflection in the connecting cable.

The thickness of the crystal is cut to $\lambda/2$ of the mechanical wavelength, based on the velocity of propagation within the crystal material. The frequency that corresponds to a half-wavelength thickness is called the *fundamental resonant frequency* of the transducer. The transducer will also resonate at odd multiples of $\lambda/2$. The radiating face of the crystal is protected by a thin layer of material which prevents direct contact of the crystal with the material being examined.

The quality factor, Q, of a transducer determines its frequency characteristics. Transducers with a high Q factor have an output which is more critically dependent on frequency than do those with a low Q. Generally,

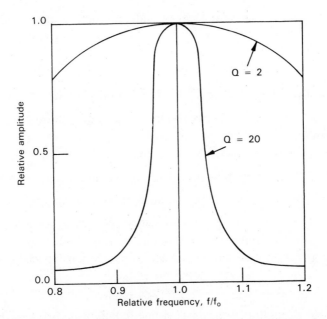

Figure 9-4 Q factor related to frequency response, for two degrees of system
damping.

the Q of the crystal, as determined by its dimensions and cut, can be
modified as required by a suitable choice of backing material. Limited
additional control over damping, which affects crystal Q, can be obtained
externally by a suitable choice of signal source impedance. The value of
the crystal Q can be measured from a frequency plot using the definition

$$Q = \frac{f_0}{f_1 - f_2} \tag{9-23}$$

where f_0 = frequency at maximum amplitude
   $f_1 = -3$-dB point below f_0
   $f_2 = -3$-dB point above f_0
See Figure 9-4.

Impedance Matching at Transducer Face

   To obtain maximum power transfer from the face of the transducer
to the tissue, the load impedance as seen by the transducer should match
that of the transducer. This is nothing more than an application of the
maximum power transfer theorem. Also, when dealing with waves, a matched
impedance means that there is no reflection of the wave at the boundary
between the two materials. Generally, we are not fortunate enough to have
the characteristic impedance of body tissue match that of the transducer.

Thus an impedance matching layer of material must be used to produce an interface that will minimize any reflections of energy at the boundary between the two materials.

CHARACTERISTIC IMPEDANCE OF MATCHING MATERIAL

We continue the analogy between ultrasound and electrical transmission lines by defining the following variables:

Z_i = input impedance of the line

Z_c = characteristic impedance of the line

Z_1 = impedance of the terminating load on the line

β = phase shift per unit distance of transmission line

From the theory of transmission lines we have the following formula for the input impedance of a transmission line with a terminating load impedance:

$$Z_i = Z_c \frac{Z_1 + jZ_c \tan \beta x}{Z_c + jZ_1 \tan \beta x} \quad (9\text{-}24)$$

If we choose a length of transmission line (which will correspond with the thickness of the matching material) to be such that $x = \pi/2$ corresponding to a wavelength of $\lambda/4$, equation (9-24) reduces to

$$Z_i = \frac{(Z_c)^2}{Z_1} \quad (9\text{-}25)$$

Solving for Z_c, the characteristic impedance of the matching line, we have

$$Z_c = \sqrt{Z_i Z_1} \quad (9\text{-}26)$$

If we make a substitution of variables in our analogy, we let

$Z_i = Z_{\text{transducer}}$

$Z_1 = Z_{\text{tissue}}$

$Z_m = Z_{\text{matching medium}}$

After making these substitutions, we have the following formula for the computation of the characteristic impedance of a quarter-wave thickness of matching material to be used on the face of the transducer.

$$Z_m = \sqrt{Z_{\text{tissue}} Z_{\text{transducer}}} \quad (9\text{-}27)$$

The same formula holds true for odd integer values of a quarter-wavelength (i.e., $n\lambda/4$, where $n = 1, 3, 5, \ldots$). Since wavelength is frequency dependent, this form of impedance matching is frequency dependent. However, transducers will normally radiate over a very narrow frequency band which allows the use of quarter-wave matching methods. Figure 9-5 illustrates how the quarter-wave section is used in an ultrasound transducer.

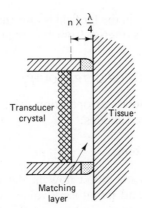

Figure 9-5 Quarter-wave impedance matching used in an ultrasound transducer.

The field generated by the excited transducer is basically cylindrical in the immediate vicinity of a disk-shaped transducer, as shown in Figure 9-6. The length of the near (Fresnel) field zone is given by

$$L = \frac{D^2}{4\lambda} \qquad (9\text{-}28)$$

where L is the length of the near field zone, D the diameter of the transducer, and λ the wavelength of the acoustic wave in the transmission medium.

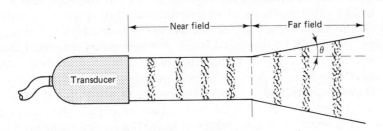

Figure 9-6 Unfocused ultrasound beam pattern.

Beyond the near field the beam diverges in a cone-shaped manner, which means that the intensity is reduced as a function of the distance from the transducer. The angle of divergence of this far (Fraunhofer) field is given by equation (9-29) and shown in Figure 9-6.

$$\theta = \tan^{-1} \frac{2\lambda}{D} \qquad (9\text{-}29)$$

From equation (9-29) it is seen that the smaller the wavelength, implying higher frequencies, in relation to the diameter of the transducer surface, the less the beam will diverge. The less divergent the beam, the better the resolution of targets that are in the beam. Most medical diagnostic work is done in the near field of a transducer, where the wavefronts act as plane disks. Divergence in the far field is often quite small, however, and meaningful data can still be gathered from this region.

DIAGNOSTIC USES OF ULTRASOUND

There are two modes of operation used with medical ultrasound systems: doppler and pulsed modes.

Doppler Systems

Information on blood flow in the vascular system of humans is of great diagnostic and monitoring value to the medical practitioner, but for a long time has been the least accessible information available. Heart rate, and later blood pressure, have been the only measurable parameters for most physicians in history. More recently, techniques have been developed for more accurate measurement of blood pressure using indwelling catheters and the measurement of blood flow using electromagnetic flowmeters. The problem with these methods is that they are invariably invasive in nature and have associated the accompanying risk, plus expense, and inconvenience. Because ultrasound can provide some information noninvasively, it has been used with steadily increasing success to measure blood flow parameters in arteries and the action of the heart in both adult and fetal hearts.

Unlike other methods of ultrasound use, this mode of operation is used for the measurement of motion within the body, as opposed to static imaging, which will be discussed later. A major use of a medical Doppler system is the detecting and recording of fetal heart movement and the determination of blood flow in different vessels of the body.

Doppler ultrasound systems work on the principle that if an incident wave of frequency f_i is reflected from a moving object, the frequency of the reflected wave, f_r, will be higher or lower than f_i depending on whether the object is moving toward or away from the transducer. Thus a Doppler system responds to motion. Its principle of operation is based on the Doppler principle, which defines the effect on the frequency of a transmitted sound caused by a change in the path length between source and receiver. The Doppler frequency shift, f, is calculated according to the classic Doppler formula

$$\Delta f = 2f_0 \frac{v}{c} \cos \theta \qquad (9\text{-}30)$$

We can now solve for the velocity in terms of the frequency shift Δf from the frequency f_0.

$$v = c \frac{\Delta f}{2f_0 \cos \theta}$$

where Δf = shift in frequency from incident frequency

f_0 = frequency of incident radiation

v = velocity of irradiated object

c = velocity of ultrasound wave in the transmission medium

θ = angle between the velocity vector and a line joining

the source of radiation and the observed point

See Figure 9-7. A block diagram of a typical Doppler system is shown in Figure 9-8.

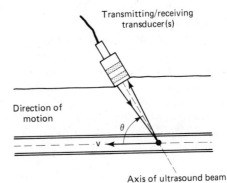

Figure 9-7 Doppler frequency shift of ultrasound reflection due to blood motion.

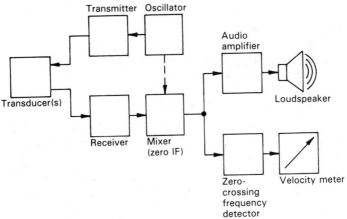

Figure 9-8 Block diagram of a typical Doppler system.

    The basic use of Doppler systems is the measurement of blood flow or the motion of heart valves or walls. It is used for the study of blood circulation to the brain and in vascular laboratories in the diagnosis of vascular disease. Doppler techniques have the potential of being able to examine a common site of flow obstruction, such as in the carotid artery bifurcation in the neck, as well as being able to determine the effects of constriction or obstructions in the arterial system.

The simplest form of Doppler system corresponding to Figure 9-8 is the continuous-wave (CW) Doppler. This type of system uses two transducers, a transmitter crystal and a receiver crystal. It is used for fetal pulse detection, blood flow analysis, and air emboli detection.

Problems with continuous Doppler systems.   Perhaps the major problem with continuous Doppler systems is that frequency changes are received from every moving reflector. Since all body tissues, especially those near large arteries or the heart, are in continual motion, the reflected signal will contain information from many structures. Thus we see that the CW Doppler systems are not capable of providing range information or determining the motion of an object at an arbitrary distance from the transmitting crystal.

There is also the problem of attenuation of the reflected wave, since attenuation is an exponential function of distance:

$$I = I_0 e^{-ax} \tag{9-31}$$

This problem is solved by using an amplifier whose gain is increased as a function of time to compensate for the decrease in amplitude of returning echoes with increasing range. This is used in pulse-echo systems, discussed in the next section. Many Doppler units are provided with controls for setting the onset of range-gain compensation and start-rate control (see Figure 9-9).

Since CW Doppler systems are not capable of providing range information or determining motion at an arbitrary distance from the transmitter probe, the pulsed Doppler or range-gated system was developed. This system is used to separate the Doppler signals generated by a number of moving targets in which the frequencies of the echoes collected at an arbitrary range are compared with that of a reference oscillator that is itself gated to provide the transmitted burst of frequency. Range gating involves shutting down the receiver/amplifier for all but a small window of time. This type of ultrasound equipment transmits a burst of ultrasound. Then the transmitting crystal is switched to the receiver/amplifiers and it becomes the receiving crystal transducer. The amplifiers are only active after a time τ_d after the beginning of the burst of frequency (see Figure 9-10). The definition of terms related to pulse Doppler systems can be seen in terms of the timing diagram in Figure 9-10.

1. Pulse repetition frequency, 1–5
2. Transmit pulse duration, 1–2
3. Range (delay, τ_d), 2–3

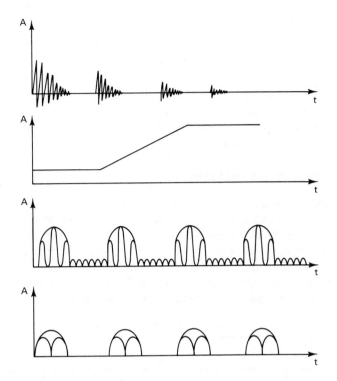

Figure 9-9   Range-gain compensation for Doppler ultrasound system.

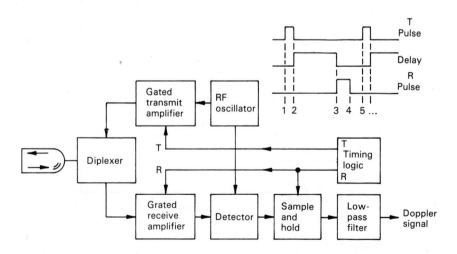

Figure 9-10   Block diagram of a range-gated Doppler system.

The range of the reflecting object can be calculated from the equation

$$\text{range} = c\,\frac{\tau_d}{2} \tag{9-32}$$

Range-gain compensation, as discussed previously, may be incorporated into the receiver amplifier. In the case of the gated Doppler system, the transducer is stimulated with a burst of RF rather than an impulse, so that the resolution is generally poorer than in the case of the pulse-echo systems that are now to be discussed.

Pulse-Echo Ultrasound Systems

A-scan systems.　　Three types of ultrasonic systems used in medicine make use of pulse-echo methods: the A-scan, B-scan, and M-mode systems. The A-scan ultrasound system is a one-dimensional detection process and is the simplest process we will take up in this section. This type of ultrasound system is essentially the same as that of radar technology. A schematic diagram of a basic A-scan system is given in Figure 9-11. A signal is emitted from the transducer and is reflected at various discontinuities (i.e., a change

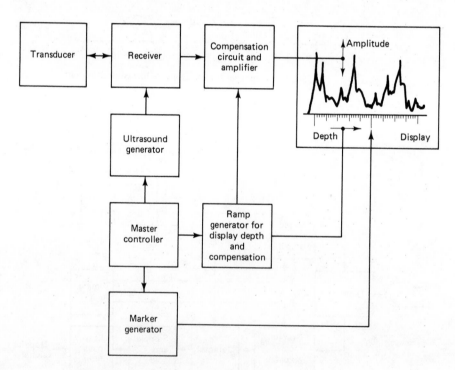

Figure 9-11　Block diagram of a basic A-scan ultrasound system.

in the characteristic impedance) back to the transducer, where the reflections are picked up and fed into an RF amplifier. As can be seen in Figure 9-9, the gain of the amplifier increases with time as the various reflections are picked up by the transducer. The time-varying gain of the RF amplifier is provided to compensate for the attenuation of the acoustic wave as it propagates through the medium in the forward and reverse (reflected) directions. This will produce echo amplitudes of approximately the same height. The signals are then fed into signal-conditioning electronics and into the vertical deflection amplifier of the CRT display for visual interpretation. The screen of the CRT is calibrated with distance markers so that the horizontal position will correspond to the depth in tissue and the detected echoes will deflect the trace in the vertical plane in proportion to the reflection amplitude. A-scan ultrasound systems are best suited for the ranging and visualization of static structures, such as in the eye and brain.

Echo encephalography, A-scan ultrasound used on the brain (see Figure 9-12) can be used in the determination of a displaced brain midline, indicating possible tumor or hemorrhage. In ophthalmology, A-scan systems are used to determine the distance of the retina from other structures of the eye, such as the lens. It can also determine whether there is a foreign body in the eye and the approximate distance behind the lens it is located. It may be noted in Figure 9-13 that for ophthalmological measurements, a water-filled tube is used. Its purpose is to act as a delay line to separate the eye lens artifact from the more important information.

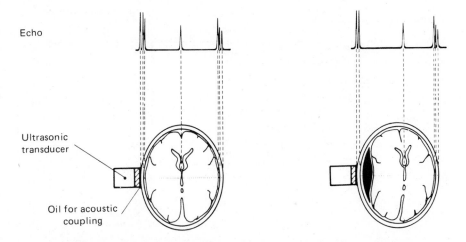

Figure 9-12 Principle for locating the midline of the brain by the A-scan. Note that the midline is displaced to the right in the right picture. (From Bertil Jacobson et al., *Medicine and Clinical Engineering,* © 1977. Reprinted by permission of Prentice-Hall, Inc., Englewood Cliffs, N.J.)

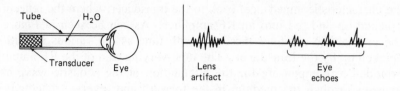

Figure 9-13   A-scan of the eye using echoes for ranging.

B-scan systems.   One of the chief disadvantages of the A-scan technique is its one-dimensionality. The display gives information only on the position of a reflecting object and provides no two-dimensional information. B-scan is a two-dimensional technique in which a cross-sectional image of the scanned region can be built up on a CRT storage system. It is based on the fundamental information provided by echoes in an A-scan mode and is used to intensity modulate the CRT electron beam instead of deflecting it vertically. The time base of the beam is made to move in the axial direction of the transducer instead of its "normal" left-to-right horizontal sweep. Figure 9-14 illustrates the basic principle of a B-scan system. The ultrasound transducer is mounted on a pantographic scanning arm, which is a system of linkages mechanically connected to the transducer. At each joint of the linkage there is a potentiometer that produces a voltage as a function of linkage angle. From knowledge of these polar angles, the relative

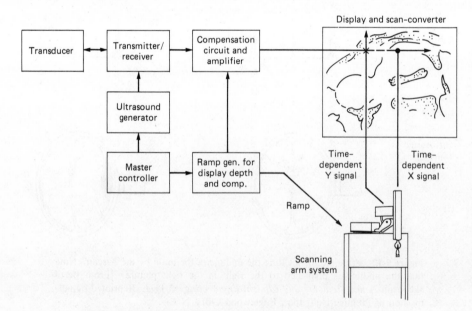

Figure 9-14   Block diagram of a static gray-scale B-mode system.

coordinate position in the scanning plane of the transducer's face can be obtained.

From Figure 9-15 we see that the X and Y coordinates of the transducer are functions of the angles α, β, and θ, and the lengths L_1, L_2, and L_3. The coordinates of the transducer can then be easily calculated from the equations

$$X = L_1 \sin \theta + L_2 \sin \beta + L_3 \sin \alpha \qquad (9\text{-}33)$$

$$Y = L_1 \cos \theta + L_2 \cos \beta + L_3 \sin \alpha \qquad (9\text{-}34)$$

Thus the scanning arm converts the actual transducer position and angle into electrical X and Y coordinate signals to correctly position the transducer line in the display screen. If the transducer is moved back and forth, new lines composed of dots from other transducer positions and angles will be added, gradually building up a complete picture due to the storage capacity of the incorporated scan converter. This device electronically adapts the X-Y signal for a normal TV video format. The picture takes from 2 to 20 s to build up and is not time varying.

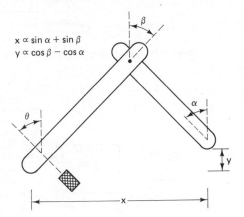

$x \propto \sin \alpha + \sin \beta$

$y \propto \cos \beta - \cos \alpha$

Figure 9-15 Schematic of pantographic scanning arm used to determine the position of the ultrasound transducer.

M-mode systems. Both A-scan and B-scan methods of ultrasonic imaging provide only static information to the user about the organs being scanned. The B-scan technique is useful for visualizing cross sections of the patient, but its use in cardiac visualization is very limited because of the rapid movement and large excursions of the heart. A thoracic B-scan would produce an incomprehensible blur in the region of the heart, to say nothing of the problems imposed by the lungs and rib cage in producing such a scan.

M-scan techniques sacrifice the ability to obtain cross sections of the scanned region and revert to the "on-axis" restriction of the A-scan method. When an object is moving, the A-scan representation will show the corresponding echoes moving in the horizontal direction. By converting these to light dots and intensity-modulating the display while slowly displacing them at a fixed velocity, a number of curves can be stored, indicating the movements of the structures under study. Figure 9-16 indicates how the time-motion (M-mode) display is built up. A typical M-mode system is illustrated schematically in Figure 9-17.

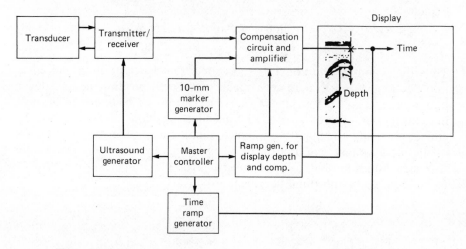

Figure 9-16   Block diagram of an M-mode, time-motion, ultrasound system.

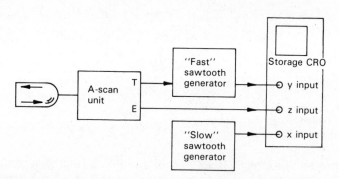

Figure 9-17   Basic components of an M-mode system used to build up the time-motion picture.

A strip-chart recorder is often substituted for (or used to augment) the oscilloscope, as in Figure 9-17. In this case, the paper-drive mechanism performs the function of the slow time base. In either case, the resulting image is a raster with a vertical line structure, intensity-modulated by the echoes returning to the transducer. The ordinate represents distance along the transducer axis, usually with the origin at the top to correspond to downward "view" into a supine patient, and the abcissa representing time. In echocardiographic applications a second trace records the electrocardiogram for comparison and time reference. Figure 9-18 illustrates in diagrammatic form the recording of simple target motion.

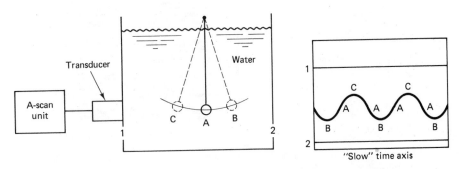

Figure 9-18 Elements of an M-mode system and display of a moving pendulum. (From Kenneth Erikson et al., "Ultrasound in medicine—a review," *IEEE Trans. on Sonics and Ultrasonics,* Vol. SU-21, No. 3, July 1974, by permission.)

Echocardiography is currently the most extensively used ultrasonic technique in cardiovascular medicine. By applying the transducer to the thorax, either at the suprasternal notch (Figure 9-19) or at the intercostal space (Figure 9-20), the motion of various heart structures may be examined.

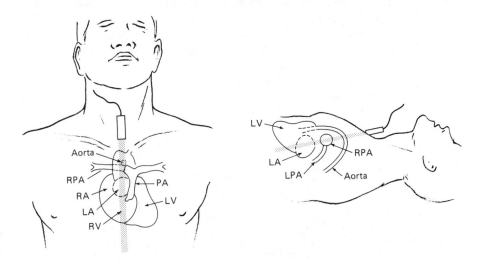

Figure 9-19 Transducer placed over suprasternal notch to "view" structures indicated.

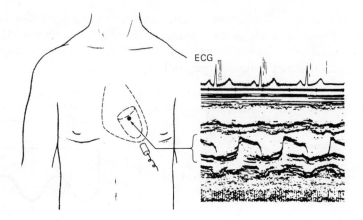

Figure 9-20 Transducer placed over an intercostal space to "view" mitral ring, mitral leaflets, and so on.

MULTIELEMENT TRANSDUCER SYSTEMS

The scanning movement of a single transducer along a straight line in an upright position can be simulated electronically by electronic switching between numerous transducer elements placed in a linear configuration. On the display each transducer has its own line trace, echoes being shown in the normal way as small intensity-modulated dots. If the transducers are placed close enough to each other, the lines on the screen tend to merge together, in a manner similar to the way that television scan lines form a TV picture. The display from the array of transducers is not perceived as discrete echo traces but as a two-dimensional picture of the structure underneath the transducer array. This technique permits a form of two-dimensional image to be formed in a very short period of time. Operated sequentially, these arrays can scan a rectangular cross-sectional plane and permit the generation of a moving image of the cross section. Figure 9-21

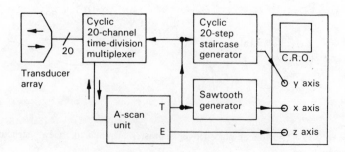

Figure 9-21 Linear array multiscan system.

shows the block diagram of a 20-transducer linear-array multiscan system, and Figure 9-22 the longitudinal and lateral cross sections.

Another form of multielement ultrasound scanning system is the phased-array electronic sector scanner. These systems are based on a well-known principle of wave mechanics called *Huygen's principle*. Huygen's principle states that a plane wavefront can be approximated as an infinite number of spherical waves propagating in the same direction. The implication of this principle in ultrasound is that the surface of any transducer may be considered to consist of small areas and that the directivity of the whole transducer can be analyzed by the addition of the contribution of each of these small transducing areas provided that phase and amplitude relations are accounted for. Huygen's principle is to plane waves what a Fourier series is to periodic functions.

As a result, a plane wave of a desired characteristic can be synthesized by an array of crystals, usually 16, whose sequence of pulsing will determine the angle of the transmitted sound beam to the transducer surface. By changing the sequence of pulsing, the ultrasound beam can be made to appear to sweep through an area that is being scanned at rates of 30 to 40 images per second. Figure 9-23 is a block diagram of a real-time sector-

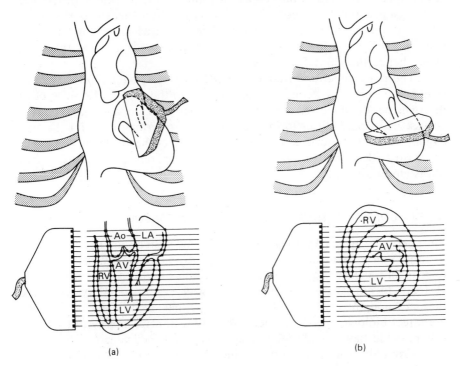

(a)

(b)

Figure 9-22 Place the transducer assembly as illustrated to visualize longitudinal cross sections (a) and lateral cross sections (b).

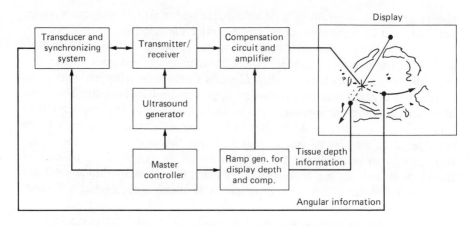

Figure 9-23 Block diagram of a real-time sector-scan ultrasound system.

scan system. This engineering system used for medical ultrasound sector scanning imaging usually requires a microprocessor to control the transmission and reception sequences of the array of transducer elements. On a sector-scan display, the traces produced by the return echoes appear to be radiating from a single point source. This point is the hypothetical position of a fictive rotating or rocking transducer of zero size which is simulated electronically. The picture is formed by a sequence of lines over the sweep angle, hence the name "sector scan."

The pulse repetition frequency (PRF) and the angular velocity of the sweep determine the actual number of lines in the sector. Because of the diverging nature of the sector-scan display, there are more lines per centimeter across the display closer to the transducer than farther away for the same angular velocity. Because of this, the resolution of the display will gradually decrease with increasing distance from the apparent point source of the ultrasound radiation.

SAFETY CONSIDERATIONS OF DIAGNOSTIC ULTRASOUND

The energy levels used in diagnostic ultrasound systems are not high enough to either ionize tissue atoms or to move electrons from a lower-energy state to a higher-energy state. The concerns one has for x-ray-type imaging systems do not carry over in the same way to ultrasound systems since they are nonionizing. This does not mean that ultrasonic radiation does not affect biological systems, because it does transfer energy to tissue and it is possible that this energy transfer could cause damage to tissue.

From all that has been determined by researchers to date, damage is a matter of energy transmitted to tissue. There are three effects produced: thermal due to molecular agitation, mechanical vibration effects, and cavitational effects. In all diagnostic ultrasound equipment, the energy levels used are at the lower end of the energy spectrum, typically 1 to 10 mW/cm^2.

Mechanical vibrations produced by very high intensity ultrasound transducers have the potential of exceeding the elastic limit of tissue, causing it to break up. Since diagnostic ultrasound intensity is very low, this does not become a factor in the safety aspects of properly designed systems.

The term *cavitation* is used by investigators to describe how gas bubbles act in ultrasonic fields. Cavitation is an effective mechanism for concentrating acoustic energy in small volumes. If the energy concentration is at a high enough level, tissue destruction is possible in the local area of the cavitation. In a continuous medium without cavitation due to gas bubbles, the same acoustic energy input would not cause any problems.

It is the belief of many investigators that the majority of effects in tissue due to ultrasound are the result of heat. It is just this heating effect that is used in therapeutic ultrasound systems for treating a variety of muscular problems, but they are used at much higher intensities than diagnostic systems. The amount of heat generated by clinical ultrasound diagnostic systems is very low due to the low level of energy applied to them in clinical use. Up to the present time, there have been no demonstrated unfavorable effects due to ultrasound. This is despite large-scale application of ultrasound in obstetrics and cardiology. Although many researchers are still looking into possible problems with the use of ultrasound, it still appears to be safe to use as long as the power levels are kept low and the transducer is not left in the same position for a long period of time. Even though ultrasound appears to be safe to use, some scientists suggest limited use of ultrasound during early pregnancy.

EQUIPMENT PERFORMANCE TESTING

Because of the large variety of ultrasound equipment on the market, it is difficult to present a sequence of performance tests that will cover all systems from single-crystal A and B scanners to multielement sector-scan systems. Because of this large variety of systems it is best for the clinical engineer to consult the manufacturer for particular test procedures and associated test equipment.

As an example of how test procedures can be performed on ultrasound equipment, let us consider a single-crystal A and B scan system. A very simple technique used to check equipment performance in the case of a scanning arm system, which does not require any special test object, is to rotate the transducer around its pivot. If the basic geometry of the equipment is correctly set, this should result in a sector of a circle on the display. If not, the shape of the circle will be distorted and give an elliptical appearance on the display. The sensitivity and compensation can then be checked against earlier results by attaching a 16-mm-thick piece of plexiglass to the sole of the transducer using scanning gel, and noting how many reflections can be seen on the A-mode display for given control settings. This number should be essentially constant with time.

A 100-mm test object, including a standard procedure for its use, was adopted as a standard at the 1974 meeting of the American Institute of Ultrasound in Medicine (AIUM) in Seattle, Washington. It consists of stainless steel rods mounted parallel to each other in a trapezoidal grid pattern. The machine parameters that can be measured according to the AIUM standards with this test object are depth calibration, axial resolution, dead-zone depth, lateral resolution, depth-compensation circuitry characteristics, and display characteristics.

AIUM-approved test objects in both open and sealed enclosures are commercially available from a number of manufacturers. Scanning oil or gel should be used with the sealed types. While the AIUM test object can

Figure 9-24 Bio-Tek ultrasound wattmeter direct-reading test instrument. (Courtesy of Bio-Tek Instruments, Inc., Burlington, Vt.)

be used to test the portions of the ultrasound system that relate to imaging, it does not measure acoustic power output from the transducer. Power measurements may be made by observing pressure produced by an ultrasound transducer against a test surface at a fixed distance from the transducer. Several companies currently offer such test devices, which provide a direct reading of the acoustic power output of the ultrasound transducer. Figure 9-24 is typical of the commercial ultrasound wattmeters available to the clinical engineer for testing the output power from an ultrasound transducer.

PROBLEMS

1. Define the following terms used in conjunction with medical ultrasound systems:
   (a) Longitudinal waves
   (b) Propagation velocity
   (c) Intensity
   (d) Absorption
   (e) Acoustic absorption coefficient
   (f) Characteristic impedance of a medium
   (g) Reflection coefficient
   (h) Piezoelectric effect
   (i) Q of an ultrasound transducer

2. What is a quarter-wave transformer, and how is it used in medical ultrasound systems?

3. What are some of the similarities and differences between Doppler and pulse-echo ultrasound systems?

4. Average brain tissue has an attenuation constant of 0.11 at 1 MHz. For this type of tissue the attenuation constant is given by

$$a = kf$$

   where k = proportionality constant
   f = frequency, MHz

   What is the attenuation in decibels at 2.25 MHz if the brain tissue is 10 cm thick?

5. For maximum power transfer, it is desired to match an ultrasound transducer having a characteristic impedance of

$$Z_t = 1.517 \times 10^6$$

   and water, which has a characteristic impedance of

$$Z_{H_2O} = 24.58 \times 10^6$$

   If an impedance matching medium has a sound velocity of 3000 m/s at 2 MHz, what is the characteristic impedance of the medium, and what is its thickness?

6. Describe the operation of a Doppler ultrasound system and draw a functional block diagram for the system.

7. Describe the operation of an A-scan ultrasound system and draw a functional block diagram for it. Explain the imaging method used for this type of ultrasound system.

8. Describe the operation of a B-scan ultrasound system and draw a functional block diagram for it. Explain the imaging method used for this type of ultrasound system.

9. What are the similarities and differences between the ultrasound systems in Problems 6, 7, and 8?

10. What are the major factors responsible for the amplitude attenuation of an ultrasound wave traveling through different mediums?

11. If the attenuation coefficient for heart muscle is 0.13 at 1.0 MHz, how far will the ultrasound wave travel before its amplitude is attenuated at 3 dB?

12. What is the reflection coefficient at the interface between the skin and skull for an ultrasound wave introduced at the head?

REFERENCES

1. Dreijer, Niels, *Diagnostic Ultrasound,* 2nd ed. (Denmark: Bruel & Kjaer, 1979).

2. Gordon, Douglas (ed.), *Ultrasound as a Diagnostic and Surgical Tool* (Baltimore: Williams & Wilkins, 1964).

3. Newhouse, Vernon, "Class notes," Purdue University, West Lafayette, Ind., 1981.

4. Reneman, Robert S. (ed.), *Cardiovascular Applications of Ultrasound* (Amsterdam: North-Holland, 1974).

5. Rose, Joseph L., and Barry B. Goldberg, *Basic Physics in Diagnostic Ultrasound* (New York: Wiley, 1979).

6. Wells, P. N. T., *Biomedical Ultrasonics* (London: Academic Press, 1977).

10

ENGINEERING PRINCIPLES OF MEDICAL X-RAY SYSTEMS

X-RAY SYSTEMS

The production of x-rays is a matter of energy conversion. The first part of this chapter is devoted to the mechanism involved with the conversion of energy into x-rays.

Definitions

The fundamental unit of work is the joule. A more useful unit is the *electron volt* (eV), or the amount of energy released when an electron falls through a potential difference of 1 volt:

$$\text{work eV} = q_e V \tag{10-1}$$

where $q_e = 1.60 \times 10^{-19}$ coulomb

$V = 1$ volt

$$1 \text{ electron volt} = (1 \text{ volt})(1.60 \times 10^{-19} \text{ coulomb})$$
$$= 1.60 \times 10^{-19} \text{ joule}$$

Since this is a small unit, a more common unit is

$$\text{keV} = 1000 \text{ eV}$$
$$\text{MeV} = 1000 \text{ keV} = 10^6 \text{ eV} = 1.60 \times 10^{-13} \text{ J}$$

Radiation. Generally, electromagnetic radiation results from a change in "orbit shells" of an electron or energy level of the neutrons and protons in the structure of the nucleus of the atom.

Electromagnetic radiation.   Radio waves, heat waves, light, ultra-violet, x-ray, and gamma rays are examples of electromagnetic radiation. They all travel at a speed of 3×10^8 m/s. As mentioned before, one source of electromagnetic radiation is the change in energy levels of either orbit electrons or particles in the nucleus. Long wavelengths are measured in meters. Short wavelengths are measured in angstroms (Å). The fundamental relation is

$$\nu\lambda = c \tag{10-2}$$

where λ = wavelength, m

    $c = 3 \times 10^8$ m/s (the velocity of light)

    ν = frequency of radiation, Hz

Quantum nature of radiation.   Although, for many purposes, we can consider all electromagnetic radiation as propagated by wave motion, when the frequency is high enough, the transmitted energy can be characterized by packets of energy called a *quantum* or *photon*. The relationship between the energy carried by the photon is dependent on the frequency of the radiation. The energy of the photon is linearly related to the frequency by the relation

$$E = h\nu \tag{10-3}$$

where E = energy, J

    h = Planck's constant (6.62×10^{-34} J-s)

    ν = frequency, s^{-1}

Example

Calculate the energy of one photon of radiation of wavelength 1 Å or 1×10^{-10} m.

Solution

$$\nu = \frac{c}{\lambda} = \frac{3 \times 10^8}{1 \times 10^{-10}} = 3 \times 10^{18} \text{ Hz}$$

$$E = h\nu = (6.62 \times 10^{-34})(3 \times 10^{18}) \text{ J}$$

$$E_{\text{MeV}} = 19.83 \times 10^{-16}/1.6 \times 10^{-13} = 12.3937 \times 10^{-3} \text{ MeV}$$

$$E_{\text{eV}} = 12,393.75 \times 10^3 = 12,400 \text{ eV}$$

A general relationship between energy of a photon is in eV and λ in Å; then

$$E = \frac{hc}{\lambda} = \frac{12,400}{\lambda(\text{Å})} \qquad \text{eV} \tag{10-4}$$

Thus a photon of wavelength 1 Å carries 12,400 eV of kinetic energy; conversely, it takes 12,400 eV of energy to produce a photon of wavelength 1 Å.

X-RAY PRODUCTION

There are two types of x-radiation that are produced when a ballistic electron interacts with target atoms (Figure 10-1): (1) continuous, and (2) characteristic.

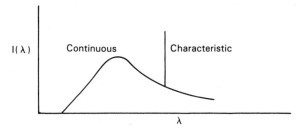

Figure 10-1   Radiation intensity as a function of wavelength.

Radiation, particularly x-radiation, can be produced when high-speed electrons are allowed to strike a target. Thus *production of x-rays is a matter of energy conversion,* that is, the conversion of the kinetic energy of the electron to that of x-radiation. The kinetic energy of an electron is altered by (1) radiation losses and (2) collision losses.

Continuous (Bremsstrahlung-Braking Radiation) or White Radiation

The intensity of x-radiation is given by the product of the energy of a photon multiplied by the number of photons at that energy in the x-ray beam.

Some points to be made about continuous x-radiation are:

1. When an electron of energy E approaches close to the nucleus of an atom, the electron will orbit partially around the nucleus due to the strong attraction of the positive nucleus to the negative electron.
2. The electron is suddenly decelerated and loses energy. In accordance with Maxwell's electromagnetic theory, the acceleration ($+$ or $-$) of a moving electric charge must be accompanied by the emission of electromagnetic radiation (Figure 10-2).

On the basis of classical electrodynamics, based on Maxwell's equation in free space we have

$$\nabla \times \mathbf{H} = \frac{4\pi\sigma}{c}\mathbf{E} + \frac{E}{c}\frac{\partial E}{\partial t} \qquad (10\text{-}5)$$

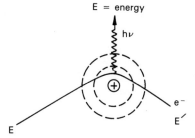

Figure 10-2   Energy loss in the form of radiation of a ballistic electron due to its change in trajectory.

where c = velocity of propagation in medium

     ε = dielectric constant of the medium

     E = electric field vector intensity

     H = magnetic field vector intensity

     σ = conductivity of the medium

If it is assumed that we have an ideal dielectric, $\sigma = 0$; then the equation above can be put in the form

$$\nabla^2 \mathbf{H} = \frac{\varepsilon\mu}{c^2}\frac{\partial^2 \mathbf{H}}{2t^2} \qquad \text{wave equation}$$

$$\nabla^2 \mathbf{E} = \frac{\varepsilon\mu}{c^2}\frac{\partial^2 \mathbf{E}}{2t^2} \qquad \mathbf{B} = \mu\mathbf{H}$$

(10-6)

where μ is the magnetic permeability. The solutions of these equations can be shown to be radiated waves **E** and **H** at right angles to each other and to the direction of propagation, which is due to the acceleration of an electric charge.

Plane-wave polarization.   For simplification consider a plane wave, that is, a linearly polarized monochromatic electromagnetic wave. The solution to Maxwell's equations in this case are

$$E = E_0 \sin\frac{2\pi}{\lambda}(x - ct) \qquad \text{electrical field intensity at any point}$$

$$H = H_0 \sin\frac{2\pi}{\lambda}(x - ct) \qquad \text{magnetic field intensity at any point}$$

(10-7)

Energy emitted by accelerated electrons.   From Maxwell's equation we have established the following:

1. When an electron is suddenly decelerated, it loses energy in the form of electromagnetic radiation provided that it does not lose energy due to collision. In this case the energy is lost due to a change in momentum.

2. The radiated energy can be considered a photon where the photon energy is, from quantum theory, (equation 10-3)

$$E = h\nu = h\frac{c}{\lambda}$$

3. On the basis of classical electrodynamics, the radiated energy from a decelerated charge also consists of orthogonal electric and magnetic waves. The magnitude of the electric and magnetic intensity, at a distance r from a charge e, is given by equation 10-8

$$E = H = \frac{ae}{rc^2}\sin\phi$$

(10-8)

where E is in esu

   H is in esu

   e is in esu

   a = acceleration of the charge

   ϕ = the angle between r and the direction of the acceleration

This relation holds when the velocity of the charge is much less than the velocity of light.

4. The amount of energy, I, which flows in a unit line through a unit area in the direction of propagation is

$$I = \frac{cE^2}{4\pi} = \frac{cH^2}{4\pi}$$

$$= \frac{c}{4\pi}\left(\frac{ue}{rc^2}\sin\phi\right) = \frac{a^2e^2}{4\pi r^2 c^3}\sin^2\phi \tag{10-9}$$

The rate at which energy is radiated from an accelerated charge can be calculated by integrating the intensity I over the surface of a sphere of radius r with the accelerated charge at its center.

Differential area:

$$dA = 2\pi r \sin\phi r\, d\phi$$

Differential energy:

$$dS = I\, dA = \frac{cH^2}{4}\pi\, 2\pi r^2 \sin^3\phi\, d\phi \tag{10-10}$$

Total energy:

$$S = \int_0^A I\, dA = \frac{1}{2}\left(\frac{a^2e^2}{c^3}\right)\int_0^\pi \sin^3\phi\, d\phi$$

$$= \frac{2}{3}\frac{a^2e^2}{c^3} \tag{10-11}$$

Thus S is the rate at which energy is radiated for the accelerated charge.

Things to note

1. Amplitude is proportional to acceleration.
2. Intensity is proportional to (acceleration)^2. For a classical interaction

$$a = \frac{F}{m} = \frac{(ze)(Ze)}{r^2 m}$$

$$I \propto a^2 = \frac{z^2 Z^2 e^4}{r^4 m^2} \tag{10-12}$$

where z = charge on particle

   Z = atomic number of nucleus

3. If Z, z, e, and r are constant,

$$I = \text{intensity of radiation is proportional to } \frac{z^2}{m^2}$$

$$I \propto \frac{1}{m^2} \tag{10-13}$$

Example

Consider an electron and a proton in their abilities to produce x-rays.

$$\text{mass ratio } \frac{m_e}{m_p} = \frac{9 \times 10^{-31} \text{ kg}}{1 \times 10^{-27} \text{ kg}} = 10^{-3}$$

$$I \propto \frac{1}{m^2} \text{ or } 10^6$$

Thus electrons are more efficient at producing x-rays than protons, by a factor of 10^6.

We can better understand now the utility of using high-energy electrons in the production of x-rays. Figure 10-1 depicts Bremsstrahlung or continuous radiation as a spectrum of intensities as a function of wavelength. This is due to the fact that a high energy electron can have a number of atomic interactions causing it to lose energy before it interacts with a nucleus to produce Bremsstrahlung. This interaction with a nucleus can take place at any energy level of the electron. Secondly, the incident electrons arrive at the target with a distribution of energies due to the nature of the emission process that produces them. This emission process will be taken up in this chapter under the topic of thermionic emission.

The point of maximum energy, λ_{min}, in Figure 10-1 can be calculated from equation (10-4). If we consider an electron accelerated through a potential of 150 keV, then the minimum wavelength of the radiation produced if all the kinetic energy of the electron is converted to radiation is given below.

$$\lambda_{min} = 12400/150000 = 0.08 \text{ Å}$$

Final notes on Bremsstrahlung.   The energy of the resulting photon of x-radiation depends on the following:

1. The original kinetic energy of the electron.
2. How close the electron comes to the nucleus.
3. The charge of the nucleus.
4. The spectrum of radiation is due to the fact that there is a velocity and a corresponding kinetic energy distribution in the electron beam and as the electron moves through the target material there are multiple interactions with other atoms before the electron expends all its energy.

5. If the electron collides with the nucleus, all the electron kinetic energy is lost as radiation.

Characteristic X-rays

Characteristic x-rays are produced when high-speed electrons interact with orbital electrons, resulting in ionization or excitation of the orbital electrons (Figure 10-3). When electrons penetrate a target, their paths are very complex and any number of atomic interactions are possible. The *probability* of any given type of interaction is a function of the energy, E, of the electron and the atomic number, Z, of the target material.

An electron with a high-enough kinetic energy may interact with an orbital electron and eject it from the atom. This can take place if the energy of the electron has more kinetic energy than the *binding energy* of the orbital electron. When an electron is ejected from an atom, the atom is in an excited state. For example, if a K-shell electron is ejected, an electron from another shell will move in, with subsequent release of monofrequency radiation $h\nu$. The types of transition that can take place are as follows:

Transition	Characteristic X-ray
$L \longrightarrow K$	K_{α}
$M \longrightarrow K$	K_{β}
$N \longrightarrow K$	K_{γ}
$M \longrightarrow L$	L_{α}
$N \longrightarrow L$	L_{β}

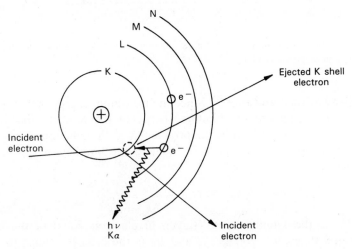

Figure 10-3 Interaction of an incident electron with an orbital electron.

Energy of emitted photon of x-ray. The energy of the radiated photon is equal to the difference in binding energies of the shells. If W_K, W_L, W_M, and W_N are the binding energies of the K, L, M, and N shells of tungsten, $W_K = 69.5$ keV, $W_L = 12.1$ keV, and $W_M = 2.81$ keV.

$$K_\alpha = W_K - W_L = 69.5 - 12.1 = 57.4 \text{ keV} \tag{10-14}$$

$$\nu_{K\alpha} = \frac{W_K - W_L}{h} \quad \text{frequency of the radiation}$$

The maximum energy of a photon cannot be greater than the binding energy of the K shell of the target material. For tungsten that would be 69.5 keV and for copper, 9 keV. The L shell for tungsten is 12.1 and the M shell for tungsten is 2.81 keV. These energies are of little use in radiology and tend to be absorbed by the target itself, transforming the energy of the incident electrons into heat. This monofrequency radiation is seen in Figure 10-4 as superimposed peaks on the Bremsstrahlung spectrum and appears at energies that are characteristic of the target material.

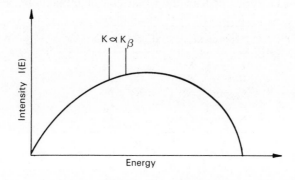

Figure 10-4 Radiation intensity as a function of energy, showing continuous and characteristic radiation.

X-ray critical voltages. Since the maximum energy of characteristic x-radiation is equal to the binding energy of the K shell, the minimum voltage across an x-ray tube to produce K-shell radiation must provide electrons whose kinetic energy is equal to the binding energy of the K-shell material.

$$W_K = H\nu_K = \frac{hc}{\lambda_K} = V_K e \tag{10-15}$$

when Ve is the kinetic energy (given in electron volts) of the electron accelerated through a voltage V, and V_K is the critical voltage to remove a K-shell electron. Increases in voltage beyond this only increase the radiation intensity.

Efficiency of x-ray production. The fraction, F, of an electron's kinetic energy, E_0, which is converted to x-ray energy is a function of the following two items:

1. The atomic number, Z, of target material
2. The electron energy, E_0

Thus the fraction of kinetic energy converted to x-rays is given in the equation (10-16).

$$F = \frac{\text{x-ray energy produced}}{\text{energy of incident electron}} = KZE_0 \qquad (10\text{-}16)$$

K has been empirically derived by various investigations and has values

$$0.7 \times 10^{-6} \leqslant K \leqslant 1.1 \times 10^{-6}$$

for E_0 expressed in keV.

Example

For tungsten $Z = 74$, voltage $= 100$ kVP, and $K = 1.1 \times 10^{-6}$. Find F.

Solution

$$F = (1.1 \times 10^{-6})(74)(100) = 8.14 \times 10^{-3} = 0.814\%$$

The heat produced is approximately equal to 99.2% of total energy.

X-RAY POWER SUPPLIES

We have seen that to produce x-rays, electrons possessing high kinetic energy must impinge on a target. Energies in the range 20 to 150 keV are the most commonly used energy levels in medical radiography. We turn our attention now to developing methods which produce ballistic electrons of an appropriate energy level that in turn will produce x-rays of the necessary energy level for the medical care intended.

To produce x-rays, we have seen that three things are required:

1. A source of free electrons
2. A means of imparting energy to the electrons
3. An appropriate target material

Force on Electron and Gain in Energy and Acceleration due to an Electric Field

Under certain conditions, electrons may escape from metal boundaries. This will be discussed later when the design of x-ray tubes is described. If electric or magnetic fields exist in the region into which they escape, and if there is a vacuum sufficiently high so that collision with gas particles is relatively rare, the field will control the subsequent movement of the electrons.

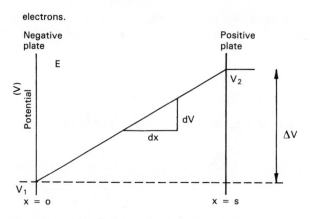

Figure 10-5 Electric field and potential between two plates.

electrons. Figure 10-5 depicts the potential gradient between two electrode plates which gives rise to the E field that accelerates electrons towards the target.

$$E = \text{electric field intensity, V/m}$$

$$\Delta V = V_2 - V_1 = \text{potential difference}$$

$$\frac{dV}{dx} = \text{potential gradient} = E$$

Consider the charge q_e of an electron moved a distance $d\mathbf{L}$. The force on q_e due to the electric field $\mathbf{E}$ is given by

$$\mathbf{F} = q_e\mathbf{E} \tag{10-17}$$

The energy expended by the electrical field in moving the charge in a direction $d\mathbf{L}$ is the work done.

$$dW = q_e\mathbf{E} \cdot d\mathbf{L} \tag{10-18}$$

$$\text{total work done } W = q_e \int \mathbf{E} \cdot d\mathbf{L} = q \int E \, dL \cos\theta$$

If the field is uniform and the electrons move in the direction of the field, $\theta = 0$ and $dL = dx$.

$$W = q_e \int_0^s E \, dx$$

but

$$W = q_e \int_0^s E \, dx = q_e \int_{V_1}^{V_2} \frac{dv}{dx} \, dx = q_e \int_{V_1}^{V_2} dv$$

$$= q_e(V_2 - V_1) \tag{10-19}$$

This illustrates that the work done on the charge q_e is equal to the charge on the electron times the difference in potential the electron has been accelerated through. This, in turn, is equal to the gain in kinetic energy of the electron. If we let

$$V_1 = 0 \text{ (ground state)}$$

then

$$V_2 = V \text{ (voltage)}$$

and the gain in kinetic energy is given by

$$W = q_e V_2 = \tfrac{1}{2} m_e v^2 \tag{10-20}$$

We see that to impart a given amount of energy to an electron it must be accelerated by an electric field through a voltage difference V. To produce accelerating potentials that are high enough for the production of x-rays, a high-voltage transformer is used. The characteristics of this device are described in the following section.

The Transformer

In the study of transformers, it is useful to introduce a device known as the *ideal transformer* (see Figures 10-6 and 10-7). It is a coupled circuit with the following properties:

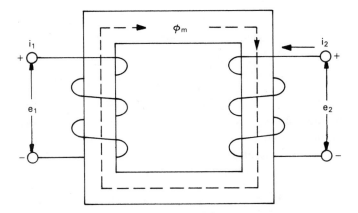

Figure 10-6 Idealized representation of a transformer.

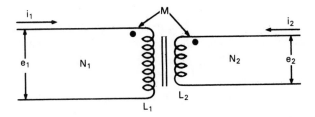

Figure 10-7 Circuit diagram for an ideal transformer.

1. Its windings have zero resistance.
2. Its core has infinite permeability.
3. Its coefficient of coupling is unity.
4. Eddy current and hysteresis losses in the core are zero.

In Figure 10-7, L_1 and L_2 are the self-inductance of the primary and secondary, and M represents mutual inductance. The core flux (Figure 10-6), which is the flux linking both windings (it may have more than two windings), is given the symbol ϕ_m, the subscript m meaning "mutual." Then, by Faraday's law, the *induced voltages* are

$$e_1 = N_1 \frac{d\phi_m}{dt} \qquad e_1(t) = L_1 \frac{di_1}{dt} - M \frac{di_2}{dt} \tag{10-21}$$

$$e_2 = N_2 \frac{d\phi_m}{dt} \qquad e_2(t) = M \frac{di_1}{dt} - L_2 \frac{di_2}{dt} \tag{10-22}$$

If we assume the core is infinitely permeable and the flux is finite, the total magnetomotive force (mmf) around the core must be zero. Thus

$$N_1 I_1 + N_2 I_2 = 0 \text{ amp-turns} \tag{10-23}$$

Equations (10-21) to (10-23) lead to two important relationships for the ideal transformer. They are

$$\frac{e_1}{e_2} = \frac{N_1}{N_2} \tag{10-24}$$

and

$$\frac{I_1}{I_2} = -\frac{N_2}{N_1} \tag{10-25}$$

We call the quantity N_1/N_2 the *turns ratio*. Equation (10-24) says that the ratio of induced voltages is proportional to the turns ratio. Equation (10-25) in words is: The ratio of currents is inversely proportional to the negative reciprocal of the turns ratio.

Referred quantities.   If the primary voltage is given, then, knowing the turns ratio, the secondary voltage is known.

$$e_2 = \frac{N_2}{N_1} e_1$$

The quantity $(N_2/N_1)e$ is called the primary voltage referred to the secondary; or the primary voltage in secondary terms. In a similar manner,

$$I_1 = -\frac{N_2}{N_1} I_2$$

This says that the quantity $(N_2/N_1)I_2$ is the secondary current *referred* to the primary.

Example

An ideal transformer has 100 turns on its primary and 200 turns on its secondary. The flux in the core is given by

$$\phi_m = 0.02 \sin 377t \qquad \text{webers}$$

By Faraday's law,

$$e_1 = N_1 \frac{d\phi_m}{dt} = 754 \cos 377t \qquad \text{volts}$$

This voltage has a root-mean-square (rms) value of $754/\sqrt{2}$ volts. We would say: The primary induced voltage is $754/\sqrt{2}$ volts, and the secondary induced voltage is the turns ratio times the primary voltage. The secondary voltage is $1508/\sqrt{2}$ volts, and this secondary voltage referred to the primary is $754/\sqrt{2}$ volts. In a similar manner, if the effective value of secondary current is 10 A,

$$I_2 = 10 \text{ A}$$

we would say: The secondary current referred to the primary is 20 A.

Referred impedances.   Consider an ideal transformer with a primary source, voltage e_1, and a resistance R_2 and inductance L_2 in series with its secondary (Figure 10-8). It supplies a current I_2 to a load. Because the secondary current is out of the dot while primary current is into the dot, equation (10-23) must be changed to $N_1 I_1 - N_2 I_2 = 0$.

For the secondary loop,

$$e_2 = v_t + I_2 R_2 + L_2 \frac{dI_2}{dt}$$

By equation (10-24),

$$e_1 = \frac{N_1}{N_2} e_2$$

$$e_1 = \frac{N_1}{N_2} v_t + \frac{N_1}{N_2} I_2 R_2 + \frac{N_1}{N_2} L_2 \frac{dI_2}{dt}$$

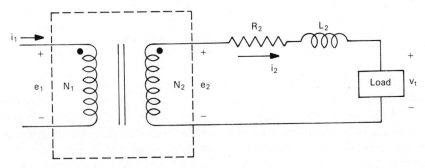

Figure 10-8 Circuit diagram of ideal transformer and equivalent secondary resistance and inductance.

Now, replace I_2 by its equivalent in primary terms:

$$I_2 = \frac{N_1}{N_2}I_1$$

$$e_1 = \frac{N_1}{N_2}v_t + R_2\left(\frac{N_1}{N_2}\right)^2 I_1 + L_2\left(\frac{N_1}{N_2}\right)^2 \frac{dI_1}{dt}$$

$$\frac{N_1}{N_2}v_t = e_1 - R_2\left(\frac{N_1}{N_2}\right)^2 I_1 - L_2\left(\frac{N_1}{N_2}\right)^2 \frac{dI_1}{dt} \qquad (10\text{-}26)$$

Note that $(N_1/N_2)v_t$ is the terminal voltage referred to the primary of the ideal transformer. The circuit represented by this equation is shown in Figure 10-9. This process has moved R_2 and L_2 *across* the ideal transformer, requiring a modification in their values. We would say "the secondary resistance referred to the primary is $(N_1/N_2)R_2$ ohms." A similar statement may be made for the inductance.

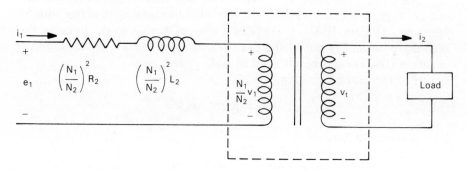

Figure 10-9 Circuit diagram using the concept of reflected impedance from the secondary to the primary circuit.

Example

An ideal transformer with $N_1 = 100$ turns and $N_2 = 200$ turns has a resistance if $3\ \Omega$ is in series with its secondary. This resistance referred to the primary side is $3 \times (100/200)^2$ or $\frac{3}{4}\ \Omega$. Thus $\frac{3}{4}\ \Omega$ placed in series with the primary *replaces* the 3-Ω resistor in the secondary insofar as *terminal* voltages are concerned.

Transformer ratings.   Transformer capacity is rated in kilovolt-amperes (kVA). An output rating for a transformer is based on the maximum current that the transformer can carry without exceeding a certain temperature rise or reaching magnetic saturation of the core material. Power in an ac circuit depends on the power factor of the load as well as the amount of current flowing. Since the load is generally unknown to the transformer manufacturer, the transformers are rated in kVA, which is independent of the power factor (PF).

Example

(a) What is the rated kilowatt output of a 5-kVA 2400/120-V transformer at 100%, 80%, and 30% power factor? (b) What is the rated current output?

Solution

  (a) The power delivered in kW is given by $P = \text{kVA} \times \text{PF}$, so

$$P = 5 \times 1 = 5 \text{ kW}$$

$$P = 5 \times 0.8 = 4 \text{ kW}$$

$$P = 5 \times 0.3 = 1.5 \text{ kW}$$

  (b) The rated current output is

$$I = \frac{\text{kVA}}{120} = 41.7 \text{ A}$$

Full rated current of 41.7 A is supplied by the transformer at three different power factors even though the kW output is different in each case.

Transformer voltage regulation. The voltage regulation of a transformer is defined as the change in output voltage (produced by a change in load current from zero to rated value) divided by the rated value:

$$\text{regulation} = \frac{\text{no-load voltage} - \text{full-load voltage}}{\text{full-load voltage}} \qquad (10\text{-}27)$$

Example

Consider the circuit shown in Figure 10-10. Let the output of a transformer be 100 kV. It will feed an x-ray tube whose equivalent impedance is 40,000 Ω. Maximum current through the load is 300 mA. Find (a) the percent regulation and (b) the primary voltage, V_p, to produce rated secondary voltage.

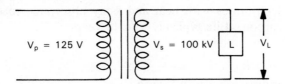

$V_p = 125$ V $V_s = 100$ kV   L V_L

Figure 10-10   Circuit diagram for the transformer.

Solution

  (a) $V_L = I_L Z_L = (0.3)(40 \text{ k}\Omega) = 12{,}000 \text{ V}$

$$\text{full-load voltage} = 100{,}000 \text{ V} - 12{,}000 \text{ V} = 88{,}000 \text{ V}$$

$$\text{regulation} = \frac{100 \text{ kV} - 88 \text{ kV}}{88 \text{ kV}} = 13.64\%$$

  (b) Turns ratio $= V_s/V_p = 100 \text{ kV}/125 = 800$ at full-load $V_s = 88$ kV, but we want 100 kV at 300 mA. We need $V_s = 100 \text{ kV} + 12 \text{ kV} = 112 \text{ kV}$:

$$V_p = V_s \frac{N_p}{N_s} = \frac{V_s}{\text{turns ratio}} = \frac{112 \times 10^3}{800} = 140 \text{ V}$$

The autotransformer. If we wish to change the output voltage of a high-voltage transformer, the voltage at the primary must be changed. We can do this with a dropping resistor in the primary, but this is a waste of power. Consider a transformer of the form shown in Figure 10-11. If $N_s < N_p$, then $V_s < V_p$. Autotransformers are used in the range 1 to 2.5 turns ratio.

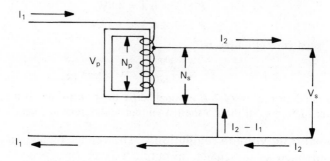

Figure 10-11 Schematic of an autotransformer.

High-voltage transformer characteristics

1. Range of output voltages: 20 to 150 kV
2. Turns ratio: approximately 500:1
3. Input voltage: 110/220 single-phase or three-phase
4. Electrical insulation: usually oil
   a. Oil has 10 to 15 times the dielectric strength of air.
   b. Oil deteriorates with heat and x-ray absorption to organic acids and carbon. Life typically 5 to 15 years.
   c. Acts as a coolant.

THEORY OF DESIGN OF X-RAY TUBES

We have seen that the three conditions necessary for the production of x-rays are:

1. A source of electrons
2. A means of accelerating the electrons
3. A target in which the kinetic energy of the electrons is converted to x-rays

The x-ray tube is constructed to satisfy these conditions (see Figure 10-12). It has three basic component parts:

1. *Cathode:* supplies electrons

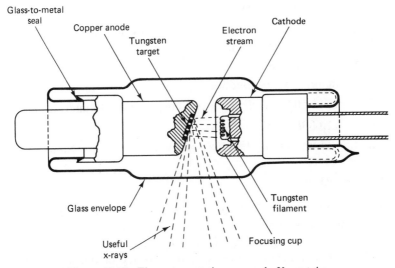

Figure 10-12 Elementary stationary anode X-ray tube.

2. *Anode:* which includes the target
3. *Glass envelope:* to maintain a vacuum; at least 10^{-6} mmHg

The Production of Electrons: Thermionic Emission

To describe properly the amount of current that flows in the situation when an electric field is placed across an anode and cathode, it is necessary to examine two things:

1. The factors determining the number of electrons that can be emitted from the cathode material
2. The effect that these emitted electrons have on each other as well as the influence of the electric field acting on the electrons

Hot cathodes are prolific producers of free electrons. This was discovered by Edison in his light-bulb experiment and studied by Richardson and Dushman. The production of electrons using heat is called *thermionic emission.* On the basis of theoretical calculations using classical kinetic theory, both investigators were able to show that *electrons could be liberated from metals provided that sufficient thermal energy was imparted to the fastest-moving electrons* so that they could overcome the surface potential barrier.

Expressed in terms of current resulting from the liberated electrons, they showed that the thermionic current was related to the temperature of the cathode by the equation

$$I_{th} = aAT^2e^{-(E_w/E_T)} = aAT^2e^{-b/T} \qquad (10\text{-}28)$$

where I_{th} = thermionic emission current, A

$\quad a$ = a constant characteristic of the material, A/m^2-(°K)^2

$\quad A$ = cathode emitter surface area

$\quad T$ = temperature, °K

$\quad E_W$ = material work function

$\quad E_T$ = $T/11{,}600$, the electron-volt equivalent of temperature, eV

$\quad b$ = $11{,}600 E_W$, °K

This equation is known as the *Richardson–Dushman equation.*

Work Function

The quantity E_W is called the work function. It represents the amount of energy that must be imparted to fast-moving electrons in a metal in order to free them from the bounds of the metal. See Figure 10-13, where E_B is the potential-energy barrier existing at the surface of the metal, and E_f is the Fermi energy level, representing the maximum kinetic energy that an electron can possess at 0°K. As the temperature is increased, E_f can become $>E_B$ and electrons escape:

$$E_W = E_B - E_f \qquad \text{eV} \qquad (10\text{-}29)$$

For a cathode emitter, E_W ranges from 1 to 4.5 eV.

Example

For a particular x-ray tube, let $a = 6.02 \times 10^5$, $b = 5.25 \times 10^{4°}$K for tungsten, $A = 1$ cm^2, $T = 2200$°K. Use the Richardson–Dushman equation (10-28) to find the thermionic current.

Solution

$$I_{th} = (6.02 \times 10^5)(10^{-4})(2200)^2 e^{-(5.25 \times 10^4)/2200}$$

$$= 12.6 \text{ mA}$$

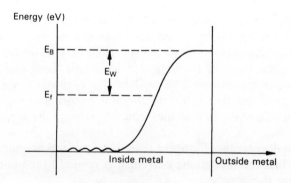

Figure 10-13   Energy states inside a metal.

Space-Charge Limiting

Figure 10-14 shows a plot of the anode current as a function of temperature for fixed values of voltage. If the anode–cathode voltage is fixed, the current becomes a function of temperature. The Richardson–Dushman equation holds quite well until a temperature is reached where the positive voltage between the cathode and anode is not large enough to remove all the emitted electrons. At this point the current becomes independent of the temperature for a fixed tube voltage and deviates from the prediction of the Richardson-Dushman equation. Therefore, for each voltage a saturation current I_s is reached. This deviation is caused by an equilibrium between the attractive force due to the anode–cathode voltage on the electrons and the mutual repulsion between all the electrons, which causes a portion of the emitted electrons to return to the hot cathode. When equilibrium is established, it is as if there were a static charge around the cathode. This effect is called the *space-charge effect*. This is a limiting effect which makes the anode current, I_a, less than the thermionic current, I_{th}, that is,

$$I_{th} > I_a$$

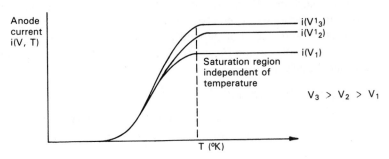

Figure 10-14 Anode current as a function of cathode temperature for constant anode–cathode voltage.

Derivation of the Child–Langmuir Law

We wish to derive the relationship between the voltage $V_{a\text{-}c}$ between the anode and cathode of an x-ray tube and the current through it, $i(V)$. Assume a simple parallel-plane configuration as shown in Figure 10-15. Also assume that the electrons start at rest; that is, their kinetic energy is zero, the cathode temperature is sufficiently high to cause the emission of all the needed electrons and that the dimension of the plates is large compared with the separation.

None of these assumptions are actually true for x-ray tubes, but laboratory measurements coincide reasonably well with the derived relation. We start with Gauss' law, which relates the electric field intensity (hence

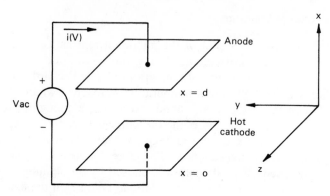

Figure 10-15 Schematic of a parallel-plane configuration with a voltage imposed between the planes.

the voltage) to the charge (hence the current):

$$\int_{surface} \mathbf{D}_n \cdot d\mathbf{S} = \int_{vol} \rho \, dv = Q$$

that is, the total electric flux density across a surface equals the charge enclosed by the surface. By virtue of rectangular symmetry of the parallel planes, we can infer that the electric field intensity, $\mathbf{E} = \mathbf{D}/\varepsilon_0$, lies only in the X direction and does not depend on Y or Z in any way.

Consider a small volume $dv = dx \, dy \, dz$, where the charge density is $\rho(x)$; then (see Figure 10-16)

$$
\begin{aligned}
\int \mathbf{D}_n \cdot d\mathbf{S} &= D(x) \, dy \, dz + [D(x + dx)] \, dy \, dz \\
&= -D(x) \, dy \, dz + [D(x) + dD(x)] \, dy \, dz \\
&= dD(x) \, dy \, dz \\
&= \frac{dD(x)}{dx} \, dx \, dy \, dz
\end{aligned}
\tag{10-30}
$$

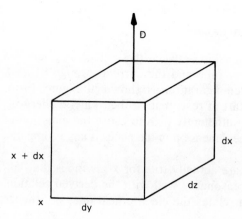

Figure 10-16 Incremental volume dv.

This equals the total charge enclosed:

$$\frac{dD(x)}{dx} dx\, dy\, dz = \rho(x)\, dx\, dy\, dz$$

or

$$\frac{dD(x)}{dx} = \rho(x) \tag{10-31}$$

$$\frac{dE(x)}{dx} = \frac{\rho(x)}{\varepsilon_0}$$

But $V(x) = -\int E(x)\, dx + \text{constant}$:

$$E(x) = -\frac{dV(x)}{dx} \tag{10-32}$$

We can relate the voltage to the charge density by

$$\frac{d^2 V(x)}{dx^2} = -\frac{\rho(x)}{\varepsilon_0} \qquad \text{Poisson's equation} \tag{10-33}$$

The charge within a volume element is not stationary but moves with a velocity $U(X)$. Thus the current density

$$J = \rho(x) U(x) \tag{10-34}$$

Under equilibrium conditions, then, J is a constant. If a charged particle q starts at rest and accelerates through a potential difference $V(x)$, it is easily shown that

$$u(x) = \sqrt{\frac{2q}{m}[-V(x)]} \tag{10-35}$$

$$\frac{d^2 V(x)}{dx^2} = -\frac{1}{\varepsilon_0}\sqrt{\frac{m}{2q}}\frac{J}{\sqrt{-V(x)}} \tag{10-36}$$

If the particles are electrons,

$$\frac{d^2 V(x)}{dx} = -1.90 \times 10^5 \frac{J}{\sqrt{V(x)}} \tag{10-37}$$

The result at this point is a nonlinear differential equation. From the theory of differential equations we can assume a solution of the form

$$V(x) = kx^n$$

Substitution produces

$$k^{1.5}(n-1)nx^{1.5n-2} = -1.90 \times 10^5 J \tag{10-38}$$

which is true for all x; thus, $n = \frac{4}{3}$ and substitution produces a value of k such that

$$V(x) = (-\tfrac{9}{4} \times 1.90 \times 10^5 J)^{2/3} x^{4/3}$$

Solving for J, we have

$$J = -\frac{4}{9}\frac{1}{1.9 \times 10^5}\frac{[V(x)]^{3/2}}{x^2} \tag{10-39}$$

where J is in the negative downward direction. For anode current given by $i(V) = -JA$, and $V = V(x)$ evaluated at $x = d$, the anode current becomes

$$i(V) = 2.34 \times 10^{-6}\frac{A}{d^2}V^{3/2} \tag{10-40}$$

This equation is known as the *Child–Langmuir law* [10]. It demonstrates that the anode current depends on the $\frac{3}{2}$ power of the anode–cathode voltage when the current is limited by the interelectrode space charge.

Child–Langmuir law.   From the development above, equation (10-40) shows that when the tube is space-charge limited, the anode current becomes a function of the cathode–anode voltage. The relationship that accounts for this is the *Child–Langmuir law* and describes the anode current when the system is space-charge limited. Equation (10-41) is a version of equation (10-40) that is applicable to x-ray tubes.

$$I_a = 2.34 \times 10^{-6}\frac{A}{d^2}V_{a\text{-}c}^{3/2} \tag{10-41}$$

where A = cathode–anode area of parallel plate, m^2

$\quad\quad\ \ d$ = anode–cathode distance, m

$\quad\quad\ \ V_{a\text{-}c}$ = anode-to-cathode voltage

Assumption in derivation

1. Electrodes are flat, parallel equipotential surfaces.
2. Only electrons having zero velocities are present in the interelectrode space.
3. For a cathode–anode voltage V, the number of electrons emitted by the cathode exceeds the number that can be received by the anode.

This is for an ideal case of parallel electrodes. In the case of an x-ray tube the cathode and anode are not parallel plates. Under these circumstances, Langmuir showed that the current collected at the anode continued to be functionally dependent solely on the $\frac{3}{2}$ power of V. The only change is a geometry factor K that depends on shape and separation. Thus equation (10-41) becomes

$$I_a = KV_{a\text{-}c}^{3/2} \tag{10-42}$$

If we plot tube current against tube voltage, we have the graph shown in Figure 10-17, where I_c is the cathode heating current. There are two regions of operation:

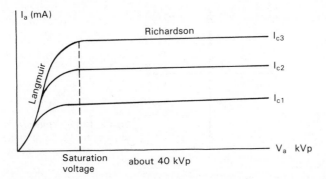

Figure 10-17   Anode current as a function of anode–cathode voltage for constant cathode current.

1. Langmuir region—where the anode current is space-charge limited
2. Richardson region—where the cathode temperature determines the anode current

Forward resistance of x-ray tube.   When operating in the space-charge-limited region (Langmuir), the forward resistance is given by the slope of the volt–ampere curve:

$$R_f = \frac{V_a}{I_a} = \frac{V_a}{KV_a^{3/2}} = \frac{1}{KV_a^{1/2}} = \frac{1}{K}\frac{1}{\sqrt{V_a}} \qquad (10\text{-}43)$$

Thus the forward resistance of the x-ray tube is proportioned to $1/\sqrt{V_a}$. By the nature of its construction an x-ray tube acts like a vacuum tube diode as pictured in Figure 10-18.

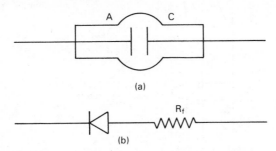

Figure 10-18   X-ray tube (a) and associated equivalent circuit (b).

Ideal characteristics and equations.   Figure 10-19 shows the characteristic of an ideal diode where

$$V = 0 \qquad i > 0$$
$$i = 0 \qquad V < 0$$
$$Vi = 0 \qquad \text{for all } i \text{ and } V \text{ (no power absorbed)}$$

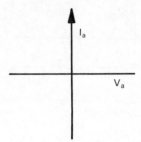

Figure 10-19 Voltage–current characteristics for the ideal diode.

With a 60-Hz high-voltage applied across the anode–cathode electrodes of the x-ray tube, there are two distinct regions of operation:

1. Langmuir region, where $I = KV_a^{3/2}$ for a constant filament current
2. Richardson region, where $I_s \propto T^2 e^{-b/T}$

Since the saturation current, I_s, occurs during the peak voltage (Figure 10-20), the filament current must be stabilized since emission changes very rapidly as a function of filament current I_{fil} because filament current determines the cathode temperature and emission current must be kept below the saturation current I_s (Figure 10-21).

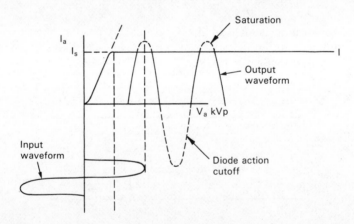

Figure 10-20 Clipped half-wave rectification due to cathode emission saturation.

To save time in heating the filaments and to prevent undue evaporation loss of tungsten from the filament, the heating time must be limited to the time necessary to achieve the desired temperature plus the time for a single exposure or a series of exposures. To keep this time to a minimum, the filament has a standby preheater which is increased to the operating value during the actual exposure.

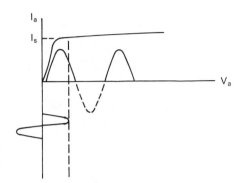

Figure 10-21 Half-wave rectification produced by an x-ray tube when the saturation level is not reached.

X-RAY TUBE ANODE DESIGN

The first requisite of an x-ray-producing machine is to produce a clear, sharp image of biological material. In order that the x-ray picture be sharp, ideally, a point source should be used at the anode. Two problems with realizing this are:

1. Since more than 99% of the input energy produces heat, the heat loading would melt the point source.
2. We cannot focus an electron beam to the size of a point source nor design a tungsten filament with a small enough surface area to produce the necessary number of electrons.

Suboptimal Solution

From a heat-loading point of view, we would want the largest area possible for dissipation of heat and a large surface area of the emitting filament to produce enough electrons for the radiation required. A method known as *line focus principle* achieves a compromise of these demands. The *focal spot* is the area of the tungsten target anode, which is bombarded by the ballistic electrons from the cathode. The size and shape of the focal spot is determined by the size and shape of the electron stream when it hits the anode. If the focal spot is of dimensions $ab \times cd$ and the target is angled by $\phi°$, the projection of the target focal spot on a plane below the x-ray tube is $cd \times ab \sin \phi$. If by design $cd = ab \sin \phi$, the apparent focal spot is $(cd)^2$.

The size of the effective focal spot affects the detail on a radiograph: the smaller it is, the better the detail. The size of the effective focal spot is determined by

1. The angle ϕ of the target
2. The size of the actual focal spot
3. The size of the cathode filament (see Figure 10-22)

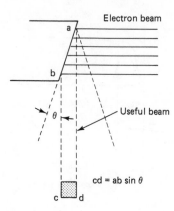

Figure 10-22   Illustration of the line focus principle.

Anode angles differ according to individual tube designs and may vary between 10° and 20° (see Figure 10-23). There is a limit to this reduction in focal spot angle which is due to a physical phenomenon known as the heel effect. Some common effective focal spot sizes are 0.3, 0.6, 1.0, and 2.0 mm.

Stationary anode.   Anode material requirements include:

1. High atomic number
2. High melting point
3. Low vapor pressure
4. Good heat conductivity

The element tungsten best satisfies these requirements. Because over 99% of the applied energy to the x-ray tube is lost as heat, heat buildup on the

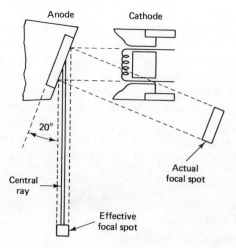

Figure 10-23   Stationary anode system illustrating the reduction in focal spot size.

anode sets the limitation on the power level and time-duration use of x-ray tubes.

Construction.   A small plate of tungsten 2 or 3 mm thick is embedded in a large mass of copper. The anode angle is usually between 15 and 20°. Tungsten must be bonded to the copper in a way that allows for the different coefficient of expansion of each of the tungsten and copper.

Heat-Loading Properties of Fixed-Anode X-Ray Tubes

For the purpose of analysis we will consider the anode as a circular cylinder. When in use, the conversion of electron kinetic energy to thermal energy in the anode can produce thermal equilibrium in a few minutes. The temperature is a maximum at the target, which is the point of impact for the electrons, and decreases moving out from the input surface or focal spot to the outside of the anode cylinder. The temperature must be kept below the *vaporization temperature of tungsten, 2200°C,* to avoid too rapid erosion of the target. If the applied voltage to the anode is alternating, using self-rectification, the anode temperature must be kept below *1200°C,* which is the temperature at which, under a reverse-polarity situation, electrons can be emitted from the anode, producing a reverse current in the x-ray tube. This condition is called *secondary emission* and will be discussed in more detail later. The junction between the tungsten and copper must be kept below the melting point of copper (800°C). For the fixed-anode x-ray tube, most of the heat produced is disposed of by means of conduction. The rate at which heat is conducted away through the anode is given by equation (10-44).

$$H = -KA\frac{dT}{dL} \qquad \text{cal/s or Btu/hr} \qquad (10\text{-}44)$$

where T = temperature
   L = length of conducting medium
   A = cross-sectional area
   K = coefficient of thermal conductivity
   dT/dL = temperature gradient

If the temperature gradient is uniform through the medium, as it is in the copper anode of the x-ray tube,

$$\frac{dT}{dL} = \frac{\Delta T}{L} = \frac{T_2 - T_1}{L} \qquad (10\text{-}45)$$

The rate of heat conduction then becomes

$$H = \frac{KA(T_2 - T_1)}{L} \qquad (10\text{-}46)$$

At temperature equilibrium, the power input in thermal energy must equal the power conducted away.

If K is in Btu/ft-hr-°F, T_1 and T_2 in °F, and L in feet, then

$$H_{\text{watts}} = \frac{0.2929KA}{L}(T_2 - T_1) \qquad (10\text{-}47)$$

If K is in cal/cm-s-°C, T_1 and T_2 in °C, and L in centimeters, then

$$H_{\text{watts}} = \frac{4.1844KA}{L}(T_2 - T_1)$$

The junction between the tungsten and copper must be kept below the melting point of copper, about 800°C. The temperature of the tungsten target must be below 2200°C. The rated power of a tube varies as the thermal area of the focal spot.

It is desirable to keep T_1, the temperature surrounding the target, as low as possible. Therefore, in some systems a liquid coolant, usually oil, is circulated through the anode.

Rotating-Anode X-Ray Tube

As the need increased in radiology for more penetrating x-rays requiring higher tube voltages and current, the limiting factor in the output of the system became the x-ray tube itself. This was due to heat generated at the anode and the metallurgical properties of the metals used in the construction of the anode.

Since the heat capacity of the anode is a function of the focal spot area, the absorbed power can be increased if the effective area of the focal spot can be increased. This is accomplished by the rotating-anode type of x-ray tube (Figure 10-24). The anode of the x-ray tube is a disk of tungsten or an alloy of tungsten and 10% rhenium. This alloy helps to reduce the changes in the anode track due to stress produced in the track as a result of rapidly changing temperatures. The anode theoretically rotates at speeds

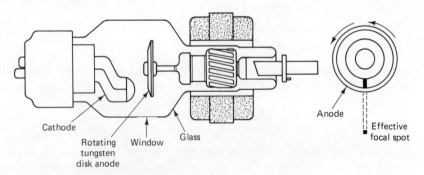

Cathode

Rotating tungsten disk anode

Window

Glass

Anode

Effective focal spot

Figure 10-24 Rotating-anode x-ray tube.

of 3600 or 10,000 rpm. The actual rate is 3000 to 3600 or 9000 to 10,000 rpm because of slip between the motor rotor and the changing magnetic fields of the induction motor used to drive the anode. The tungsten disk that represents the anode has a beveled edge that may vary between 10° and 20° depending on the use of design of the particular type. Typical angles are between 15° and 16.5°, in keeping with the line focus principle. Thus the sole purpose of the rotating anode is to spread the heat produced during an exposure over a larger area of the anode, increasing the heat loading capacity of the tube and allowing higher power levels to be used, which produces more intense x-radiation.

Engineering design considerations.   Because of the greater amount of power absorbed by a rotating-anode x-ray tube system, this also presents problems in the dissipation of this heat.

Problem

In the rotating-anode tube, absorption of heat by the anode assembly is an undesirable characteristic because the heat will be transferred partially to the bearing assembly of the rotor, causing it to expand and bind.

Solution.   The anode stem that connects the anode disk target to the rotor of the induction motor is made of molybdenum because it has a high melting point (2600°C) and is a poor conductor of heat. Thus this molybdenum provides a partial barrier between the target and the bearings of the rotor assembly.

The length of the molybdenum stem must be considered. From thermal considerations we want the stem to be long in order to increase the thermal resistance it represents. On the other hand, the longer the stem, the more the load on the bearings due to added cantilevering of the anode target producing larger moments at the bearings. There is also the increased rotational inertia which increases the time it takes for the rotating anode to reach final speed before an exposure can be taken. Typical startup time is about 1 s.

Problem

Because the anode rotation system is a high-speed system, the bearings must be lubricated. The lubricant must withstand temperatures of several hundred degrees Celsius and have a very low vapor pressure to operate in the vacuum of the x-ray tube.

Solution.   Each x-ray tube manufacturer will select the particular lubricant used. The following are some of the bearing lubricants which are used: lead, gold, graphite, silver, and molybdenum.

To extend the life of the bearings, the rotation of the anode is braked following an exposure. This is accomplished by passing a dc current through the starter winding or slowing it down using 60-Hz line frequency to bring the anode speed down to about 3000 rpm.

Maximum heat loading of rotating-anode x-ray tubes.   Since the rotating anode is essentially thermally insulated from the rotor, the primary means of heat dissipation is by radiation through the vacuum to the wall of the glass envelope, then into the surrounding oil and tube housing.

In 1879, Joseph Stephan developed the mathematical relation between the rate of emission of radiant energy per unit area and the temperature of radiating body. This relation is the *Stephan–Boltzmann law,* given as

$$P_R = e\sigma T^4 \qquad (10\text{-}48)$$

where P_R = rate of emission of radiant energy per unit area

e = emissivity of the surface (accounts for nonblack body radiation)

σ = 5.67 × 10^{-8} (Stephan–Boltzmann constant)

T = °K

The significance of the emissivity, e, is twofold:

1. It is the ratio of power radiated by a nonperfect radiator, a gray body, to that radiated by a perfect blackbody radiator.
2. It is the fraction of incident radiation that is absorbed by a gray surface, this fraction being unity for a black surface.

Thus the Stephan–Boltzmann law takes the form

$$P_R = 5.67 \times 10^{-8}\sigma T^4 = 56,700\sigma\left(\frac{T}{1000}\right)^4 \qquad (10\text{-}49)$$

in watts per square meter for a theoretical black surface.

Since any "warm" body can transfer heat by radiation, the surrounding glass envelope at temperature T_g will radiate heat energy back to the x-ray tube anode. The net energy radiated is then given by

$$P_R = 5.67 \times 10^{-8}e\sigma(T_a^4 - T_g^4) \qquad (10\text{-}50)$$

where T_a is the temperature of the anode. Now, if $T_a^4 \gg T_g^4$, the total power radiated by the anode to its surrounding is approximated reasonably well by equation (10-51).

$$P = 5.67 \times 10^{-8}eAT_a^4 \qquad \text{watts} \qquad (10\text{-}51)$$

where A is the area of the emitting surface in square meters.

Example

Consider a rotating anode that is 50 mm in diameter. The filament and focusing cup produce an electron beam of 7 mm × 2 mm. The area of bombardment of the target is 7 × 2 = 14 mm². The angle of the beveled edge is 16.5°, producing an effective focal spot of 2 mm × 2 mm. If the anode were stationary, the entire heat load would be delivered to this small 14-mm² area. If the target rotates at 3600 rpm and the average radius is *40 mm*, the bombardment area is 2π(40 mm)(7 mm) = 1759.3 mm². Thus the thermal capacity is increased due to an increase in area of 125 times.

Consider that the maximum temperature to which tungsten can be safely raised is 3000°C. Above this level, considerable vaporization of tungsten occurs. At this temperature, what is the radiate power?

Solution

$$P = 5.67 \times 10^{-8} eAT_a^4$$

$$T = 3000 + 273 = 3273°K$$

$$e = 0.35 \text{ for tungsten}$$

$$A = 2 \times \pi(0.05)^2 = 1.57 \times 10^{-2} \text{ m}^2 \text{ (both sides of anode radiant)}$$

$$P = 5.67 \times 10^{-8}(0.35)(1.57 \times 10^{-2})(3273)^4 = 35,754.8 \text{ W}$$

power radiated = 3575.8 W or J/s

This is the maximum power that can be put into the rotating anode safely on a continuous basis.

Thermal capacity of a rotating anode. If

$$m = \text{mass of anode}$$

$$c = \text{specific heat of anode material}$$

$$t_e = \text{exposure time}$$

$$T_0 = \text{initial temperature}$$

$$\Delta t = \text{time between exposures}$$

then

$$mc(T - T_0) = \sum_{i=1}^{N} W_i t_{ei} - \sum_{i=1}^{N} W_{Ri} \Delta t_i \qquad (10\text{-}52)$$

where $W_i t_{ei}$ is the energy input into the anode and $W_{Ri} \Delta t_i$ is the energy radiated between exposures. The thermal capacity of a rotating-anode x-ray tube is the energy, in joules, which the anode can accumulate as heat before reaching its temperature limit T_{max}.

Example

mass of anode = 500 g

total energy = 100 kV × 100 mA × 10 s = 100,000 J

$$C_{Tung} = 0.184 \text{ J/g}$$

$$T_{max} = T_0 = \frac{100,000}{500 \times 0.184} = 1087°C \text{ rise in temperature}$$

Heat transfer considerations. The power that can be applied to an x-ray tube is determined primarily by three factors:

1. The surface area of bombardment (focal spot size)
2. The length of exposure (exposure time)
3. The type of applied tube voltage (single-phase or three-phase)

The basic principles controlling permissible heat loads for stationary and rotating anode tubes are the same, but it is obvious that the rating of the stationary anode tube will be lower.

Since anode temperature is the limiting factor to input power, then in the design and rating of x-ray tubes, the thermal limits must be taken into consideration. From the previously developed relationship on anode thermal capacity, we see that there is a definite relationship between power input, exposure time, and anode temperature. All these factors are related by the manufacturer in the x-ray tube rating charts.

Maximum loading calculation of rotating anode.   If D is the average diameter of the track, the surface area of the thermal target track is

$$S = \frac{\pi f D}{\sin \phi} \tag{10-53}$$

where ϕ = bevel angle of the anode

$\quad\ f$ = cathode filament length, mm

If n equals the number of revolutions per second that the anode makes, then the time for one revolution, t_0, is $1/n$. If P_0 is the maximum power that can be applied to the anode during time t_0, the energy deposited per revolution of the anode is

$$Q_0 = P_0 t_0 \qquad \text{J or W-s}$$

The loading per mm^2 is

$$Q = \frac{P_0 t_0}{S} = \frac{P_0 t_0 \sin \phi}{\pi f D} \qquad \text{J/mm}^2 \tag{10-54}$$

Maximum power ratings.   The maximum power rating, P_0, is the power that can be applied during the shortest possible exposure time, t_e, without damage to the anode; that is, anode temperature does not exceed its specified safe limit. For a rotating anode, this is the power that can be applied during 1 revolution. At 3600 rpm,

$$t_e = \tfrac{1}{60} \text{ s}$$

At 9000 rpm,

$$t_e = \tfrac{1}{150} \text{ s}$$

$$\text{power} = \text{rate of doing work or an energy rate } \frac{dE}{dt}$$

$$\text{total energy} = \int_0^{t_e} \frac{dE}{dt}\, dt = \int_0^{t_e} P(t)\, dt = E_{t_e} \tag{10-55}$$

$$t_e = \text{exposure time}$$

If the power is constant,

$$E_{t_e} = P \int_0^{t_e} dt = Pt_e \qquad (10\text{-}56)$$

But for an x-ray tube system,

$$P = \text{kVp} \times \text{mA}$$
$$E_t = \text{kVp} \times \text{mA} \times t_e$$

$$\text{kVp} = \text{kilovolts peak}$$

This product is commonly called "heat units," although in reality these are the units *watt-seconds or joules.*

In considering x-ray tube power ratings, three different modes of operation must be considered when determining the input power:

1. Single-exposure operation
2. Multiple rapid exposure (as in angiography)
3. Multiple exposures during several hours of heavy use

HEAT-LOADING CHARACTERISTICS OF X-RAY TUBES AS GIVEN BY TUBE RATING CHARTS

From the heat transfer considerations developed in the preceding section, it is seen that the safe operation of the x-ray tube is temperature limited. The rise in anode temperature is a function of the rate of energy input to the anode and the rate of total heat energy dissipation by means of conduction and radiation to the tubes surrounding.

The manufacturers of x-ray tubes consolidate all the heat-loading characteristics and limitations of their product in rating charts. X-ray tubes must be operated within the limitations set by the manufacturer or permanent damage to the tube may result due to excessive heating of the anode. Since x-ray tubes and the subsequent cost of repairs are very costly, a complete understanding of tube rating charts is essential when the x-ray machine is called upon to operate at maximum energy or for a series of exposures. Thus the x-ray tube rating charts are a summation of all the anode and cathode design information, such as anode voltage and current, focal spot size, diameter, mass, conductivity, emissivity, and speed of rotation needed to produce maximum life from the tube.

Three factors must be considered when determining permissible exposures for rotating-anode x-ray tubes:

1. The maximum permissible energy value of a *single exposure*, determined from the radiographic rating charts for the focal spot used and type of voltage supplied to the anode
2. The maximum rate at which a *series of exposures*, less than 7 to 8

minutes apart, may be safely made, determined with the aid of the anode cooling chart and calculations

3. The effect of heavy exposure schedule on the *tube housing temperature,* determined from the house cooling chart

Single Exposures

Radiographic rating charts for single exposures are based on the assumptions that the anode, at the beginning of the exposure, is at room temperature and is rotating at normal speed.

Rating charts for tubes are arranged so that one may be selected for the particular type of x-ray generator and combination of focal spot sizes to be used and the other charts discarded to avoid confusion. Note that the charts apply to generators differing in the following characteristics:

1. 50- or 60-Hz frequency
2. Single- or three-phase power supply

Use of radiographic rating charts. To use the radiographic rating charts, find the intersection of two of the three exposure factors (i.e., kVp, mA, time) and see that the third is less than the value which also intersects at that point. For example, suppose that a chart like the one shown in Figure 10-25 (2-mm-focal spot and a full-wave 60-Hz, single-phase power) is being used.

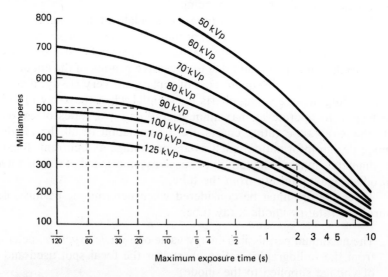

Figure 10-25 X-ray tube rating chart for the determination of maximum exposure parameters for one exposure.

If 300 mA for 2 s is selected, their point of intersection corresponds with 80 kVp, *the maximum kVp to be used with those factors*. On the other hand, if 500 mA at 90 kVp is to be used, *the maximum exposure time is slightly over 1/20 s*. Similarly, if 1/60 s at 110 kVp is to be used, *the maximum is 450*. (Usually, 400 mA would be used and the time adjusted accordingly.)

Series of Exposures

When a number of exposures are to be made within 7 to 8 minutes, such as for cine angiography, chest survey, stereo, and spot-film techniques, the amount of heat that the anode can safely accumulate must be considered. This anode heat storage capacity is defined in heat units: the product of kVp × mA × seconds, which is dimensionally watt-seconds. Sometimes the notation MAS is used. This is the product of milliamperes times seconds (MAS = mA × s).

Note. The heat units derived from the product of kVp × mA × seconds apply only to tubes operating from generators using single-phase power. For three-phase generators producing 6 pulses per cycle, the heat unit product must be multiplied by 1.35 (35% *more* heating is produced); for those producing 12 pulses per cycle, multiply by 1.40. The rationale for this will be taken up in the section on x-ray generators.

Use of anode cooling charts.   The anode cooling chart (Figure 10-26) is used to determine whether a series of exposures are permissible within a certain time period.

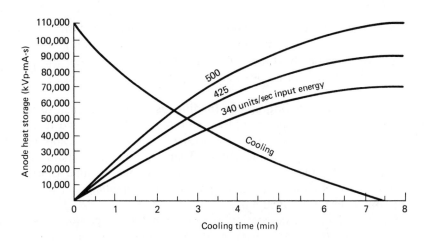

Figure 10-26   Anode heat storage and cooling chart.

Example

Assume that there have been exposures totaling 80,000 heat units on a tube with 110,000-unit anode heat storage capacity. Two minutes later, an exposure at 82 kVp and 300 MAS is made. What waiting time (if any) is required before making an exposure at 75 kVp and 150 MAS?

Solution.   Refer to the anode cooling curve for the tube (Figure 10-26) and note that 80,000 units is at a cooling time of $1\frac{1}{4}$ minutes.

Two minutes later (or $3\frac{1}{4}$ minutes total) the anode will have cooled to 45,000 units. The second exposure equals 24,600 heat units (82 × 300), so that after this exposure the total anode heat will be 69,600 units (45,000 + 24,600). The third exposure requires 11,250 heat units (75 × 150), so that the total anode heat storage will be 80,850 units (69,600 + 11,250). Since this is less than the 110,000-heat-unit capacity of the tube, the exposure can be made without any waiting time. Had this not been the case, the user would have to wait an appropriate length of time for the anode to cool down to a sufficient level so that the energy imparted to the anode from the third exposure does not cause the anode to exceed its temperature design limit.

Fluoroscopy.   Fluoroscopic exposure time is also limited in some instances by the total heat that can be safely absorbed by the anode of the x-ray tube. This limit and the cooling rate of the anode are shown on the anode cooling charts. Each curve shows three characteristics of the anodes:

1. The maximum heat storage capacity of the anode in heat units
2. The time it takes for the anode to cool from one energy level to another
3. The amount of heat that will accumulate during a fluoroscopic examination

Permissible fluoroscopic exposures are presented in terms of input energy curves, which account for the cooling that occurs during exposure. Several typical values have been plotted on the anode cooling charts. Interpolation is needed for values between plotted values.

Use of anode cooling charts for fluoroscope.   The heat to be dissipated by the tube is determined by multiplying the kVp and mA, then selecting the input energy curve corresponding to this value (or its approximate position) on the anode cooling chart. Then by finding the point on the curve that corresponds with the estimated examination time (horizontal axis), the corresponding number of heat units accumulated in the anode can be found on the vertical axis.

For example, an examination at 85 kVp and 4 mA produces 340 heat units per second (85 × 4). From the corresponding curve in Figure 10-26, it can be seen that this amount of exposure adds about 50,000 heat units to the anode after 4 minutes. Usually, this amount of heat is handled easily by the tube. Notice that the curve representing 425 heat units (H.U.) per second reaches 90,000 H.U. in approximately 7 minutes. From this point on, the number of heat units stored in the anode remains constant at a level below the maximum design level of 110,000 H.U. for this tube. The significance of this is that the anode is in thermal equilibrium with the energy input of the bombarding electrons. Thus at the temperature where equilibrium takes place, the energy input to the anode equals the energy dissipated mostly by radiation and whatever thermal conduction takes place. Thus we have the equilibrium relation

$$mc(T - T_0) = \text{kVp} \times \text{mA} - 5.67 \times 10^{-8} eAT^4 \qquad (10\text{-}57)$$

where m = mass of the anode assembly

c = specific heat of the anode assembly

T = anode temperature, °K

e = emissivity of the anode assembly

A = area of anode assembly

T_0 = ambient temperature

This means that exposure at this energy level can be continued indefinitely without damage to the anode.

For higher input levels, say 10 kVp at 5 mA (500 H.U. per second), the anode heat capacity (in this instance 110,000 units) is reached after $7\frac{1}{2}$ minutes of examination. Continued operation could damage the tube!

To find the time required for the anode to cool a given amount, locate the intersection of the original heat units on the cooling curve given in Figure 10-26, then note the equivalent time. Repeat this procedure at the lower heat unit value and determine the time difference. In summary, the input energy curves provide information on the rate of heat energy that is added to the anode during fluoroscopy and the cooling curve provides information on the rate that heat is dissipated after exposure is terminated.

Tube Housing Temperature

The third factor that may have to be considered in determining permissible exposure schedules is the total heat storage capacity of the housing (usually 1,250,000 units). This rating must be followed to prevent the oil in the housing from becoming too hot to remove the heat radiated from the anode, and to prevent damage to the interior parts of the tube assembly, such as the port, expansion diaphragm, and cable insulators.

Use of the housing cooling chart. To illustrate the use of the housing cooling chart (heat storage capacity and cooling rate), assume that 70-mm chest survey films will be made using an average exposure of 90 kVp, 200 mA, $\frac{2}{5}$ s (80 MAS) at the rate of 4 per minute.

From the radiographic rating chart (Figure 10-27) it will be noted that

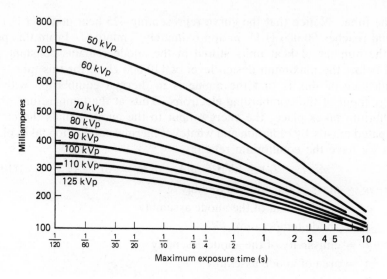

Figure 10-27 X-ray tube rating chart.

the individual exposure is well within the rating (1½ s). The total anode heat storage is 7200 units per exposure (90 × 80), and 28,800 units per minute (7200 × 4) at the rate of 4 exposures per minute. Since the anode cooling rate is generally at least 30,000 units per minute (note the thermal rating on the tube data sheets), this technique is permissible.

However, after about 44 minutes (1,250,000/28,800) a total housing heat storage in excess of 1,250,000 units would be reached. (Most tube housings have a housing heat storage capacity of 1,250,000 units; note the thermal rating on the tube data sheets.)

When this maximum is reached, it is then necessary to reduce the rate of heat input to the maximum cooling rate of the housing (note thermal rating) or less, by adding air circulation, or by reducing the number of exposures. Assuming a maximum overall cooling rate of 15,000 units (which is correct for most tube housings without air circulation), and since in this case the heat units per exposure are 7200, two exposures per minute (2 × 7200 = 14,400 units) could be made safely.

Or, assume that after operating for 44 minutes at the original rate the tube is permitted to cool for 30 minutes, which from the housing cooling chart (Figure 10-28) would reduce the heat still in the housing to about 800,000 units. It would then be possible to resume operation for 15 minutes [(1,250,000 − 800,000)/28,800 = 15]. After this time, the housing heat would reach the maximum permissible, and it would be necessary to reduce the exposures to the rate described previously.

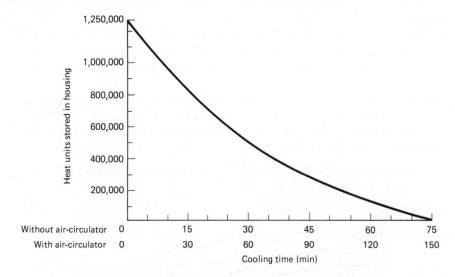

Figure 10-28   Cooling chart for x-ray tube housing.

Special Techniques

Following are the methods for calculating the duration for cineradiography and angiography exposures from a full-wave-rectified single-phase-system rating chart.

Note: All mA values referred to are full-wave, single-phase values as indicated on the tube charts and mA meter during normal continuous exposure. They are not readings made during intermittent or pulsed exposures.

Cineradiography

Method of Solution Summary

1. Determine the exposure (kVp and MAS) required for proper film density.
2. Using these factors (kVp and MA), refer to the appropriate radiographic rating chart and determine the longest single exposure that is permissible.

Note: For focal spots of 0.77 mm and larger, the kVp and mA values selected will usually permit exposure times longer than the maximum time on the rating chart. If so. the following formula is used to determine the time of film run.

maximum time for film run (seconds)

$$= \frac{\text{anode heat storage capacity (heat units)}}{\text{(heat units per exposure)} \times \text{(cine film rate)}} \qquad (10\text{-}58)$$

Example

To illustrate, assume exposure factors of 100 kVp, 20 mA, 1/60 s, and 30 frames per second using a tube having the radiographic rating chart given in Figure 10-29 and an anode heat storage capacity of 135,000 units. Calculate the maximum continuous film run.

Solution. Referring to the radiographic rating chart, Figure 10-29, note that the permissible continuous exposure for the assumed factors is over 40 s. The heat units per exposure = $100 \times 20 \times \frac{1}{60} = 33\frac{1}{3}$ heat units. Therefore, the formula for maximum film run time applies.

$$\text{maximum time for film run} = \frac{135,000}{33\frac{1}{3} \times 30} = 135 \text{ s of continuous film run}$$

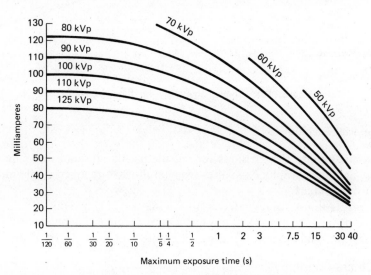

Figure 10-29   X-ray tube rating chart.

Exposure Cycle Factor.   For a given kVp and mA, the full-wave rating chart will give the longest permissible continuous exposure time, which would also be the film run time for a continuous exposure. However, the number of seconds of film run can be increased if the exposures are turned off between frames during film transport.

The *exposure cycle factor* is a number expressing the relationship between a continuous and an intermittent exposure sequence—the factor by which the film run can be increased for a sequence of *intermittent* exposures. It is expressed by the formula

$$\text{exposure cycle factor} = \frac{1}{(\text{exposure time}) \times (\text{exposure rate})} \tag{10-59}$$

Exposure cycle factors for typical cine film runs are given in Table 10-1.

TABLE 10-1   Exposure Cycle Factors Used in Cineradiography

Exposure Time (s)	Exposure Rate Per Second	Exposure Cycle Factor
1/120	60	2
1/120	30	4
1/120	15	8
1/120	$7\frac{1}{2}$	16
1/60	30	2
1/30	15	2
1/15	$7\frac{1}{2}$	2

Example

To illustrate the use of the exposure cycle factor, assume that for a tube with a smaller focal spot having the radiographic rating chart shown in Figure 10-30 for this size focal spot, the longest continuous exposure is 22 s (for 100 kVp, 20 mA). What is the maximum number of seconds of film run?

Solution.   Note from Table 10-1 that the exposure cycle factor is 2 for 30 frames per second and 1/60-s exposure. The maximum number of seconds of film run is thus $22 \times 2 = 44$ s.

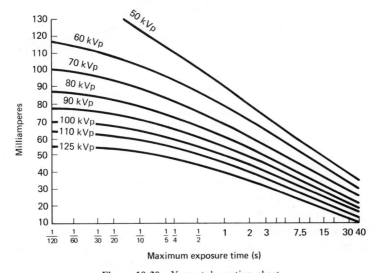

Figure 10-30   X-ray tube rating chart.

Method of Solution Summary

1. Determine the exposure cycle factor as related to a single continuous exposure.
2. To obtain the maximum number of seconds of film run, multiply the longest permissible continuous exposure by the exposure cycle factor.
3. If the film run time is too short, recalculate on the basis of a slower frame rate or larger focal spot.

Angiography

Example

To illustrate, assume a technique of 100 kVp, 400 mA, and 1/20 s exposures at the rate of 6 per second using a tube having the radiographic rating chart in Figure 10-31. Can this technique be used?

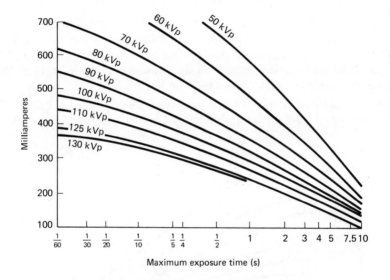

Figure 10-31   X-ray tube rating chart.

Solution.

1. From the tube rating, the maximum exposure time at 100 kVp, 400 mA is $\frac{3}{20}$ s, so the $\frac{1}{20}$-s exposure is permissible.
2. The total time for each film sequence would be 1 s (6 films).
3. The maximum allowable heat input is obtained by multiplying kVp $\times$ mA $\times$ time for any intersection at the total time (from step 2) on the rating chart. In this case either 70 kVp $\times$ 400 mA $\times$ 1 s or 100 kVp $\times$ 280 mA $\times$ 1 s could be used—28,000 heat units.

4. The heat units for the desired technique = 100 kVp × 400 mA × 1/20 s × 6 exposures/second = 12,000 units and is less than the 28,000 units permitted, so this technique could be used.

The angiographic rating charts may be used to determine the maximum number of exposures directly if desired.

Method of Solution Summary

1. Determine that the individual exposure factors are within the radiographic rating for the type of tube and focal spot employed.
2. Determine the total time for each sequence ("on" time plus "off" time).
3. Find the maximum number of heat units which can be applied to the tube for the total time determined in step 2.
4. Compare this maximum heat input with the heat input of the desired technique.

RECTIFICATION: THE APPROXIMATION OF CONSTANT POTENTIAL

With the high-voltage transformer connected directly to the x-ray tube, the tube acts as a type of vacuum-tube diode. From previous descriptions of the conditions for current flow in the x-ray tube, current will flow only when the anode is positive with respect to the cathode. If the anode becomes negative with respect to the cathode, emitted electrons from the cathode filament will be attracted back to the cathode and there will be no current flow across the tube. Thus the x-ray tube acts as a diode with respect to the flow of current through the tube. This simple high-voltage circuit for x-ray emission is given in Figure 10-32. In this circuit configuration we see that the current is unidirectional and that x-rays are produced only during the time when there is current conduction in the tube.

Mathematics of Rectified Circuits

For a periodic function as we have in the voltage and current in an x-ray system we have the following relationship: If $f(t)$ is periodic in T, then $f(t) = f(t + nT)$, $n = 1, 2, \ldots$.

Average value of a function

$$f_{ave} = \frac{1}{T} \int_0^T f(t) \, dt \qquad (10\text{-}60)$$

For half-wave rectified current the average value is

$$I_{ave} = \frac{1}{2\pi}\int_0^{2\pi} I(t)\, dt = \frac{1}{2\pi}\int_0^{\pi} I_m \sin \omega t \, d\omega t$$

$$= \frac{1}{2\pi}\left(-I_m \cos \omega t \Big|_0^{\pi}\right) = \frac{I_m}{2\pi}(2) = \frac{I_m}{\pi} \qquad (10\text{-}61)$$

$$= \frac{I_m}{\pi}$$

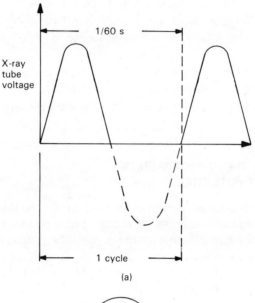

(a)

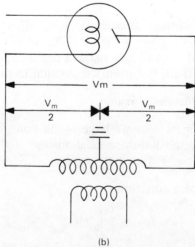

(b)

Figure 10-32 Circuit showing x-ray tube acting as a half-wave rectifier, self-rectified, one pulse per cycle.

Rms or effective value.   The rms value of a function is

$$\sqrt{\frac{1}{T} \int_0^t f^2(t) \, dt} = f_{\text{rms}} \qquad (10\text{-}62)$$

For voltage and current the effective value of each is the same as above, with $V(t)$ or $I(t)$ replacing $f(t)$.

For half-wave rectification,

$$I_{\text{eff}} = \frac{I_m}{2} \quad \text{and} \quad V_{\text{eff}} = \frac{V_m}{\sqrt{2}}$$

$$\text{power} = V_{\text{eff}} I_{\text{eff}} = \frac{V_m I_m}{2\sqrt{2}} = 1.11 \, V_m I_{\text{ave}} \qquad (10\text{-}63)$$

in a half-wave-rectified system.

Self-Rectified X-Ray System Deficiencies

The self-rectified circuit is the simplest to design, as the x-ray tube is connected directly to the secondary of the high-voltage transformer. There are, however, problems that occur when the high-voltage portion is configured in the self-rectification mode.

Secondary emission of electrons.   If the surface of the x-ray tube anode becomes hot enough, some electrons in the anode may have enough kinetic energy to overcome the work function of the material and escape from the surface. The number of these secondary electrons emitted by the anode per impinging electrons depends on the work function of the anode material and the kinetic energy of the cathode-emitted electrons.

The emission of secondary electrons poses little problem during that portion of the applied tube voltage when the anode is positive with respect to the cathode. It is during the portion of the cycle when the applied tube potential is reversed (i.e., when the anode is negative with respect to cathode) that the secondary-emission electrons are attracted to the cathode and gain kinetic energy as they move toward the cathode due to the force of the electric field that is exerted on the electrons emanating from the anode. This produces current in the reverse direction. This reverse current tends to heat the cathode to a higher temperature, which causes it to emit more electrons.

This corresponds to increased cathode emission and increased tube current. The increased tube current increases anode heating. The increased kinetic energy of the anode electrons produces an increase in secondary emission electrons, which increases reverse tube current, which further increases the temperature of the cathode.

It is seen that this is an unstable condition and will soon result in the destruction of the cathode filament. Thus the use of the self-rectified circuit is limited to use in small, low-power, portable machines, but there is now a tendency to abandon this type of circuit. When used, though, the self-rectified system is limited to approximately one-third the rating allowed on single-phase rectified power.

Use of in-line rectifiers (half-wave-rectification circuit).   By inserting diodes, either vacuum-tube or solid-state types, into the high-voltage circuit in the manner illustrated in Figure 10-33b, a half-wave-rectifying circuit for the tube current is obtained which does not suffer from the problem of tube damage due to secondary emission from the anode of the x-ray tube. In the case where vacuum diodes are used in the rectification circuit, the plate (anode) of the vacuum-tube diode is kept at a temperature below that where secondary emission will take place. This is accomplished by providing a large plate area for the emitted electrons of the diode to impinge on, and the anode-to-cathode voltage is kept relatively low so as not to impart too much kinetic energy to the electrons in the diode. Thus the electrons do not have sufficient energy to heat the anode to any significant degree. This circuit configuration produces half-wave current rectification, but the addition of the in-line rectifiers in the circuit prevents current flow in the reverse direction due to secondary emission, which in turn protects the x-ray tube. Half-wave-rectified x-ray circuits are not employed to any great extent in diagnostic radiology but find use in low-power portable x-ray systems.

Full-wave-rectification x-ray circuit.   A major deficiency of the half-wave-rectified x-ray circuit is the fact that x-rays are produced only during the half-cycle of the applied voltage, when the anode is positive with respect to the cathode. This means that the x-rays are produced in a burst for this half-cycle and a considerable amount of exposure time is lost during the half-cycle when the x-ray tube is not conducting (see Figure 10-34). In Figure 10-35 two diodes are added to make a bridge rectifier circuit. The bridge rectifier circuit produces full-wave rectification where each half-cycle of the applied sinusoidal 60-Hz voltage is applied across the anode and cathode in such a way that the anode is positive with respect to the cathode over both half-cycles. From Figure 10-35b we note the following conduction sequence. If the applied voltage to the tube is

$$V(t) = V_m \sin 120 \, \pi t$$

then for $0 \leqslant t \leqslant 1/20$ s, point A is positive with respect to B. Thus diodes 1 and 3 are forward biased (conducting) and diodes 2 and 4 are reversed biased (nonconducting). For $1/20 < t \leqslant 1/60$ s, B is positive with respect to A and diodes 4 and 2 are forward biased and 1 and 3 are reversed biased. This sequence of events produces the voltage waveform of Figure 10-35a,

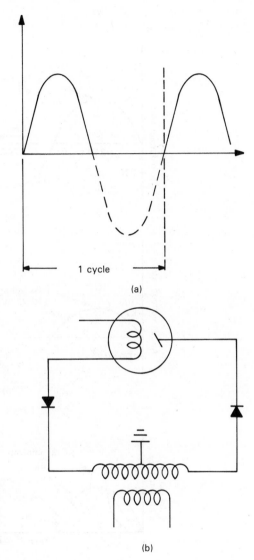

Figure 10-33 Circuit illustrating placement of external diodes in a single-phase circuit to prevent secondary emission current flow in the x-ray tube, half-wave rectified, two rectifiers, one pulse per cycle.

which is typical of full-wave rectification. An x-ray tube connected to a full-wave rectification circuit of the type illustrated above utilizes both halves of the applied sinusoidal voltage supplied by the high-voltage transformer. This means that x-rays are produced during each half-cycle of the applied voltage reducing the exposure time for a given radiation exposure. Full-wave-rectified circuits are used in medium- and high-capacity x-ray units and are the circuits most commonly employed for diagnostic x-ray examination.

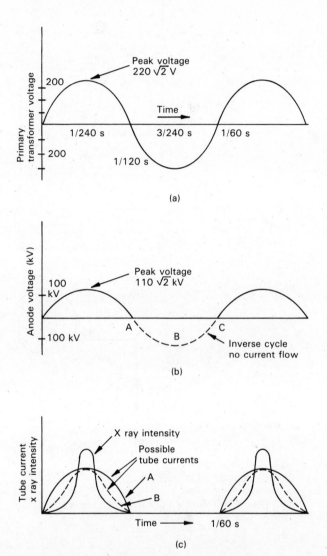

Figure 10-34 Graphs illustrating the variation with time of the voltage developed by the transformer, the x-ray tube voltage, the x-ray tube current, and the x-ray intensity for the circuit of Figure 10-35b. The tube current A or B will depend on the construction of the cathode assembly of the x-ray tube. If the filament is placed well below the surface of the cathode, electrons will flow to the anode only near the peak of the cycle, as in curve B. (From H. E. Johns and J. R. Cunningham, *The Physics of Radiology*, 3rd ed., Charles C Thomas, Publisher, Springfield, Ill., 1977.)

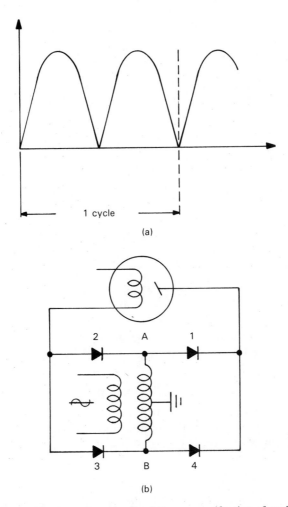

Figure 10-35 Circuit configuration for full-wave rectification of applied anode voltage using a four-rectifier, center-tapped transformer, two pulses per cycle.

VOLTAGE, CURRENT, AND POWER RELATIONSHIPS AT THE X-RAY TUBE

X-ray tubes operated in the self-rectified mode are generally not operated in the saturation regions of their operating characteristics. Referring to Figure 10-36, this means that the x-ray tube is operated in the Langmuir region, where the anode current is a function of anode-to-cathode voltage. Thus the anode current is described by the equation

$$I_a = KV_{\text{a-c}}^{3/2}$$

In this region of operation, the x-ray tube is space-charge limited. The

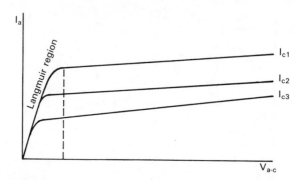

Figure 10-36 Anode current as a function of anode–cathode voltage for constant cathode current.

voltage across the tube is simply the sinusoidal voltage of the secondary at 60 Hz:

$$V(t) = V_m \sin 120 \, \pi t$$

$$V_{\text{eff}} = \frac{V_m}{\sqrt{2}} = 0.707 V_m$$

V_m is the amplitude of the applied sine wave. Since the anode current is a half-wave-rectified form, the average value of the current is given by

$$I_{\text{ave}} = \frac{I_m}{\pi}$$

where I_m is the maximum anode current.

The effective or rms value of the anode current in the half-wave-rectified mode of operation is given as

$$I_{\text{eff}} = I_{\text{rms}} = \frac{I_m}{2} = \frac{\pi I_{\text{ave}}}{2}$$

To calculate the power absorbed by the x-ray tube, we have that

$$P_T = V_{\text{eff}} I_{\text{eff}} = \frac{\pi V_m I_{\text{ave}}}{2 \sqrt{2}} = 1.11 V_m I_{\text{ave}} \tag{10-64}$$

The average value of current is used here because the current meter in the x-ray system usually reads average value and the voltmeter is set up to read peak or maximum voltage.

During the portion of the applied tube voltage cycle when the x-ray tube is not conducting, the voltage applied to the tube in the inverse direction is the "no-load" voltage of the transformer. Depending on the transformer regulation supplying the high voltage, the no-load voltage can be excessively high and could damage the x-ray tube or the high-voltage insulation of the cables if dielectric breakdown were to occur because of these excessive voltages (see Figure 10-37).

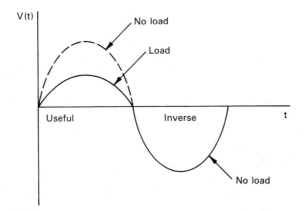

Figure 10-37 Plot of anode voltage illustrating the effect of no inverse reducer in the circuit.

To limit this excessive no-load voltage in the secondary, a power resistor in parallel with a diode is put in series with the primary of the transformer. This diode–resistor combination is sometimes called a *transient suppressor* or *inverse reducer*. The primary current is conducted by the rectifier during the conducting half-cycle and by the resistor during the nonconducting half-cycle of the x-ray tube. This is illustrated in Figure 10-38. If it is desired to have an inverse reducer load in the primary during the nonconducting portion of the cycle be the same as seen by the primary circuit in the conducting phase, the load resistor must look to the primary circuit like the load of the conducting x-ray tube. The resistance the tube presents to the transformer is computed from the inverse slope of Langmuir region of operation of the x-ray tube.

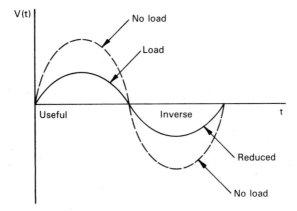

Figure 10-38 Plot of anode voltage illustrating the reduction in anode voltage due to the presence of an inverse reducer.

$$R_T = \frac{\Delta V_{\text{a-c}}}{\Delta I_a}$$

This resistance must be reflected into the primary circuit, making use of the turns ratio. Thus the value of resistance in the primary is

$$R_R = \left(\frac{N_P}{N_S}\right)^2 R_T$$

Full-Wave-Rectified X-Ray Systems

For a full-wave-rectified system, we have four diodes configured in a bridge network. They are used in conjunction with a center-tapped transformer which produces anode-to-cathode voltage that is of the form typical of a full-wave-rectified ac voltage. See Figure 10-39, where

$$0 \leq V_{\text{a-c}}(t) \leq V_m$$
$$0 \leq I(t) \leq I_m$$

Full-wave-rectified systems employ rotating-anode tubes and are usually run in the saturation region of operation.

For full-wave rectification,

$$I_{\text{rms}} = \sqrt{\frac{1}{2\pi}\int_0^{2\pi}(I_m \sin \omega t)^2\, d\omega t} = I_m\sqrt{\frac{\pi}{2\pi}} = \frac{I_m}{\sqrt{2}}$$

$$I_{\text{ave}} = \frac{1}{\pi}\int_0^{\pi} I_m \sin \omega t\, d\omega t = \frac{2I_m}{\pi} = 0.637 I_m$$

Using the foregoing methods of computation the rms and average values for the tube anode voltage are given as

$$V_{\text{rms}} = \frac{V_m}{\sqrt{2}} = 0.707 V_m \tag{10-65}$$

$$V_{\text{ave}} = \frac{2V_m}{\pi} = 0.637 V_m \tag{10-66}$$

The power absorbed by the x-ray tube anode is given by

$$P_a = V_{\text{rms}}I_{\text{rms}} = \frac{V_m}{\sqrt{2}}\frac{\pi I_{\text{ave}}}{2\sqrt{2}} = 0.785 V_m I_{\text{ave}} \tag{10-67}$$

This gives the power absorbed by the x-ray tube in terms of the peak voltage and average current, which are the meter readings on the x-ray generator.

Example

Calculate the power delivered to a full-wave-rectified x-ray tube by a generator set at 100 kVp and 400 mA.

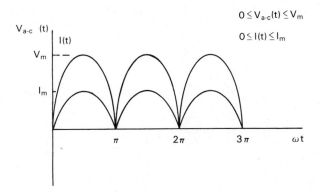

$$0 \leq V_{\text{a-c}}(t) \leq V_m$$

$$0 \leq I(t) \leq I_m$$

Figure 10-39   Voltage and current waveform for a full-wave rectified x-ray tube circuit.

Solution

$$P = 0.785 V_m I_{\text{ave}} = 0.785(100)(400) = 31,400 \text{ W}$$

HIGH-VOLTAGE CABLES IN X-RAY SYSTEMS

The primary function of high-voltage cables is to conduct the output of the high-voltage transformer to the x-ray tube. Because a typical x-ray machine can employ voltages in excess of 100 kV, the design and subsequent effect of the cable capacitance on the applied voltage to the x-ray tube must be considered.

The cable consists of an inner conductor of radius a surrounded by electrical insulating material of a dielectric constant, ε, and an outer metal braid cover over the dielectric material (see Figure 10-40). This outer metal braid is connected to the center tap of the high-voltage transformer, which is grounded. The entire cable is then covered with a protective cover of vinyl or other plastic. By using a center-tapped secondary in the high-voltage transformer (Figure 10-41), the voltage on each cable is reduced by 50% of x-ray tube voltage relative to ground. This reduces the amount of dielectric required in the high-voltage cables, making them smaller in diameter. The grounded metal braid also serves to reduce electromagnetic

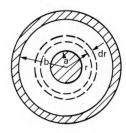

Figure 10-40   Cross-sectional view of coaxial cable.

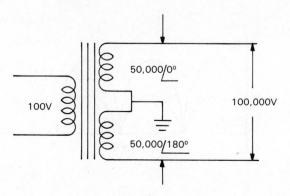

Figure 10-41 High-voltage transformer with center-tapped secondary.

radiation from the cables and serves as a safety path to ground for the high voltage should there be a breakdown in the dielectric material for any reason.

Because the high-voltage cable contains an inner conductor and an outer conductor separated by dielectric material, we can consider the cable as a tubular electrical capacitor.

Referring to Figure 10-40, the electric field strength in an annular space is given by

$$E = \frac{q}{2\pi r \varepsilon} \qquad \text{V/m} \tag{10-68}$$

where q = charge, coulombs per meter of length

$\quad r$ = radial distance, meters from the line charge

$\quad \varepsilon$ = dielectric constant of the insulating material around the center conductor

The potential difference, V, between the inner and outer conductors of the cable is the integral of the electric field strength over the distance between the two conductors: that is,

$$V = \int_a^b \mathbf{E} \cdot d\mathbf{l} = \int_a^b \frac{q}{2\pi r \varepsilon} \, dr$$

$$= \frac{q}{2\pi \varepsilon} \ln \frac{b}{a} \tag{10-69}$$

Since capacitance is defined by

$$C = \frac{q}{V}$$

the capacitance per unit length is given by

$$C = \frac{q}{(q/2\pi \varepsilon)\ln(b/a)} = \frac{2\pi \varepsilon}{\ln(b/a)} \qquad \text{F/m} \tag{10-70}$$

Thus for a high-voltage cable of length L, the total capacitance of the cable is

$$C_{\text{total}} = \frac{2\pi\varepsilon L}{\ln(b/a)} \qquad (10\text{-}71)$$

which illustrates that the cable capacitance is proportional to the length of the cable. A typical cable capacitance of high-voltage cables is 40 to 70 pF/ft or 131 to 230 pF/m.

Effect of High-Voltage Cable Capacitance on X-Ray Tube Voltage

From what has been developed previously, the power absorbed by the x-ray tube when full-wave rectification is used is given by

$$P_a = 0.785 V_m I_{\text{ave}}$$

The capacitance of the high-voltage cables in a full-wave-rectified system appears in parallel with the x-ray tube. Since the shielding on the cables is grounded and tied to the center tap of the secondary of the high-voltage transformer, the ground point is only a common node for the capacitance of each cable. The equivalent circuits for this configuration are given in Figure 10-42, where R_{tube} is the equivalent load resistant of the x-ray tube and C is the equivalent cable capacitance of both cables.

The net effect of the added cable capacitance is that energy is stored during the conduction period of the diode network and energy is delivered to the tube during the nonconducting period. This changes the average value of the current and voltage across the tube, which increases the power delivered to the x-ray tube. Figure 10-43 is typical of the voltage waveform across the tube at low-milliampere levels.

Analysis of x-ray tube voltage. In Figure 10-43 the waveform is assumed to be in the steady state. At points t_1 and t_3 the applied transformer voltage becomes greater than the voltage on the cables at this time, thus forward biasing the diodes in the rectifier circuit. This places the full transformer voltage directly across the x-ray tube and the distributed capacitance of the high-voltage cables. Assuming relatively small voltage drops across the diodes in the rectifier circuit, the voltage at the tube becomes

$$V(t) = V_m \sin \omega t \qquad t_1 \leqslant t \leqslant t_2 \qquad (10\text{-}72)$$

During the interval between t_2 and t_3 and the corresponding points on subsequent waveforms to follow, the diodes of the bridge network are not conducting since the transformer voltage is less than the stored voltage on the cables, so the diodes are reversed biased. During this interval the stored voltage on the cables discharges through the x-ray tube. If the tube is operated in the saturation region, where the tube current is a function

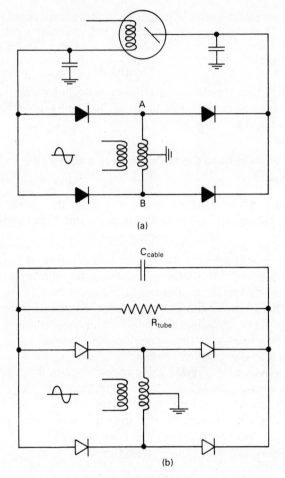

Figure 10-42 (a) Illustration of equivalent tube cable capacitance produced by the high-voltage cables, full-wave rectified; (b) equivalent circuit showing how the high-voltage cable capacitance is in parallel with the x-ray tube.

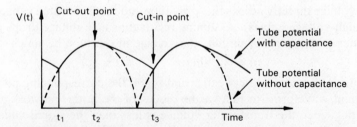

Figure 10-43 Effect of high-voltage cable capacitance on the anode voltage in a full-wave rectified circuit.

of cathode emission and is independent of anode voltage, the anode current remains constant as the anode voltage changes. In the interval between the cut-out point, t_2, and the cut-in point, t_3, the anode voltage is given by

$$\frac{dV_a(t)}{dt} = \frac{dV_c(t)}{dt} = -\frac{1}{C}I_a = \text{constant} \qquad t_2 < t \le t_3 \qquad (10\text{-}73)$$

The expression for the anode voltage is obtained by integrating the first-order differential equation using separation of variables.

$$dV_a(t) = -\frac{1}{C}I_a \, dt \qquad t_2 < t \le t_3$$

$$V_a(t) = -\frac{1}{C}I_a \int_{t_2}^{t_3} dt + K$$

$$V_a(t) = \left[-\frac{1}{C}I_a(t - t_2) + K \right]_{t=t_2}$$

$$V_a(t) = V_m - \frac{1}{C}I_a(t - t_2) \qquad t_2 < t \le t_3 \qquad (10\text{-}74)$$

Equation (10-74) defines the anode–cathode voltage over the interval when this voltage is supplied by the distributed capacitance of the high-voltage cables. This equation holds as long as the tube current is low enough so that the time rate of change of the anode voltage remains less than the maximum rate of change of the transformer voltage. Thus

$$\frac{dV_a(t)}{dt} = -\frac{1}{C}I_a < \omega V_m \qquad (10\text{-}75)$$

where V_m = value of the applied anode voltage

ω = $2\pi f$

f = frequency of the applied anode voltage

Figure 10-44 illustrates the effect of tube current on the applied anode voltage.

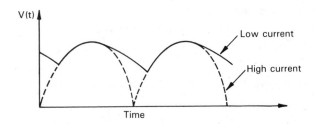

V(t)

Low current

High current

Time

Figure 10-44   Effect of current drain on the anode voltage waveform.

Effect of High-Voltage Cable Capacitance on Anode Heat Loading

In Figure 10-44 we see that for low-current work, the average value of the anode current will be higher due to the added cable capacitance in the x-ray tube circuit. This in turn changes the power dissipated in the anode and must be accounted for in the heat-loading calculation for the x-ray tube. A table of correction factors is usually supplied by the tube manufacturer to simplify the task of accounting for the added energy supplied to the x-ray tube due to cable capacity. Table 10-2 lists the typical heat input correction factors that would be used in low-milliampere fluoroscopy work.

Where low-mA, high-kVp, and relatively long high-voltage cables are used, the heat input correction factors can be as large as 1.39 (39% more energy supplied to the anode). This means that the anode heat storage capacity will be reached more quickly, due to the stored energy in the cable capacitance, than it would if there were no cable capacitance.

Example: Correction Factor Usage in Fluoroscopy and Spot Film

To illustrate the application of the factors involved in determining permissible exposures when fluoroscopy is combined with spot films, consider the following problem. Assume a stomach fluoroscopic examination for 5 minutes at 85 kVp, 4 mA combined with eight spot films at 90 kVp, 40 MAS. The anode cooling chart for this problem is given in Figure 10-26. Also assume that the length of the cables is 36 ft. Under these conditions, is the procedure permissible without exceeding the maximum loading of the x-ray tube anode?

Solution. The input energy rate for the first procedure is

$$85 \text{ kVp} \times 4 \text{ mA} = 340 \text{ H.U./s}$$

From the anode heat storage curve it is seen that the heat storage for 5 minutes is about 59,000 H.U. From the table of heat input correction factors for shockproof cable capacity for 36 ft and 4 mA, and interpolating linearly for the factor representing 85 kVp, we have a factor of 1.20. Multiplying this into the chart value, we have

$$1.20 \times 59,000 = 70,800 \text{ H.U.}$$

The total heat from the eight spot-film exposures will be

$$90 \times 40 \times 8 = 28,800 \text{ H.U.}$$

The total heat accumulated in the anode is then

$$70,800 + 28,800 = 99,600 \text{ H.U.}$$

which is within the heat storage capacity of this particular x-ray tube.

Example

How long a waiting period is required, if any, until a similar procedure can be performed?

Solution. From the anode cooling curve in Figure 10-26, it is going to take 6.1 minutes for the tube to cool to 10,000 H.U. (6.4 min − 0.3 min) before it is safe to add another 99,600 H.U. to the anode.

TABLE 10-2 Heat Correction Factors for Shockproof Cable Capacity

kVp	For 24-Ft Cables			For 36-Ft Cables			For 48-Ft Cables			For 60-Ft Cables		
	1 mA	3 mA	5 mA	1 mA	3 mA	4 mA	1 mA	3 mA	4 mA	1 mA	3 mA	4 mA
75	1.32	1.17	1.06	1.35	1.24	1.18	1.36	1.27	1.24	1.37	1.30	1.27
100	1.34	1.22	1.11	1.36	1.28	1.24	1.37	1.31	1.28	1.38	1.33	1.30
150	1.36	1.28	1.18	1.37	1.32	1.29	1.38	1.34	1.32	1.39	1.35	1.33

Source: Picker International, Cleveland, Ohio.

THREE-PHASE POWER FOR X-RAY GENERATION

All the x-ray circuits that have been described so far provide a pulsating voltage to the anode. This type of voltage waveform, when used to accelerate electrons in the x-ray tube, has the following disadvantages:

1. During the portion of the cycle when the tube voltage is appreciably lower than V_m, the x-rays produced are of low energy and the electron kinetic energy is transformed mostly into heat at the anode.
2. Most of the low-energy radiation produced during this portion of the cycle is absorbed by the filter, tube housing, or patient.
3. The timing of short radiographic exposures requires synchronizing with the line voltage so as to terminate the exposure at a zero crossing to minimize the high-voltage transient.
4. The intensity of the radiation produced is lower because no radiation is generated during a large portion of the exposure time.

Through the use of three-phase power in x-ray machines, the deficiencies of the single-phase system can be overcome. Three-phase equipment can supply steady power to the x-ray tube instead of pulsating power. It also tends to be more efficient than single-phase equipment of the same rated capacity.

Three-Phase Voltage and Current Relationships

A balanced, three-phase voltage system is composed of three single-phase voltages having the same amplitude and frequency but time displaced from one another by 120°. Figure 10-45 is a representation of the three-phase voltage representation.

The three single-phase voltages are generated by a rotating magnetic flux field that is common to the three windings of the generator, which are 120° displaced from one another but share a common housing. Figure 10-46 depicts this type of generator configuration.

How the field coils of the generator are wired will determine the voltage and current relationship for the three-phase system. If one end of each coil is connected to form a common neutral terminal, a wye (Y) connection results. This is shown schematically in Figure 10-47. When one coil is connected to the next coil in a manner illustrated in Figure 10-48, a delta (Δ) connection results.

Voltage and Current Relations for Y Connections

Every balanced three-phase system can be analyzed in terms of the procedures that apply to a single-phase circuit. Define:

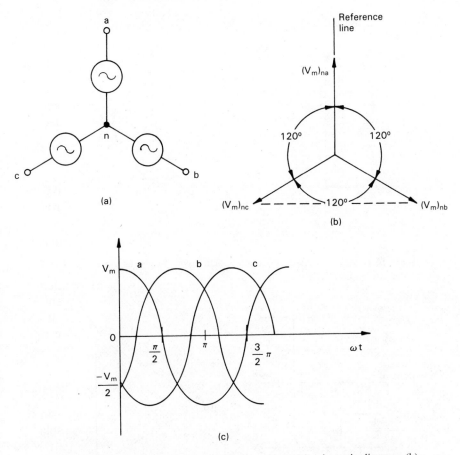

Figure 10-45 Balanced three-phase voltage system: (a) schematic diagram; (b) phasor diagram in terms of the maximum value of each phasor; (c) time diagram.

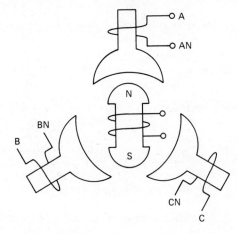

Figure 10-46 Schematic for a three-phase generator.

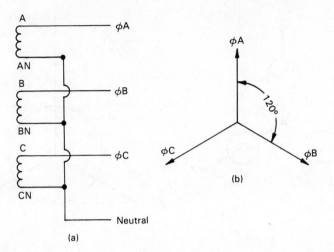

Figure 10-47 (a) Y-connected generator field coils; (b) phasor diagram for Y-connected field coils.

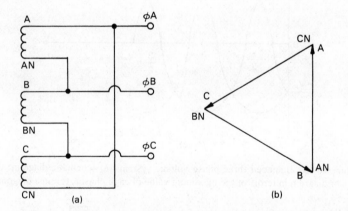

Figure 10-48 (a) Delta-connected field coils; (b) voltage phasor diagram for Δ-connected field coils.

Line voltage:

$$\mathbf{V}_{ab}, \mathbf{V}_{bc}, \mathbf{V}_{ca}$$

Phase voltages:

$$\mathbf{V}_{na}, \mathbf{V}_{nb}, \mathbf{V}_{nc}$$

Thus

$$V_{na} = V_{nb} = V_{nc} = 120\text{V rms} = V$$

Calculations of Line Voltage

The line voltage, $\mathbf{V}_{ab}$, can be represented as the sum of two phasors.

Thus

$$\mathbf{V}_{ab} = \mathbf{V}_{an} + \mathbf{V}_{nb} = -\mathbf{V}_{na} + \mathbf{V}_{nb} \qquad (10\text{-}76)$$

For a balanced system $V_{na} = V_{nb} = V_{nc} = V_p$, the effective value of the phase of the voltage.

$$\mathbf{V}_{ab} = -V_p \angle 0 + V_p \angle -120°$$

$$= -V_p + V_p(\cos 120° - j \sin 120°)$$

$$= V_p\left(-\frac{3}{2} - j\frac{\sqrt{3}}{2}\right) = \sqrt{3}V_p - 150° \qquad (10\text{-}77)$$

Similarly, we obtain

$$\mathbf{V}_{bc} = \sqrt{3}V_p \angle -270°$$

$$\mathbf{V}_{ca} = \sqrt{3}V_p \angle -30°$$

This shows that the *line voltage* is greater than the phase voltage by a factor of $\sqrt{3}$.

$$V_L = \sqrt{3}V_p \qquad (10\text{-}78)$$

valid for Y connection, and if I_L is the line current and I_p the phase current then

$$I_L = I_p$$

since the current that flows out of the point A (Figure 10-47) must also be the same as the phase current.

Current and Voltage in a Delta-Connected Three-Phase System

It can be shown, using a similar mode of analysis, that for a delta connection as illustrated in Figure 10-49 the following relations between phase and line voltage and current is true.

$$V_L = V_p \qquad \text{and} \qquad I_L = \sqrt{3}I_p \qquad (10\text{-}79)$$

Power Calculation in Three-Phase Systems

For a single-phase circuit, the instantaneous power as a function of time, $P(t)$, is given in terms of effective values.

$$P(t) = VI \cos \theta - VI \cos (2\omega t - \theta) \qquad (10\text{-}80)$$

where θ is the power factor angle. The total average three-phase power is given by

$$P_{3\phi\text{ave}} = \frac{1}{T}\int_0^T [P_a(t) + P_b(t) + P_c(t)] \, dt$$

$$P = 3V_p I_p \cos \theta \qquad (10\text{-}81)$$

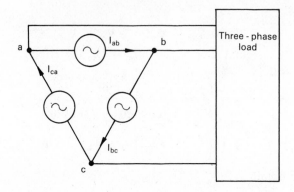

Figure 10-49 Schematic of a three-phase generator in the Δ-connected mode and its associated load.

For a Y-connected system,

$$P = 3\frac{V_L}{\sqrt{3}}I_L \cos \theta = \sqrt{3}V_L I_L \cos \theta \qquad (10\text{-}82\text{a})$$

For a Δ-connected system,

$$P = 3V_L\frac{I_L}{\sqrt{3}} \cos \theta = \sqrt{3}V_L I_L \cos \theta \qquad (10\text{-}82\text{b})$$

Note:

1. θ = angle of load impedance per phase, not the angle between V_L and I_L.
2. The total three-phase power for the system is a constant, whereas for single-phase circuits, the power pulsates at twice the time frequency.

Example

A Y-connected load of 15 Ω resistance is connected to a 240-V three-phase line (Figure 10-50). Find (a) the current through each resistance, (b) the line current, and (c) the power absorbed by the load.

Solution

(a) The phase voltage or voltage across each resistance is

$$V_p = \frac{V_L}{\sqrt{3}} = 138.6 \text{ V}$$

(b) It can be seen that $I_L = I_p$:

$$I_L = I_p = \frac{V_p}{Z_p} = \frac{138.6}{15} = 9.24 \text{ A}$$

(c) $P = 3(V_p I_p \cos \theta) = 3(138.6)(9.24) = 3841 \text{ W}.$

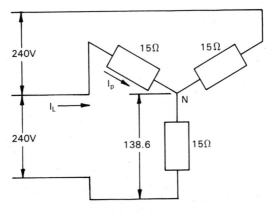

Figure 10-50

THREE-PHASE RECTIFICATION SYSTEM

As with the single-phase x-ray system where we had one pulse or two pulses of applied voltage per cycle depending on the type of rectification used, a similar situation exists with a three-phase system. Using different types of three-phase transformers and associated rectifier configurations, we can obtain 6 pulses or 12 pulses of applied anode voltage.

Rectification of Three-Phase Voltage Systems: Basic Circuit

As a starting point let us consider a three-phase full-wave bridge rectifier circuit using a Y-connected secondary as given in Figure 10-51. The transformer secondaries, a, b, and c, apply the following voltages to the x-ray tube and rectifiers in series: $(V_{na} - V_{nb})$, $(V_{nb} - V_{nc})$, and $(V_{nc} - V_{na})$. These three voltages are all in parallel as far as the x-ray tube, which acts as the load, is concerned. Whichever voltage is highest will carry the entire load of the x-ray tube, as the rest of the diodes in the bridge network will be biased off. Consider the time interval from a to b (Figure 10-51). During this time the voltage given by $(V_{na} - V_{nb})$ is the largest and therefore, during this 60° interval, the entire x-ray tube current flows out rectifier 1 and back through rectifier 5. The value of the applied tube voltage varies between $1.53V_p$ and $1.732V_p$. It should be noted that V_p is the phase voltage of the secondary winding and that

$$V_{na} = V_{nb} = V_{nc} = V_p = \text{phase voltage}$$

$$V_{na} - V_{nb} = V_L = \text{line voltage between } a \text{ and } b$$

During the next 60° interval from b to c in Figure 10-51, the voltage $(V_{nb} - V_{na})$ is the largest voltage and is negative. Thus the tube current

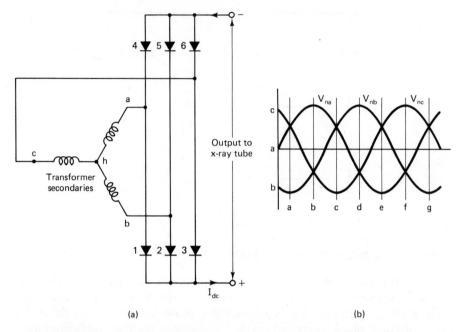

Figure 10-51 (a) Three-phase, full-wave bridge rectifier circuit using six diodes; (b) associated, per phase voltage.

will flow out diode 1 and back through diode 6. Once again the output voltage across the tube will vary from $1.53V_p$ to $1.732V_p$. At the end of this interval from b to c, the voltage $(V_{nb} - V_{nc})$ becomes the largest and is positive, and the tube current will now flow out diode 2 and back through diode 6. It is now apparent that a voltage maximum will occur during each 60° interval, producing six pulses per cycle or 360°.

The applied tube voltage from the rectifier outputs must account for the drop in voltage across the rectifiers. The actual voltage becomes, using interval a to b, for example,

$$V_T = (V_{na} - V_{nb}) - 2V_R \qquad (10\text{-}83)$$

where V_R is the voltage drop across one rectifier.

Delta-Wye/Six-Rectifier Circuit

Consider the simple circuit consisting of a Δ-connected primary with a Y-connected secondary which is connected to a six-rectifier network for full-wave rectification of the three-phase transformer output voltage as given in Figure 10-52. The rectification process takes place, producing the 6-pulse voltage waveform as discussed previously. The theoretical variation of the tube voltage, the voltage ripple, is about 13% of the maximum value, as compared to 100% for the single-phase full-wave-rectified voltage system.

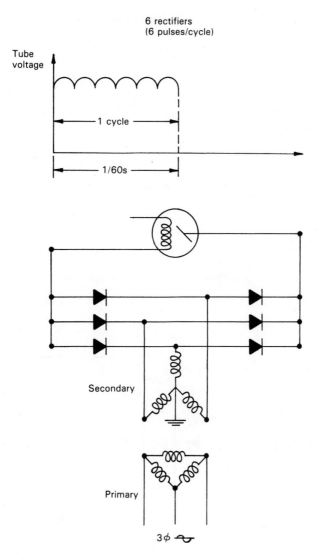

Figure 10-52   Six rectifiers, 6 pulses/cycle, three-phase anode voltage of x-ray tube.

This particular three-phase circuit configuration illustrates how three-phase rectification takes place and the improvement in anode voltage that is obtained using three-phase voltage.   In reality, this circuit suffers from the fact that at different portions of the cycle, the voltages in the transformer are not symmetrical.   This asymmetry in voltage takes place 180 times per second at each pole of the transformer.   This limits the permissible high-voltage output of the transformer.

12 Rectifiers–6 Pulses/Cycle Configuration

The shortcomings of the previous circuit are avoided by making use of two Y-connected secondaries whose rectified outputs are connected in series with the midpoint grounded as shown in Figure 10-53. In this circuit configuration the secondary voltages are balanced with respect to ground. The output voltage to the x-ray tube is still 6 pulses per cycle with a 13% ripple in the tube voltage.

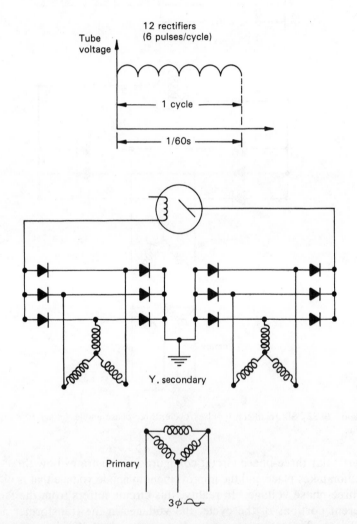

Figure 10-53 Twelve rectifiers, 6 pulses/cycle configuration for balanced voltages with respect to ground.

12 Rectifiers–12 Pulses/Cycle Configuration

In the previously described six pulses/cycle system, the two secondaries were both configured in the same way (i.e., both Y-connected). A similar result could have been achieved if the secondaries were Δ-connected. In order to reduce the voltage ripple still further and increase the effective value of the voltage, which increases the x-ray-producing efficiency of the x-ray tube, consider a transformer configuration with one secondary Y-connected and the secondary Δ-connected, as pictured in Figure 10-54. The primary of the transformer could also be Y-connected for this transformer configuration without changing any of the results to be developed.

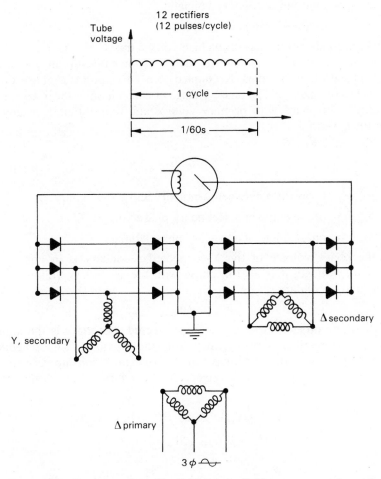

Figure 10-54 Twelve rectifiers, 12 pulses/cycle applied between x-ray tube anode and cathode.

VOLTAGE–PHASE RELATIONSHIPS IN THE SECONDARIES
OF THE TRANSFORMER

Consider first the Δ-connected primary and Δ-connected secondary of the transformer. Since they are both Δ-connected the line voltage output of the secondary is in phase with the voltage of the primary. If transformer losses are neglected, the relationship of the voltage and phase is

$$V_{\text{sec}} = n_{\Delta\text{-}\Delta}V_{\text{pri}} \qquad (10\text{-}84a)$$

$$\theta_{\text{sec-}\Delta} = \theta_{\text{pri}} \qquad (10\text{-}84b)$$

where $n_{\Delta\text{-}\Delta}$ = turns ratio Δ connection

$\theta_{\text{sec-}\Delta}$ = phase of secondary voltage

θ_{pri} = phase of primary voltage

The other secondary windings of the high-voltage transformer is Y-connected. The transformer coils are wound in such a manner that the magnetic flux of each phase of the primary is coupled to a corresponding phase of the Y-connected secondary. The phase voltage of each phase of the Y-connected secondary is in phase with primary voltage and they differ in magnitude by the turns ratio.

$$V_{\text{sec}} = n_{\Delta\text{-Y}}V_{\text{pri}} \qquad (10\text{-}85a)$$

$$\theta_{\text{sec-Y}} = \theta_{\text{pri}} \qquad (10\text{-}85b)$$

where $n_{\Delta\text{-Y}}$ = turns ratio between Δ-primary and Y-secondary

$\theta_{\text{sec-Y}}$ = phase angle of a secondary phase voltage

θ_{pri} = phase angle of a primary phase voltage

Since the phase voltage of the Y-connected secondary is in phase with the primary voltage, it is also in phase with the secondary voltage of the Δ-connected secondary.

$$\theta_{\text{sec-Y}} = \theta_{\text{sec-}\Delta}$$

Since the line voltage output of the Y-connected secondary is the applied voltage to the rectifier network, then from the previous developments of the Y-connected generators we have the following relationship between the line and phase voltage of the Y-connected secondary. From Figure 10-47 we have

$$\mathbf{V}_{ab} = \mathbf{V}_{an} + \mathbf{V}_{nb} = -V_p \,\angle 0° + V_p \,\angle -120°$$

$$= -V_p + V_p(\cos 120° - j\sin 120°) \qquad (10\text{-}86)$$

$$= V_p\left(-\frac{3}{2} - j\frac{\sqrt{3}}{2}\right) = \sqrt{3}V_p \,\angle -150°$$

In a similar fashion,

$$\mathbf{V}_{bc} = \sqrt{3}V_p \ \angle -270°$$

$$\mathbf{V}_{ca} = \sqrt{3}V_p \ \angle -30°$$

A phase diagram of the relationship between the line voltage and phase voltage of the Y-connected secondary is given in Figure 10-55. From this we see that the line voltage is shifted by 30° from the phase voltage of the secondary. This also means that the output voltage of the Y-connected secondary is shifted 30° relative to line voltage of the Δ-connected secondary. By using a set of six diodes to rectify the output of each secondary connection and the fact that the line voltage output of the Y is shifted by 30° relative to the line voltage of the Δ connection, an output of 12 pulses per cycle is obtained, as shown in Figure 10-54.

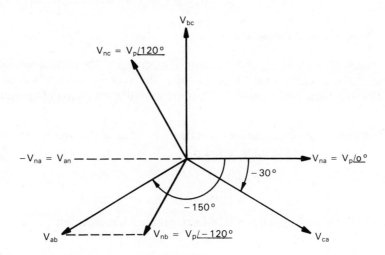

Figure 10-55 Phase and magnitude relations between the phase and line voltages of a Y-connected secondary.

For the 12-pulse three-phase rectification system the percent ripple is only 3.4%, compared to 13.5% for the 6-pulse system and 100% for the full-wave-rectified single-phase system.

POWER CALCULATION AND HEAT LOADING
IN A THREE-PHASE SYSTEM

For the sake of comparison, let us look at the heat-loading and power requirements for a single-phase full-wave-rectified system and a three-phase, 6-pulse and 12-pulse rectified x-ray system that are needed to produce the same radiographically equivalent exposures. Consider that 100 kVp and 400 mA are used in each case and that maximum heat loading of the x-ray tube anode is not exceeded.

Single-Phase System

As derived in equation (10-67), the power in a single-phase full-wave system is given by

$$P = 0.785V_m I_{ave}$$

The power level at the anode is then

$$P = 0.785(100)(400) = 31,400 \text{ W}$$

If an exposure time of 0.1 s is used, the energy into the anode is

$$\text{energy} = 31,400 \times 0.1 = 3140 \text{ W-s}$$

Three-Phase, 6 Pulses/Cycle

Since three-phase systems are usually operated in the current saturation region of the operation for the x-ray tube, the average current is equal to the effective value of the current. Thus in the saturation region of operation,

$$I_{ave} = I_{eff}$$

For the 6-pulse system, the rms or effective value of the anode voltage is

$$V_{eff} = 0.956V_m$$

$$P = V_{eff} I_{eff}$$

$$P_a = 0.956(100)(400) = 38,240 \text{ W}$$

$$\text{exposure time} = 0.064 \text{ s to produce a radiograph}$$
$$\text{equivalent to that of}$$
$$\text{the single-phase system}$$
$$\text{energy} = 38,240 \times 0.064 = 2447.36 \text{ W-s} \qquad (10\text{-}87)$$

Three-Phase, 12 Pulses/Cycle

For this configuration we have

$$V_{eff} = 0.989V_m$$

$$P = 0.989(100)(400) = 39,560 \text{ W}$$

$$\text{exposure time} = 0.05 \text{ s to produce an equivalent}$$
$$\text{radiograph}$$
$$\text{energy} = 39,560 \times 0.05 = 1978 \text{ W-s} \qquad (10\text{-}88)$$

From these calculations, it is seen that by going from 2 pulses per cycle to 12 pulses per cycle, the heat loading on the x-ray tube is reduced by 35%, which indicates increased efficiency of the x-ray generating system and allows for a higher work-load capacity for the same heat generated.

Instead of using the engineering units of watt-seconds, the use of the arbitrary measure of *heat units* is a more common practice. Since the power levels are different depending on whether the system is single-phase or three-phase, a multiplying factor must be used in conjunction with the standard definition of heat units (i.e., the product of kVp × mA × seconds). This is because the product of these factors does not accurately reflect the heat units for a three-phase system. To determine heat units in three-phase equipment, multiply the product of the exposure factors by 1.35 for 6-pulse configurations and 1.41 for 12-pulse configurations.

Example

Consider an x-ray system operated at 75 kVp at 100 mA for 1/2 s. Calculate the heat units for (a) a *single-phase system,* (b) a *three-phase 6-pulse system,* and (c) a *three-phase 12-pulse system.*

Solution

(a) H.U. = 75 × 100 × 0.5 = 3750 heat units

(b) H.U. = 75 × 100 × 0.5 × 1.35 = 5062 heat units

(c) H.U. = 75 × 100 × 0.5 × 1.41 = 5288 heat units

COMPARATIVE ADVANTAGES OF THREE-PHASE AND SINGLE-PHASE ANODE VOLTAGES

One of the essential differences between single-phase and three-phase rectified power systems used in the generation of x-rays is the reduction in anode voltage ripple from 100% in the case of single-phase systems to 13% for 6-pulse/cycle systems and 3.4% for 12-pulse/cycle systems. Voltage ripple affects certain aspects of tube operation. One factor that it influences is the speed of the electrons across the x-ray tube. With the 2-pulse wave the voltage value rises to a peak and then falls to zero (100% ripple). This rise and fall of the voltage causes the kinetic energy of the electrons across the tube to vary accordingly. Consequently, if 100 kVp were applied to the x-ray tube, the kinetic energy of the electrons would theoretically range from zero when the voltage value is at zero, to 100 keV when the voltage value is at its peak. The 12-pulse wave from a three-phase generator does not drop to zero. It only drops 3.4% below peak value. As a result, the kinetic energy of the electrons only drops 3.4% below peak kV value. In other words, the kinetic energy of electrons (at 100 kVp) in a 12-pulse system theoretically will range from 96.6 keV to 100 keV. This means that the average kinetic energy imparted to the electrons is much higher in the 12-pulse system than in the 2-pulse system.

When low-speed electrons strike the target in an x-ray tube, their energy is either converted into heat or into low-energy photons. Consequently, since a single-phase system produces comparatively more low-energy electrons, it also produces a greater proportion of low-energy x-ray photons than does a three-phase system. The greater proportion means that the average photon energy is less in a single-phase system. Low-energy x-ray photons are either absorbed by filtration or by the patient. In either case, they usually serve no useful purpose in diagnostic radiology since they do not reach the film to contribute to the exposure. (With certain examinations such as mammography, some low-energy photons are indeed useful since they contribute to the exposure. However, in this type of examination special techniques and equipment are used to effect the image-forming, low-energy radiation.) The penetrating power or quality of a beam of x-rays is governed by photon energy; in the three-phase beam the average photon energy is much greater. This, of course, is assuming that both systems are operated at equal peak kilovoltages. Notice that the qualitative difference in the two beams is in average photon energy. Both systems do in fact produce low-energy photons. The difference is in a proportion of low-energy photons, which, as stated before, is greater in single-phase systems. The intensity of an x-ray beam is greater with a three-phase system than with a single-phase system for a given tube current. Therefore, the 12-pulse system will produce a given amount of radiation in a much shorter period of time than that required for the 2-pulse system.

Figure 10-56 illustrates how the three-phase system produces more x-rays per cycle than does a single-phase system. Notice that image-forming radiation is produced only at certain times with a 2-pulse system. At other times, either no radiation is produced (when the sine wave is at zero value) or radiation is produced that has insufficient energy to reach the film. The 12-pulse wave continuously produces image-forming radiation because of its nearly constant voltage level. Thus the three-phase system is more efficient. The average energy level of the beam of radiation produced by a three-phase unit is higher than that produced by a single-phase unit when both are adjusted for the same peak kV. Therefore, to produce radiographs with the same general scale of contrast it would be necessary to use higher kVp with the single-phase unit. For a given mA, it would require approximately twice as much exposure time for a single-phase unit as for a three-phase (12-pulse).

X-ray tube capacity is greater in a three-phase system for short exposures. One reason for the increased tube capacity is because the heat is spread over a larger area on the target. The increased thermal capacity for x-ray tubes operated on three-phase is increased only for exposures less than 1/2 s. From 1/2 to 1 s the ratings are approximately the same

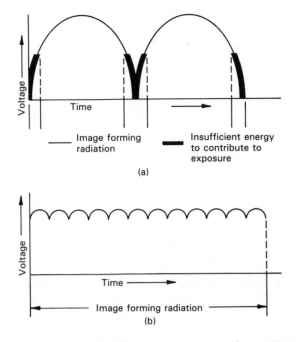

Voltage

Time

——— Image forming
radiation

█ Insufficient energy
to contribute to
exposure

(a)

Voltage

Time ——▶

◀——————— Image forming radiation ——————▶
(b)

Figure 10-56 (a) Relationship between anode voltage and time, illustrating the time in the two-pulse cycle where useful x-rays are produced; (b) twelve pulses/cycle produces useful x-radiation during the entire cycle.

as for single-phase. Above 1 s the ratings can be greater for x-ray tubes operated on single-phase.

The use of three-phase circuits does have some disadvantages compared with single-phase systems. First, the adjustment of the primary voltage of a three-phase system does require the use of an autotransformer per phase. Also, there is a problem with inductive transients that are generated in starting and terminating an x-ray exposure. To avoid these transients, switching is done at a point where the voltage is zero. For a single-phase system, switching the voltage on or off can be done every 1/120 s, which is when the voltage passes through zero. In three-phase systems, the zero-crossing switching is employed sequentially, which tends to reduce the magnitude of the transient.

In summary, then, three-phase x-ray systems offer the following advantages over single-phase systems:

1. The power supplied to the tube by a three-phase generator supplies more usable voltage for the production of x-rays per unit time, thus allowing for shorter exposures.

2. Because the tube voltage for a three-phase system is no lower than 13% of the maximum voltage for a 6-pulse system and 3.4% of maximum voltage for a 12-pulse system, the intensity of the x-ray radiation produced for a given tube current is much higher than that of a single-phase system.

3. With the anode voltage remaining high throughout the exposure interval, for three-phase systems, the radiographic quality of the x-rays produced is better than that of single-phase systems. This is due to the fact that less "soft" radiation is produced.

4. A three-phase tube voltage is more efficient in utilizing the anode heat storage capacity, as the target is not subjected to low-energy electrons, which produce heat in the anode but little or no x-radiation.

X-RAY SWITCHES AND TIMING METHODS FOR EXPOSURE CONTROL

Exposure control that is related to the time or duration the x-ray tube is on, once kVp and mA are determined, can be broken down into two components:

1. Switching circuits that turn the x-ray tube on and off
2. Timing circuits that activate the switching circuits

In the early days of x-rays, the fixed-anode tubes were operated at low currents because of the lack of sufficient heat storage capacity. Consequently, exposure times were long and simple timing devices were adequate for exposure control.

With the advent of the rotating-anode x-ray tube and the use of much higher kVp and mA, there was a need for more accurate times and faster timing circuits and switches. Still further sophistication was required of the exposure times with the introduction of cineradiography—thus the topic of timing and switching circuits for x-ray exposure control.

Switching Circuits

The prime function of an x-ray switch is to start and stop the current in the x-ray tube. At the present time, the three common means of initiating and terminating an exposure are with mechanical switches or electronic switches consisting of either a thyratron tube or a silicon-controlled rectifier (SCR).

Mechanical switches. The simplest switches are mechanical relay contactors. These solenoid-operated electromagnetic switches get their coil

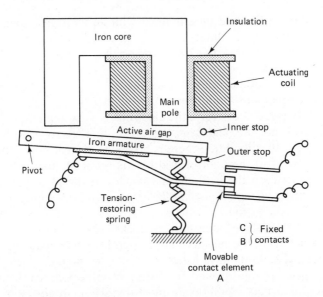

Figure 10-57 Basic construction of an elementary relay contactor.

current from the exposure timer. A relay consists of the following components (Figure 10-57):

1. A magnetic circuit made up of a fixed core, movable armature, and an air gap
2. One or more actuating coils which establish the magnetic flux between the pole piece and the armature
3. One or more sets of contact elements or points, one set on the armature and the other set fixed
4. A restoring spring which maintains the armature at maximum air-gap position

The size of the points or contactor is proportional to the amount of current they must carry. Because of high-voltage arcing problems, relays are always used in the primary or low-voltage side of the high-voltage transformer. Electromagnetic relays work well with small currents and relatively long exposures.

The maximum accuracy of magnetic relays is one half-cycle of a full-wave-rectified system, $\frac{1}{120}$ s. Due to mechanical inertia, relays have a cycle time which limits them to two or three exposures per second.

Electronic switches: thermionic type—thyratron. Consider now a device that looks like a vacuum tube and has an anode and an indirectly heated cathode. Also, the tube is filled with an inert gas, such as argon, neon, or mercury vapor. Interposed between the cathode and anode, but close to the cathode, is a third electrode called a *grid*. This device is a gas-filled triode tube.

When the cathode is indirectly heated by the cathode heater, electron emission takes place in a manner similar to that of the x-ray tube. If the control grid is biased negatively, the electrons are dominated by the electric field of the grid, not the field from the anode, and are repelled back to the cathode. Under this condition there is no anode current flow. If the anode is made positive and the control grid bias is reduced, some electrons will start to flow between the cathode and anode. If the plate voltage is well above the ionizing voltage of the particular gas used in the tube, the electrons will cause the gas to ionize and large currents can now flow due to the ionized gas in the tube. Once the gas in the tube is ionized, the positive ions are attracted by the negative control grid, further reducing the grid's negative potential, so the grid loses control of the flow of electrons due to a positive space charge around it. These positive ions cause current to flow in the grid circuit. As a result, most thyratron circuits have a resistor inserted in series to limit the grid current to a safe value. Once the thyratron has started to conduct, the anode voltage must be reduced below the ionizing potential of the gas before the control grid can regain control. The "dot" in Figure 10-58 indicates that the tube is gas filled.

There are a number of advantages to the use of thyratrons in x-ray switching circuits. First, thyratrons can conduct large currents with only a small voltage drop across them. This current can be used to drive a relay in the primary circuit of the high-voltage system. Since there are no moving parts in a thyratron, its response time is almost instantaneous. This property is used in single-phase systems to supply current for x-ray tube operation during the portion of the cycle where the x-ray intensity is the highest. This reduces the anode heat loading of the x-ray tube and the exposure of

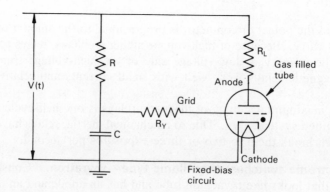

Figure 10-58 Thyratron tube circuit used as a switch to cut off anode current.

the patient to nondiagnostic radiation. Exposure times as short as 1 ms are obtained using a thyratron.

With the advent of solid-state high-current switching devices in recent years, such as the silicon-controlled rectifier (SCR), the use of thyratron tubes is now diminishing because of their large size, cost, heat production, and difficulty of maintenance.

Electronic switches: solid state—SCR. The SCR is the solid-state equivalent of the thyratron in terms of its terminal characteristics. It is a four-layer semiconductor material, a *P-N-P-N* system. Figure 10-59a is the four-layer representation and Figure 10-59b is the circuit symbol for the SCR. In order to interpret the action of this device, it is handy to visualize the SCR as two transistors, *NPN* and *PNP*, configured as in Figure 10-59c and d. If the bias on the gate, which is visualized as the base of the *NPN* transistor, maintains the transistor in the cutoff state, no current will flow in this transistor and so the *PNP* transistor will also not conduct. The only current flowing under this condition is the minority current. In this state the anode-to-cathode impedance is very high.

If the pulse is applied to the gate, which now drives the *NPN* transistor into conduction, the collector current of this transistor supplies base current to the *PNP* transistor. The collector current in the *PNP* transistor is fed back to the base of the *NPN* transistor. This positive feedback configuration is inherently unstable and both transistors are driven quickly into saturation regardless of the gate potential just so long as the anode potential remains positive. In this saturated condition the impedance between the anode and cathode is very low. The SCR is shut off when the anode voltage becomes zero or negative with respect to the cathode. This resets the SCR in a nonconducting state to await the next gate pulse. Because of the voltage ratings on the SCR, they are usually used in the primary circuit and controlled by a set of system logic circuits.

Exposure Timing Systems

X-ray exposure timers have developed through numerous stages, starting with simple mechanical timers and ending with microprocessor electronic timers. With the variety and age of the x-ray equipment in use today, almost all of the various types of exposure times are still in use. Basically, the timer is a device which initiates and terminates the x-ray exposure. The timer controls the x-ray contactor, which in turn controls the voltage to the primary of the high-voltage transformer.

Mechanical timers: spring driven. The simplest of the timing devices is one that is a spring-driven, escapement mechanism similar to that of a wind-up stopwatch, clock, or kitchen timer. The timer is set by turning a

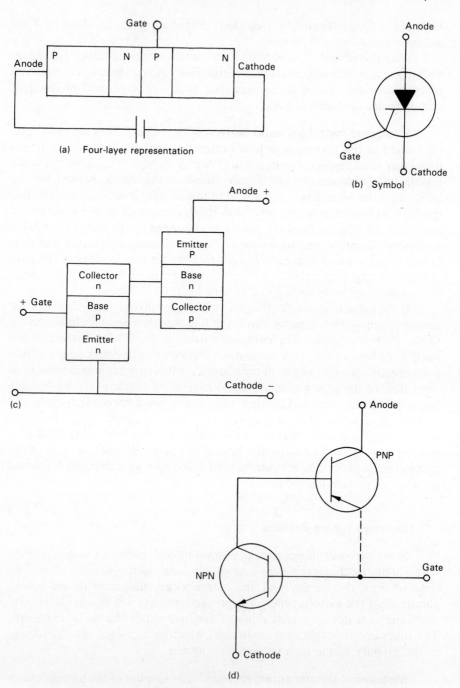

Figure 10-59   Representation of a silicon-controlled rectifier (SCR) and its equivalent discrete transistor representation.

dial to the desired time. Usually, a button is pushed, which starts the clock mechanism and energizes a relay in the primary of the high-voltage transformer to initiate the exposure. At the end of the exposure time the timer terminates the electrical supply to the primary relay, which in turn opens the primary circuit of the high-voltage transformer. Mechanical timers are usually used in low-power portable or dental x-ray machines. The timer range is normally from 1/20 s to 8 or 12 s. The accuracy is somewhat uncertain below 1/4 s.

Mechanical timers: synchronous.   The synchronous timer is similar to that of the spring-driven timer but instead of a spring-driven escapement, a small constant-speed synchronous motor such as the type found in an electric clock is used. In this type of device, the synchronous motor, whose speed is determined by the power-line frequency, is connected through a drive unit and gear train to an electrical switch which energizes the relay in the primary of the high-voltage transformer. Due to inertia and play between gears, this type of timer is not accurate for short exposure times, the shortest interval being about 1/20 s. Synchronous timers have an accuracy of about 0.1 s; thus these timers are used for long exposures, as in x-ray therapy. Figure 10-60 is a schematic diagram of a synchronous timer. When the switch is on, line voltage is applied to the timer motor. With the pushbutton depressed, current flows from point c through d, energizing the clutch relay. When the clutch relay is turned on, the contactor relay in the transformer primary is also energized and the exposure begins. At the same time the clutch rotates with the synchronous motor and after the preselected time has elapsed, an arm (dog) on the clutch opens the primary relay coil circuit, terminating the exposure.

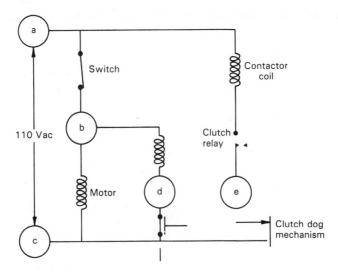

Figure 10-60   Mechanical synchronous timer used to control exposure time.

Electronic timers. As the power of x-ray generators increased, exposure times became too short to be controlled accurately with mechanical timers. To meet the demand of accurate short-term timing, the electronic timer was developed. There are three classifications of electronic timers:

1. *RC* time-constant types
2. Impulse timer
3. Digital timing circuits

RC Timing Circuits. The basic principle of operation of this class of timing circuits is that the length of the x-ray exposure is determined by the time constant of a simple *RC* circuit. Consider the circuit shown in Figure 10-61. The switch *S* is closed between exposures. At the beginning of an exposure switch *S* is opened and a relay in the primary high-voltage transformer circuit is closed, which begins the exposure.

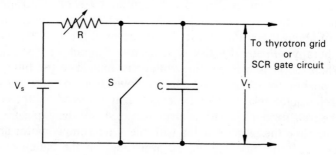

Figure 10-61 *RC* timing circuit used to control a thyratron tuber or SCR.

The voltage on the capacitor, V_t, is given by the equation

$$V_t(t) = V_s(1 - e^{-t/RC})$$

If *C* is constant and *R* is a variable resistor, the *RC* time constant can be changed to meet exposure needs. Figure 10-62 illustrates the capacitor voltage as a function of time for two different time constants.

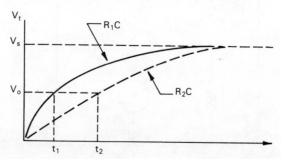

Figure 10-62 Timer *RC* curves. Voltage rises as a function of the resistance setting R_1 or R_2.

The voltage across the capacitor is used to overcome the negative bias on a thyratron tube and at a firing voltage V_o, which is a function of the thyratron-biased voltage, the thyratron will fire, which will open the x-ray circuit switch and terminates the exposure. The voltage on the capacitor can also be used to fire an SCR to accomplish the same function of terminating the exposure. This type of time circuit has long been used. They are not as accurate as impulse or digital timers but are superior to synchronous motor timers for short exposures.

Digital Timers. Digital timers are commonplace today and are found in x-ray machines, microwave ovens, and digital computers. The principles of digital timing are straightforward and make use of a reference oscillator, a counter, and input logic. The generation of a time period T is based on counting out N cycles of a precise frequency F such as those produced by a quartz crystal oscillator. Thus we have the relation

$$N = T \times F$$

If we start with a predetermined counter as shown in Figure 10-63, the reference oscillator generates the precise frequency of F counts per second that is used by the counter. The value of N is programmed into the counter logic. The initiation of the time period is controlled by the set–reset flip-flop, which is set to "1" by a "start" signal. Through the use of an AND gate the output of the oscillator is put into the counter. When N counts have been made corresponding to a time interval T, a pulse is issued that resets the flip-flop and disconnects the oscillator from the counter. The counter pulse output can be used to trigger an SCR and terminate the x-ray exposure.

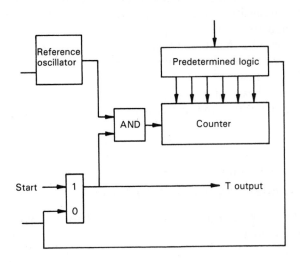

Figure 10-63 Time-period generation using digital counters.

X-RAY COMPUTED TOMOGRAPHY

History

The reconstruction of the internal structure of an object from its projections is neither new nor novel. As early as 1917, mathematical investigations applicable to the theory of reconstruction were conducted by Radon in a paper "On the Determination of Function from Their Integrals Along Certain Manifolds." In 1956, Bracewell [2], a radioastronomer, wished to identify regions of the sun that emitted microwave radiation. Microwave antennas on earth could not focus on points but made measurements of the total radiation from ribbonlike strips. By using a series of these "strip sums" in different directions, Bracewell was able to reconstruct a map of the sun's microwave emissions.

There were many other investigators in the area of image reconstruction of x-ray projections, but it was not until 1970 that Godfrey Hounsfield, a British engineer working for EMI Ltd. in England, produced the first x-ray scanning machine, the EMI brain scanner. For his work in the development of the CT scanner, Hounsfield was awarded the Nobel Prize for Medicine in 1979. It has been said that there has been no comparable discovery of this magnitude in radiology since Roentgen discovered x-rays in 1895.

CT Scanning Techniques: The Reconstruction Process

An ordinary x-ray system takes pictures by passing x-rays through the body and detecting them with photographic film. The different tissue and bone structure of the body attenuate the x-ray beam differently and thus vary the intensity of the x-radiation as a function of location. These changes in intensity are what the photographic film responds to. The CAT scanner, on the other hand, consists of an x-ray tube, x-ray detectors with photomultipliers tubes, A/D converters, computer systems, and a video display (Figure 10-64).

Consider a beam of x-rays passing through the body. The beam is attenuated by absorption and scattering. The amount of absorption depends on the following factors:

1. Physical density
2. Atomic composition
3. Photon energy spectrums of x-ray beams

Notation: An (x, y) coordinate system is used to describe points in the layer under examination.

Rays = dashed lines,

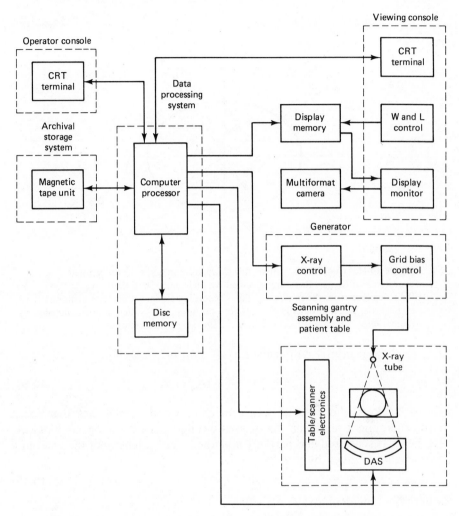

Figure 10-64 CT scanner block diagrams: DAS, data acquisition system; CRT, cathode ray tube; W and L, window and level. (From *Introduction to Computed Tomography,* General Electric Company, Milwaukee, Wis., 1976, by permission.)

ϕ = angle of ray with respect to the y axis, and

r = ray distance from the origin.

The contribution of each point toward the detected signal is denoted by the density function $f(x, y)$ used simply in a mathematical sense (see Figure 10-65). In x-ray tomography, the function $f(x, y)$ actually represents the linear attenuation coefficient, μ.

Let all points on the ray path be described by (r, ϕ). The integral of

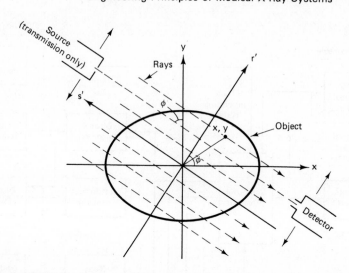

Figure 10-65 Coordinate systems. The density function is described with x, y coordinates, $f(x, y)$. Rays are specified by their angle with the y axis, ϕ, and their distance from the origin, r'. The s' coordinate denotes distance along the ray. (From R. A. Brooks et al., "Theory of image reconstruction in computed tomography," *Radiology*, Vol. 117, Dec. 1965, by permission.)

$f(x, y)$ along a ray (r, ϕ) is called a *ray sum* or *projection p*

$$p(r, \phi) = \int_{r,\phi} f(x, y)\, ds \qquad (10\text{-}89)$$

Going back to the case of x-ray imaging, we have the fact that for a monoenergetic x-ray beam that passes through a small homogenous object, the beam is scattered and partially absorbed. The intensity of the emerging x-ray beam is given by

$$I = I_0 e^{-\mu s} \qquad (10\text{-}90)$$

where I_0 = incident x-ray intensity

I = transmitted x-ray intensity

s = path length

μ = linear attenuation coefficient

(see Figure 10-66)

If the medium through which the x-ray beam passes is nonhomogenous, the linear attenuation coefficient becomes a function of position, which is given by

$$\mu = \mu(x, y)$$

The relationship between the incident and transmitted x-ray intensity now becomes

$$I = I_0 e^{-\int \mu(x,y)ds} \qquad (10\text{-}91)$$

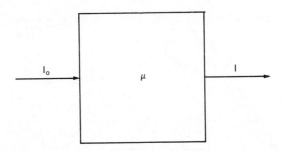

Figure 10-66 Illustration of x-ray beam intensity attenuation.

By letting the density function $f(x, y) = \mu(x, y)$, the ray sum or projection becomes

$$p(r, \phi) = -\ln\frac{I}{I_0} = \int_{r,\phi} \mu(x, y)\, ds \qquad (10\text{-}92)$$

If the transmitted x-ray intensity is detected by the detector, the ray sum is proportional to the logarithm of the detector output.

Projection or Profile. This is a complete set of ray sums at a given angle. The reconstruction problem can be stated as: Given

$$p(r, \phi) = \int_{r,\phi} f(x, y)\, ds \qquad (10\text{-}93)$$

as the measured projection, p, find the density function $f(x, y)$.

Reconstruction methods

Back-Projection. The first attempt at reconstruction tomography is a crude method known as *back-projection*. The attraction of this method is that it is easily implemented without a digital computer. Kuhl and Edwards [9] were the first to produce reconstructed tomograms of living patients. The basic problem with the method is that the reconstruction is crude. In Figures 10-67 and 10-68, two projections of a rectangular object of uniform absorption are shown. In practice, however, many more would be required.

Reconstruction is performed by assuming that a given signal intensity corresponding to a ray sum is applied to all points that make up the ray. When this is done for all projections, an approximation of the original object is produced.

Mathematically, the process can be described as follows:

$r = x \cos \phi + y \sin \phi$

$\hat{f}(x)$ = approximation of the density function produced by back-projection

Then

$$\hat{f} = \sum_{j=1}^{m} p(x \cos \phi_j + y \sin \phi_j, \phi_j)\, \Delta\phi \qquad (10\text{-}94)$$

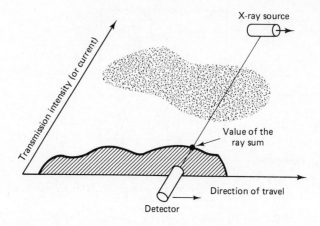

Figure 10-67 Single view or profile obtained in one scan plane. (From *Introduction to Computed Tomography*, General Electric Company, Milwaukee, Wis., 1976, by permission.)

where ϕ_j = jth projection angle

$\Delta\phi$ = angular distance between projections

Note: The expression $x \cos \phi_j + y \sin \phi_j$ essentially represents the rays passing through the point (x, y), so the back-projected density at each point (x, y) is simply the sum of all ray sums passing through the point.

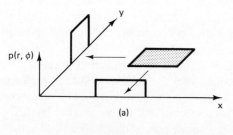

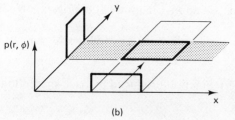

Figure 10-68 Back-projection. Profiles of a rectangular object (a) are back-projected across the plane (b) and superimposed to form an approximation of the object. (From R. A. Brooks et al., "Theory of image reconstruction in computed tomography," *Radiology*, Vol. 117, Dec. 1965, by permission.)

Example: Back-Projection Method

Consider a four-pixel array with relative attenuations of

0	2
1	3

which are the values we wish to determine.

1. Horizontal irradiation—first estimate:

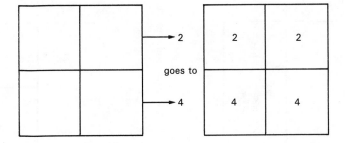

2. Vertical irradiation—add the ray sums to the second estimate:

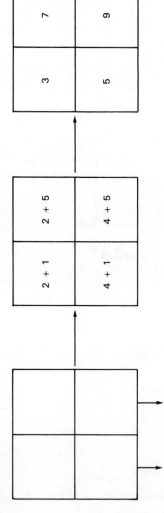

3. Diagonal irradiation up to the right; add the ray sums to the above:

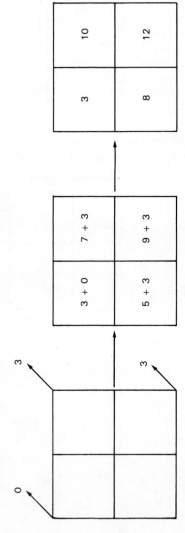

4. Diagonal irradiation up to the left; add the ray sums to the above:

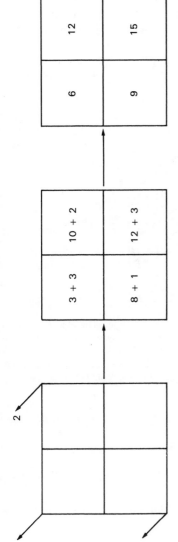

5. If we subtract 6 from each pixel (background) and divide by 3 to reduce the pixels to their simplest ratio, we get

The drawback of the back-projection method is that it does not produce a good reconstructed image because each ray sum is applied not only to points of high density, but to all points along the ray. This artifact shows up as a "star pattern."

Iterative Reconstruction.   Consider now that a grid is imposed on the object as in Figure 10-69, and that the attenuation or density function is uniform in the ith cell and is called f_i. Let w be the width of a cell. Then for a diameter D,

$$n = \frac{D}{w} \text{ total cells in domain } N \simeq \frac{\pi n^2 w^2}{4} \text{ diameter of target} \qquad (10\text{-}95)$$

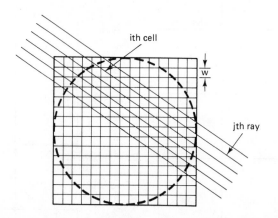

Figure 10-69 Ray geometry for iterative reconstruction. The object, bounded by the dashed circle, is reconstructed on an $n \times n$ array of cells. The rays are actually strips of finite width. The contribution of the ith cell to the jth ray is referred to as w_{ij} (heavy outline). (From R. A. Brooks et al., "Theory of image reconstruction in computed tomography," *Radiology*, Vol. 117, Dec. 1965, by permission.)

We can consider that this is n rays of width w. The problem now becomes

$$p_j = \sum_{i=1}^{N} w_{ij} f_i \qquad (10\text{-}96)$$

w_{ij} is a weighting factor that represents the contribution of the ith cell to the jth ray sum, p_j.

This equation represents a set of M equations in N unknowns, which, in principle, can be solved by matrix inversion:

$$\mathbf{p} = \mathbf{Wf} \qquad \mathbf{W} = \mathbf{M} \times \mathbf{N}$$
$$\mathbf{f} = \mathbf{W}^{-1}\mathbf{p} \qquad \mathbf{p} \text{ and } \mathbf{f} = N \text{ vectors} \qquad (10\text{-}97)$$

If $M < N$, many solutions are available.

Iterative Method of Reconstruction.   This is a technique similar to matrix inversion by iteration, where an arbitrary starting point for each pixel is chosen.

Example

Consider a simple case of four volume elements represented by the 2 × 2 matrix of pixels below.

1. The x-ray beam passes through each volume element to obtain the ray sum, horizontal and vertical:

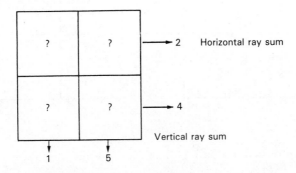

2. As a first estimate of pixel values, the average value of the ray sum is assigned to each pixel (6/4 = 1.5):

1.5	1.5
1.5	1.5

3. The new ray sum in the horizontal direction is calculated and compared to the old ray sum:

		Old	New
1.5	1.5	2	3
1.5	1.5	4	3

4. Corrections are made by adding the difference between old and new ray sums divided by the number of pixels. See steps 5 and 6.

5. First correction (horizontal):

$$\text{top row: } \frac{2-3}{2} = -0.5 \qquad \text{bottom row: } \frac{4-3}{2} = +0.5$$

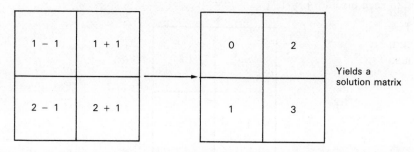

| | Old | 1 | 5 |
| | New | 3 | 3 |

6. Second correction (vertical):

$$\text{left column: } \frac{1-3}{2} = -1 \qquad \text{right column: } \frac{5-3}{2} = 1$$

Yields a solution matrix

The strong similarity of the iterative method to the back-projection method should be noted. To begin with, if the starting point is a blank screen, the first iteration is equivalent to back-projection, since the calculated projection in this case is zero. In succeeding iterations, a correction factor is back-projected.

Two-Dimensional Fourier Transforms

Much of the material to be presented on the analytic reconstruction of an image from data obtained from projections is based on the mathematics of the two-dimensional Fourier transform. This transform can be considered as a space-to-frequency transformation. Given the function $f(x, y)$ with a number of mathematical constraints put on it, the two-dimensional Fourier transform of the function is given by

$$F(k_x, k_y) = \int_{-\infty}^{\infty} \int_{-\infty}^{\infty} f(x, y) \exp\left[-j2\pi(k_x x + k_y y)\right] dx\, dy \qquad (10\text{-}98)$$

where k_x and k_y are called spatial frequencies and $F(k_x, k_y)$ is the Fourier transform of $f(x, y)$ and may be real or complex.

The inverse Fourier transform is given by the relation

$$f(x, y) = \int_{-\infty}^{\infty} \int_{-\infty}^{\infty} F(k_x, k_y) \exp\left[j2\pi(k_x x + k_y y)\right] dk_x \, dk_y \qquad (10\text{-}99)$$

Using properties of the double-integration process, the two-dimensional Fourier transforms can be viewed as a succession of one-dimensional transforms. Thus, with $G(k_x, y)$ denoting the one-dimensional Fourier transform of $f(x, y)$, we have

$$G(k_x, y) = \int_{-\infty}^{\infty} f(x, y) e^{-j2\pi(k_x x)} \, dx \qquad (10\text{-}100)$$

$$F(k_x, k_y) = \int_{-\infty}^{\infty} G(k_x, y) e^{-j2\pi(k_y y)} \, dy \qquad (10\text{-}101)$$

By the same reasoning, the inverse two-dimensional Fourier transform can be thought of as a decomposition of $f(x, y)$ into elementary complex sinusoids of infinitesimal amplitude $F(k_x, k_y) \, dk_x \, dk_y$. Looking at this fact from a different perspective, we can think of the function $f(x, y)$ as being built up by coherent (phase-preserving) addition of a very large number of complex sinusoids of different amplitudes and spatial frequencies.

Two-Dimensional Fourier Reconstruction

The reconstruction of an image from the measured projection of the image using analytic methods is based on direct solution of the equation

$$p(r, \phi) = \int_{r, \phi} f(x, y) \, ds$$

Bracewell [2] was the first to use two-dimensional Fourier methods for reconstruction problems in radioastronomy. His contribution was the use of a two-dimensional Fourier transform in image reconstruction, but at the time the method was only of theoretical importance. The numerical methods of the time and the speed at which digital computers worked proved to be too slow and inefficient to make his method applicable.

With the advent of the fast Fourier transform algorithm by Cooley and Tukey [4], the numerical computation of the Fourier transform became feasible in the methods developed by Bracewell [2].

Consider the representation of the density function $f(x, y)$ as a two-dimensional Fourier integral of the form

$$f(x, y) = \int_{-\infty}^{\infty} \int_{-\infty}^{\infty} F(k_x, k_y) \exp\left[j2\pi(k_x x + k_y y)\right] dk_x \, dk_y$$

The parameters k_x and k_y are the wave numbers in the x and y directions. The Fourier coefficients $F(k_x, k_y)$ are given by the inverse Fourier transform:

$$F(k_x, k_y) = \int_{-\infty}^{\infty} \int_{-\infty}^{\infty} f(x, y) \exp\left[-j2\pi(k_x x + k_y y)\right] dx\, dy$$

This equation can be simplified by transforming to the rotated coordinates (r, s). The angle of rotation is given by

$$\phi = \tan^{-1}\frac{k_y}{k_x} \tag{10-102}$$

Let

$$k = (k_x^2 + k_y^2)^{1/2}$$

We obtain the new expression

$$F(k_x, k_y) = \int_{-\infty}^{\infty} \int_{-\infty}^{\infty} f(x, y) \exp\left[-j2\pi kr\right] dr\, ds \tag{10-103a}$$

$$F(k_x, k_y) = \int_{-\infty}^{\infty} \left[\int_{-\infty}^{\infty} f(x, y)\, ds\right] \exp\left[-j2\pi kr\right] dr \tag{10-103b}$$

By interchanging the order of integration, we see that the expression in brackets is nothing more than the ray projection $P(r, \phi)$, that is,

$$p(r, \phi) = \int_{r,\phi} f(x, y)\, ds \tag{10-104}$$

Substituting this back into the integral expressions for the Fourier coefficients, we have that

$$F(k_x, k_y) = \int_{-\infty}^{\infty} p(r, \phi) \exp\left[-j\pi kr\right] dr$$

But this is nothing more than the Fourier transform of $P(r, \phi)$ with respect to r, that is,

$$P(k, \phi) = \int_{-\infty}^{\infty} p(r, \phi) \exp\left[-j2\pi kr\right] dr = F(k_x, k_y) \tag{10-105}$$

This conclusion is of fundamental importance in image reconstruction; that is, each Fourier coefficient or wave amplitude of the density function is equal to the corresponding Fourier coefficients of the projection taken at the same angle as the Fourier wave. Then the Fourier coefficients of the image can easily be obtained from those of the projections and the picture of the image can be synthesized (Figure 10-70).

The results of the foregoing development can be stated as the Fourier transform of a one-dimensional projection of an object is identical to the corresponding central section of the two-dimensional Fourier transform of the object. This means, in terms of x-ray scanning systems, that the Fourier transform of the x-ray profile from a scan pass at a given angle is the same as the two-dimensional Fourier transform density function $f(x, y)$. The reconstruction algorithm can be stated in the following way:

1. Take the one-dimensional Fourier transform of the projection at angle ϕ, $0 \leq \phi \leq 180°$ (usually done using a discrete Fourier transform computed using the fast Fourier transform algorithm).
2. Equate to the two-dimensional Fourier coefficients at the corresponding angle in the Fourier plane.
3. Use interpolation methods to provide a two-dimensional array of Fourier coefficients.
4. The image is reconstructed by taking the inverse two-dimensional transform of these coefficients.

Computationally, the inverse transform is obtained by summing up the sine and cosine waves whose amplitudes are given by the Fourier coefficients, $F(k_x, k_y)$. This provides a reconstruction of the object or density function.

The Filtered Back-Projection Method

We wish to relate the approximate method of back-projection reconstruction to a more accurate method that will eliminate the artifacts that were produced by this method. In the back-projection method, the reconstruction is performed by back-projecting each profile across the (x, y) plane. This means that the magnitude of each ray sum is equated to all points in the grid matrix that make up the ray.

If we let $\hat{f}(x, y)$ be the reconstructed density function which is used as an approximation to the true density function $f(x, y)$, the back-projection method can be put in the form

$$\hat{f}(x, y) = \sum_{j=1}^{m} p(x \cos \phi_j + y \sin \phi_j, \phi_j) \Delta\phi$$

where ϕ_j = jth projection angle

$\Delta\phi$ = angle distance between projection ($\Delta\phi = \pi/m$)

m = number of projection equally spaced for $\phi = 0$ to $180°$

The interpretation of the factor $x \cos \phi + y \sin \phi_j$ is that it represents a ray which passes through the point (x, y), so that the back-projection density at each point is the sum of all the m ray sums passing through the point.

It has been shown that the back-projection method produces artifacts due to the projecting of values along the entire ray. To compensate for

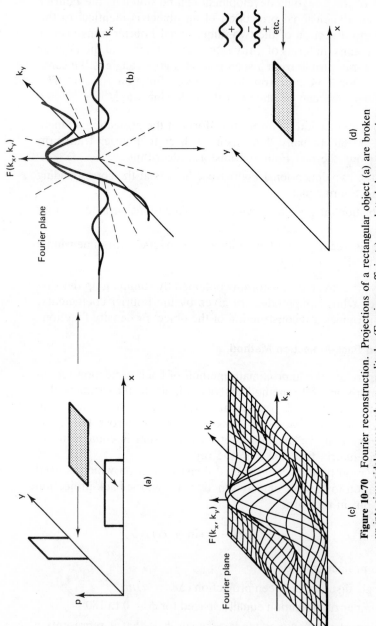

Figure 10-70 Fourier reconstruction. Projections of a rectangular object (a) are broken up into sinusoidal waves, whose amplitudes (Fourier coefficients) are plotted along appropriate lines in the frequency plane (b). Dashed lines suggest the presence of other transformed projections. A rectangular array of Fourier coefficients is then obtained through interpolation (c). Finally, the image is reconstructed by adding together sinusoidal waves (d) with amplitudes given by the Fourier coefficients. In summary, (b) and (c) show the amplitudes of sinusoidal waves of various frequencies which when combined together produce the spatial functions (a) and (d). (From R. A. Brooks et al., "Theory of image reconstruction in computed tomography," *Radiology*, Vol. 117, Dec. 1965, by permission.)

this production of artifacts by back-projection, the method of filtered back-projection was developed.

First we write the back-projection equation in integral form:

$$\hat{f}(x, y) = \int_0^{\pi} p(x \cos \phi + y \sin \phi, \phi) \, d\phi \qquad (10\text{-}106)$$

Let $P(k, \phi)$ be the one-dimensional Fourier transform of $p(r, \phi)$; then

$$p(r, \phi) = F^{-1}[P(k, \phi)]$$

$$p(r, \phi) = \int P(k, \phi) \exp [j2\pi kr] \, dk \qquad (10\text{-}107)$$

which is the inverse Fourier transform of $P(k, \phi)$. By substitution we have the relation that

$$\hat{f}(x, y) = \int_0^{\pi} \int_{-\infty}^{\infty} \frac{P(k, \phi)}{|k|} \exp [j2\pi k(x \cos \phi + y \sin \phi)] |k| \, dk \, d\phi \qquad (10\text{-}108)$$

The integrand here is in the form of a two-dimensional Fourier integral in polar coordinates. By taking the two-dimensional Fourier transform of equation (10-108), we have

$$\hat{f}(k_x, k_y) = \frac{P(k, \phi)}{|k|} = \frac{F(k_x, k_y)}{|K|} \qquad (10\text{-}109)$$

This relationship illustrates that the back-projected reconstructed image is equal to the true image where the Fourier amplitudes are scaled by the magnitude of the spatial frequencies.

If we wish the exact reconstruction, we use the Fourier integral expression for $f(x, y)$, putting the expression in polar form:

$$f(x, y) = \int_0^{\pi} \int_{-\infty}^{\infty} F(k_x, k_y) \exp [j2\pi k(x \cos \phi + y \sin \phi)] |k| \, dk \, d\phi \qquad (10\text{-}110)$$

The inner integral has the form of a one-dimensional Fourier transform. Substituting $F(k_x, k_y)$ by $P(k, \phi)$ as derived previously, we have that the modified profile function can be given as

$$p^*(r, \phi) = \int_{-\infty}^{\infty} |k| P(k, \phi) \exp (j2\pi kr) \, dk \qquad (10\text{-}111)$$

Then

$$f(x, y) = \int_0^{\pi} p^*(x \cos \phi + y \sin \phi, \phi) \, d\phi \qquad (10\text{-}112)$$

For computer use the integral is replaced by a summation:

$$f(x, y) = \sum_{j=1}^{m} p^*(x \cos \phi_j + y \sin \phi_j, \phi_j) \, \Delta \phi \qquad (10\text{-}113)$$

where m = number of projections

$\Delta\phi$ = interval between projections

This is basically the same as the earlier back-projected method, but now p^* is back-projected. Since p^* is modified in the Fourier domain, the operation $P(k, \phi)$ is in this domain and then the transformation back-projecting p^* is a filtering operation in which the high-frequency components are increased in proportion to their wave number. Thus the filtered back-projection method algorithm can be given as:

1. Take the Fourier transform of a projection.
2. Multiply each coefficient by $|k|$ and take the inverse transform.
3. Back-project onto the image plane.
4. Repeat for all projections. [3]

PERFORMANCE STANDARDS FOR MEDICAL X-RAY EQUIPMENT

X-ray equipment standards that are set up by the federal and state governments are designed to reduce patient exposure to ionizing radiation. The need for these standards and the clinical engineering determination of equipment compliance arises from the fact that 130 million people are estimated to receive some kind of x-ray examination each year in this country. X-ray equipment standards are issued by the Food and Drug Administration under authority of the Radiation Control for Health and Safety Act (Public Law 90-602). The Act requires the Secretary of Health and Human Resources to conduct a radiation control program to reduce human exposure to electronically produced radiation. The basic philosophy of these regulations is to provide better diagnostic information with a reduced level of radiation.

There are two major sources of unnecessary patient exposure: (1) x-ray equipment that does not meet its own performance standards, and (2) physicians or other medical practitioners who order x-rays for a patient when they may not be called for. With regard to the first item, the clinical engineer should be actively involved with a radiation safety program in the hospital and should perform routine performance checks on the hospital's x-ray systems as he or she would for any patient-oriented equipment in the hospital. When x-ray equipment does not meet its performance specifications, this always results in overexposure of the patient to x-radiation. If equipment does not produce enough radiation for the given exposure time to obtain a usable radiograph by the radiologist, a second exposure is usually required. Thus the patient has received more radiation than was intended. Obviously, nothing need be said about the case where more radiation is produced than is required to produce a usable x-ray picture on film. Hopefully, with the use of video output and x-ray detectors beginning to take the place of the standard x-ray film, radiation to the patient may be reduced even more.

In the second case, only the education of medical personnel will help them understand when x-rays are called for and when they will be useful diagnostically. Taking an x-ray of a patient is not like taking their photograph, as there is danger in exposing patients to any form of ionizing radiation, regardless of how little. Although there are many opinions on the subject of "safe" levels of x-radiation dosage, x-rays are not safe and should be prescribed for a patient in the same way as a dangerous drug.

There are two things that should be noted about x-ray equipment standards. First, they are equipment performance standards and are not intended to regulate how the equipment is designed, as it is up to the manufacturer to determine how to achieve levels of equipment performance as required by the standards. Second, equipment standards do not regulate the x-ray equipment user. They neither require health professionals to practice radiology in a certain way nor prohibit them from using x-ray equipment for their desired purpose—thus the need for the education of these practitioners on the appropriate use of diagnostic x-rays.

The scope of medical professionals includes dental practitioners. All too often, dental x-rays are taken when they are not required. It has become a routine procedure when a patient comes in for an examination to "look" for cavities using x-rays. It is a question of whether the risks in this case do not outweigh the benefits. Although most dental practitioners will say that the dose of x-rays is "low," there is also the other side of the coin, which says that there is no safe level of radiation.

Governmental standards for x-ray equipment components and systems apply to units manufactured after August 1, 1974. The components covered by the standard include x-ray tube housing assemblies, x-ray controls, high-voltage generators, spot-film devices, and image intensifiers, tables, cradles, film changers, cassette holders, and beam limitation devices. It is required that all components manufactured after the August 1, 1974, effective date of the standard be certified by the manufacturer for compliance. These components must bear permanent certification labels which are readily visible after component assembly.

Positive X-Ray Beam Limitation

One of the major causes of unnecessary x-ray exposure to parts of the body that are not under study is the practice of using an x-ray beam larger than needed to produce a diagnostically acceptable radiograph. Standards now require that all diagnostic x-ray equipment be able to limit the usable x-ray beam to the size of the film or detector being used. At one time it was common practice to leave the beam unadjusted just as long as the film area was covered. The fact that other parts of the body were being radiated for no diagnostic reason did not seem to be meaningful to practitioners at that time. Thus manufacturers are required to provide automatic beam limitation so that the area covered by the x-ray beam is no larger than the film cassette or detector being used. Once the equipment is in place in the hospital, the clinical engineer should check to see that the positive beam limitation system is in proper working order and that there is good correspondence between the working area as described by the light beam in the tube head and that actually covered by the x-ray beam itself.

Beam Reproducibility and Linearity

The reproducibility and linearity of the x-ray system are important to the performance of the system and to the ultimate reduction of radiation exposure of the patient. Reproducibility pertains to the ability of the x-ray equipment to duplicate certain radiation exposures for given voltage, current, and time settings. These parameters correspond to a "dose" of radiation and thus when medical personnel call for an exposure using these parameter settings, they are assuming that this relates to a corresponding dose of radiation. As with any piece of patient-oriented equipment, it is the responsibility of the clinical engineer to determine whether the x-ray system can meet this performance criterion with respect to reproducibility. Linearity has to do with the fact that the x-ray beam output intensity is linearly proportional to the tube current, or putting it another way, proportional to the MAS of the machine.

X-Ray Beam Quality

The standards now prescribe a certain acceptable level of beam quality, which can be achieved through appropriate filtration. Systems that have removable filtration, such as general-purpose systems used in mammography, must be provided with a positive means of assuring the presence of at least the added filtration needed to achieve the minimum beam quality required. In some units where the amount of filtration is a function of the examination, positive filtration may be accomplished by a filter system interlock with the voltage selector so that the system is not functional above a set voltage unless the required filtration is in place.

Other aspects of the radiation standards which dictate what features are required to be present and functional on an x-ray system are beam-limiting devices that must be between the x-ray source and the patient. This will reduce patient exposure, since in many cases the practice in the past has been to use shielding directly in front of the film to confine exposure to the desired portion of the film, instead of obtaining the same results by limiting the beam before it enters the patient's body.

Automatic Exposure Control

It is also required to have a backup timer for automatic exposure control systems to ensure that the equipment cuts off at a maximum exposure time even if the automatic timer fails. This is another of the areas of performance that the clinical engineer should check during a performance survey of the x-ray equipment. Also, all radiographic systems are required to give visible warnings when the x-ray beam is on and an audible signal when the exposure has been terminated.

Equipment manufacturers are required to furnish purchasers with a schedule of maintenance necessary to keep the system in compliance with government regulations as well as instructions with respect to procedures and precautions that should be followed in case of any unique features of the equipment. It is the hospital's responsibility and thus the clinical engineer's duty to have the x-ray equipment maintained according to the schedule furnished by the manufacturer to ensure compliance with the standard for the life of the equipment. This can be done by the clinical engineering department of the hospital, or it can be accomplished through maintenance contracts with the equipment vendor or with outside equipment maintenance

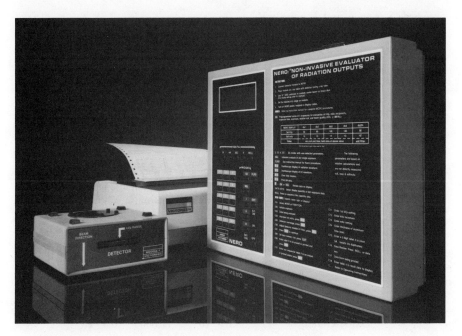

Figure 10-71 Victoreen NERO noninvasive radiation output test instrument. (Courtesy of Victoreen, Inc., Cleveland, Ohio.)

groups or clinical engineering shared services. In the case where equipment maintenance is performed under a service contract, it is important that the contract include a provision to keep the x-ray system in compliance with prevailing codes and standards by strict adherence to the maintenance schedule.

If the clinical engineering department is supervising the performance of the x-ray equipment, there are a number of electronic test devices that can be used to monitor the performance of an x-ray system. One such device is shown in Figure 10-71. This device will determine the kVp, timer accuracy, x-ray reproducibility, mA linearity, half-value layer for beam-quality determination, and real-time fluoro rate and kVp. All of this can be done noninvasively, that is, without having to remove the high-voltage cables from the x-ray head.

PROBLEMS

1. Derive the following expression relating the wavelength of an x-ray photon to its energy:

$$\lambda = \frac{12.4}{E}$$

where λ is in Å and E is in keV.

2. Calculate the energy in keV and the frequency in Hz of electromagnetic photons of wavelengths of 1, 0.5, 0.1, and 0.05 Å.

3. What is the fraction of energy that would appear as heat when 75-keV electrons are used to produce x-rays in a target made of the following materials:
   (a) Aluminum
   (b) Copper
   (c) Tungsten
   (d) Gold
   Let $k = 1 \times 10^{-6}$.

4. Maxwell's equations relate electric and magnetic fields, and the time variations and the solutions to these equations give rise to mathematical descriptions of radiated EM waves. Explain the phenomenon of Bremsstrahlung in terms of Maxwell's equations.

5. Three filaments each have the same dimensions. One is made of tungsten, the second is made of thoriated tungsten, and the third is a barium-coated metal. The physical constants for these filaments are:

	a	b	E_w
Tungsten	60.2×10^4	57,400	4.57 eV
Thoriated tungsten	3×10^4	30,500	2.63 eV
Oxide-coated	1×10^4	11,600	0.1 eV

Assuming that each is operated at a temperature of 600°K, determine which filament yields the greatest electron emission.

6. The thermionic emission and space-charge-limited characteristics of an x-ray tube are described, respectively, by the expressions

$$I_{th} = \frac{T^2}{500} e^{-16,000/T}$$

$$I_a = \frac{(V_{a\text{-}c}^{3/2})}{500}$$

If the cathode is assumed operated at 2000°K, determine the threshold value of the anode voltage beyond which operation is in the temperature-limited region.

7. Good anode design for x-ray tubes requires that the anode material have the following properties:
   (a) High melting point
   (b) High atomic number
   (c) Low vapor pressure
   (d) Good heat conductivity
   Discuss in detail, using mathematical models when needed, to show how each of these properties is necessary in the selection of x-ray tube anode materials.

8. Assume a rotating anode x-ray tube that has a maximum heat storage capacity of 110,000 heat units (watt-seconds). If the anode is 50 mm in diameter and is made of tungsten of 550 g mass, determine the cooling-curve equation for the anode if its maximum temperature can be 3000°C. Plot a curve of heat units (watt-seconds) as a function of time in minutes.

9. The high-voltage transformer for a medical x-ray machine has a 500:1 turns ratio. The primary is excited by 110 V rms at 60 Hz. Find the rms and peak voltage across the secondary of the transformer.

10. Find the power dissipated in a diagnostic x-ray tube operating at 80 kV at 400 mA. If the tube is operated for 2 s at this rating, calculate the number of joules and calories imparted to the anode. If the anode is made mostly of copper and weighs 5 lb, what is the rise in temperature of the anode over room temperature if radiation losses are neglected?

11. Define the term "power factor" and explain what it means when applied to transformers.

12. A 10-kVA 60-Hz 1000/500-V transformer supplies a 0.5 leading power factor load. All transformer impedances referred to the secondary are $R = 1\ \Omega$ and $X = 3\ \Omega$. Find the voltage regulation of the transformer.

13. For a focal spot size of 1 mm, an exposure requiring 125 kVp at 300 mA for 1/20 s, would you proceed with this exposure? Use the tube chart (Figure 10-27) to help in the determination.

14. Assume that 100,000 H.U. of exposure has been taken. Two minutes later an exposure of 90 kVp at 300 MAS is taken. Using Figures 10-25 to 10-28, determine what waiting time, if any, is required before making an exposure of 100 kVp at 150 MAS.

15. For cineradiography, exposure parameters of 150 kVp at 20 mA for 1/60 s at 30 frames/s are to be used. Calculate the maximum continuous film run for a focal spot size of 1 mm.

16. For the three-phase load in Figure 10-72 where there are three equal impedances of $Z = 10\ \underline{/45°}$ ohms connected in a Y configuration to a line voltage of 220 V:

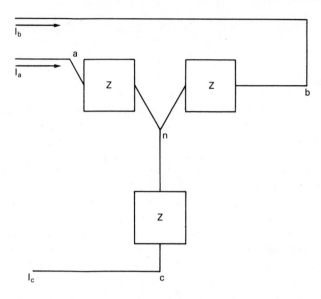

Figure 10-72

(a) Determine the phase voltages.

(b) Determine the phase and line currents.

(c) Determine the phase and total power absorbed by the load.

(d) Draw a phasor diagram showing clearly the line voltages, phase voltages, and line currents.

17. Given a three-phase Δ primary and Y secondary transformer with a rated full-load output of 100 kV with 15% voltage regulation and a primary line voltage of 208 V, what is the turns ratio?

18. Derive the rms voltage values for a 6-pulse and a 12-pulse anode voltage. Derive an expression for the power dissipated in the anode of a 6-pulse and a 12-pulse three-phase x-ray tube with a rotating anode. Assume that the applied voltage is such that the tube operates in the Richardson region.

19. Calculate the percent ripple for a three-phase, 6-pulse and 12-pulse anode voltage.

20. The high-voltage cables of an x-ray machine must withstand 100 kVp without having a dielectric breakdown. If the capacitance is to be no greater than 60 pF/ft, with the cable to have a diameter of 1 in. with an inner conductor diameter of 1/16 in., what dielectric constant must the cable insulation have?

21. For a single-phase full-wave rectified x-ray generator system which develops 100 kVp at 5 mA:

(a) Plot the waveform applied to the tube anode.

(b) Derive the equation for the applied anode voltage.

(c) What is the tube current that produces a voltage "cut-in" point just as the applied voltage to the rectifiers begins to increase from zero [i.e., when the time is equal to $n(1/120)$ s for $n = 1, 2, 3, \ldots$].

22. Define the following terms used in computed tomography:
(a) Ray sum
(b) Projection
(c) Pixel
(d) Reconstruction
(e) Gray scale
(f) Hounsfield number

23. Given the four-pixel matrix with six ray sums in Figure 10-73, find the associated values of the density function for each pixel using iterative reconstruction and back-projection methods.

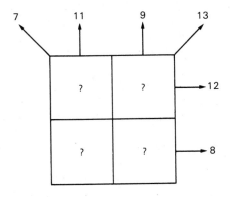

Figure 10-73

24. Let the density function $f(x, y) = \delta(x, y)$, the delta function. Show that the simple back-projection method produces a reconstruction density function given as

$$\hat{f}(x, y) = \frac{1}{\sqrt{x^2 + y^2}}$$

REFERENCES

1. Bogardus, Carl R., *Clinical Application of Physics of Radiology and Nuclear Medicine* (St. Louis, Mo.: Warren H. Green, 1964).

2. Bracewell, R. N., "Strip integration in radioastronomy," *Aust. J. Phys.*, Vol. 9, 1956, pp. 198–217.

3. Brooks, R. A., and G. Di Chiro, "Principles of computer assisted tomography (CAT) in radiographic and radioisotopic imaging," *Phys. Med. Biol.*, Vol. 21, 1976, pp. 689–732.

4. Cooley, J. W., and J. W. Tukey, "An algorithm for the machine calculations of complex Fourier series," *Math. Comput.*, Vol. 19, Apr. 1965, p. 297.

5. General Electric Co., *Introduction to Computed Tomography* (Milwaukee, Wis.: GE, 1976).

6. Hayt, William H., and George Hughes, *Introduction to Electrical Engineering* (New York: McGraw-Hill, 1968).

7. Hendee, William R., *Medical Radiation Physics, Roentgenology, Nuclear Medicine and Ultrasound,* 2nd ed. (Chicago: Year Book Medical Publishers, 1979).

8. Johns, Harold Elford, and John Robert Cunningham, *The Physics of Radiology,* 3rd ed. (Springfield, Ill.: Charles C. Thomas, 1977).

9. Kuhl, D. E., and R. Q. Edwards, "Image separation radioisotopic scanning," *Radiology,* Vol. 80, 1963, pp. 653–661.

10. Langmuir, J., "The effect of space charge and residual gases on thermic currents in high vacuum," *Phys. Rev.,* Vol. 2, 1913, p. 450.

11. Ter-Pogossian, Michel, *The Physical Aspects of Diagnostic Radiology* (New York: Harper & Row, 1967).

12. Trout, E. D., *Course Manual for Machine Sources of X-rays,* (Rockville, Md.: U.S. Department of Health, Education, and Welfare, 1977).

11

ENGINEERING ASPECTS
OF RADIATION

INTERACTION OF RADIATION WITH MATTER

In order for the engineer working in the field of x-radiation to understand the basic principles of x-ray detection measurement, exposure, absorbed dose, and the theory of radiation shielding, an understanding of the basic interactions of x-rays with matter is essential. When x-rays interact with matter such as biological material, for instance, the interaction involves mainly the transfer of energy from the radiation to the matter being radiated. Since matter is composed of atomic nuclei and orbital electrons, the interaction of x-radiation with matter constitutes the manner in which energy is transferred to either or both of these components of matter. This energy transfer to the absorbing material results in excitation and ionization of the absorbing medium, which ultimately is dissipated as heat. The absorbing materials of primary concern to the engineer are tissue and radiation shielding materials. When living matter is irradiated, the predominant interaction in the sequence of atomic events leading to biological damage is either ionization or excitation of the tissue atoms. The ionization potential of an element is the amount of energy necessary to remove the least tightly bound electron in an atom of the element. A photon collision with sufficient energy to produce ionization dissipates its energy by imparting some of its energy over that needed for ionization to kinetic energy imparted to the ejected electron and to the positive ion, which recoils under the impact of the collision. This inelastic collision takes place when the ballistic photon possesses sufficient energy to meet the requirement of the law of conservation of momentum. When this is not the case, an elastic collision with the atom as a whole will occur.

The objective of the clinical engineer or radiological engineer is to minimize radiation exposure to the patient and medical personnel. To do this, it becomes necessary to introduce quantities for the assessment of radiation dose and for establishing limits to prevent excessive exposure to x-radiation. The clinical or radiological engineer does this in two ways:

1. By facilities design, when the radiology area is shielded properly for the safety of medical and nonmedical personnel in the general area
2. By performance check of the x-ray machine itself, to assure that exposures called for at the x-ray generator are what is produced by the machine

From any point of view, a patient is bound to be overexposed to x-radiation if the x-ray machine does not conform to its performance specifications. If for a given kVp and mA setting too little radiation results, the radiograph may be underexposed and of little value to the radiologist, resulting in a retake of the x-rays and thus subjecting the patient to excessive radiation. The same scenario is true with overexposure of the radiograph. In this case not only is the patient overexposed initially but may have to be re-exposed because of the poor-quality radiographs produced. Thus, after an x-ray facility is designed properly, the primary objective of the engineer in radiology is the proper performance of the x-ray equipment.

As one approach to the measurement of x-radiation and its interaction with biological and shielding material, we define several groups into which the quantities to be defined will fall. These groups are:

1. Those describing the radiation field
2. Those describing the initial interaction of the radiation with matter
3. Those describing the absorption of energy
4. Those describing biological effects or describing risk

DEFINITIONS OF INTERACTIONS

Radiant energy in the form of x-rays when passing into an absorbing material such as body tissue or shielding material interacts with the material by losing energy or changing the direction of travel. In the process of passing into and through material, photon energy is either absorbed, scattered, or unattenuated. In the absorption process, the energy of the photon is given up in atomic processes that will be discussed later. When a photon is scattered, its direction of travel is changed. There are two forms of scattering that can take place:

1. Elastic scattering, in which no energy is lost
2. Inelastic scattering, where there is a loss of energy

Another mechanism of the transfer of energy is that of ionization. Ions are free electrons or atoms which carry an electric charge. Ionization is the process by which electrons are removed from an atom producing negative and positive ion pairs. Ionization may occur when an x-ray photon collides with an orbital electron. This mechanism is of importance to the engineer in understanding the interaction of x-rays with matter, as this is the means by which energy is transferred from radiation to matter. When living matter is irradiated, the sequence of events at the molecular level leading to biological damage is either excitation or ionization.

MECHANISMS OF RADIATION INTERACTION WITH MATTER

In the absorption of x-ray energy there are three basic interactions that occur in the absorption of photon energy. They are:

1. The photoelectric effect
2. The Compton effect
3. Pair production

Photoelectric Effect

When an incident x-ray photon whose energy is great enough to ionize an atom collides with a tightly bound orbital electron, the photon disappears and the electron is ejected from the atom forming an ion pair. If W is the atomic binding energy and $h\nu$ is the energy of the incident photon, the kinetic energy of the ejected electron is given by

$$(KE)_{e^-} = h\nu - W = \tfrac{1}{2} m v_{e^-}^2 \qquad (11\text{-}1)$$

where m = mass of electron

v_{e^-} = velocity of the ejected electron

When an electron is ejected in this manner, the process is called a *photoelectric interaction* and the ejected high-speed electron is called a *photoelectron*. Under these conditions, all the energy of the photon is absorbed and the atom emits characteristic radiation corresponding to the transitions made by the remaining orbital electrons to fill the vacancy left by the ejected electron (Figure 11-1). The photoelectron has sufficient energy, in general,

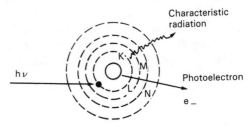

Figure 11-1 How incident radiant energy, $h\nu$, is converted into a photoelectron and characteristic radiation.

to produce many hundreds of additional ion pairs as it travels through matter expending energy.

The probability of a photoelectric interaction depends on the energy $h\nu$ of the incident photon and the binding energy W of the orbital electrons of the material. The probability of photoelectric interaction is maximum when the energy of the photon equals the binding energy of the electron (i.e., $h\nu = W$). The probability of interaction decreases as the energy of the photon increases over that of the binding energy of the atom. This probability varies with the fourth power of the atomic number Z and inversely with the cube of the photon energy, that is,

$$\text{probability of P.E. interaction} \propto \frac{Z^4}{(h\nu)^3} \qquad (11\text{-}2)$$

Thus the probability of photoelectric interaction is low for low-Z material and decreases rapidly with increasing photon energy, whereas the converse is true for low-energy photons and high-Z material. For tissue the photoelectric effect becomes negligible for x-ray photon energies above 100 keV.

Compton Scattering

Compton scattering is an elastic collision between a photon and a "free" electron, an electron where binding energy is small compared to the energy of the incident photon. In this type of interaction some of the energy of the incident photon is imparted to the free electron, the rest remaining with the now-scattered photon. Since the scattered photon is of lower energy, its frequency changes in proportion to the energy decrement. To calculate the change in energy of the incident photon, we can use Figure 11-2 as a model.

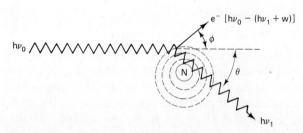

Figure 11-2 Radiation attenuation due to the Compton effect.

The amount of energy transferred in a collision can be computed by invoking the laws of conservation of momentum and energy. An energy balance equation describing the model shown in Figure 11-2 is given by

$$\frac{hc}{\lambda} + m_0 c^2 = \frac{hc}{\lambda'} + mc^2 \qquad (11\text{-}3)$$

Writing equations for the conservation of momentum in the horizontal and vertical directions, we have that

$$\frac{h}{\lambda} = \frac{h}{\lambda'} \cos \theta + mv \cos \phi \qquad (11\text{-}4)$$

and

$$0 = \frac{h}{\lambda'} \sin \theta \, mv \sin \phi \qquad (11\text{-}5)$$

Solving for the change in wavelength of the photon yields

$$\lambda' - \lambda = \Delta\lambda = \frac{h}{m_0 c^2} (1 - \cos \theta) \qquad cm \qquad (11\text{-}6)$$

where $m_0 c^2$ is the rest energy of the electron. Substituting for all of the numerical constants and converting centimeters to angstrom units, we have that

$$\lambda' - \lambda = \Delta\lambda = 0.0242(1 - \cos \theta) \qquad \text{Å} \qquad (11\text{-}7)$$

We can also write equation (11-3) in terms of the energy of the scattered photon as a function of the initial photon energy, $h\nu_0$, and the angle of scatter is given in the equation

$$h\nu_1 = \frac{h\nu_0}{1 + (h\nu_0/m_0 c^2)(1 - \cos \theta)} \qquad (11\text{-}8)$$

$m_0 c^2$ = rest energy of electron = 0.511 MeV when the photon energy is expressed in MeV. The energy of the Compton or recoil electron given in terms of the photon scatter angle θ is given by

$$E = h\nu_0 \frac{(h\nu_0/m_0 c^2)(1 - \cos \theta)}{1 + (h\nu_0/m_0 c^2)(1 - \cos \theta)} \qquad (11\text{-}9)$$

If the incident x-ray photon impinges on an electron such that the ejected electron travels in the same direction as the x-ray photon and absorbs maximum energy from the photon, the scattered photon will travel in the reverse direction, $\theta = 180°$, with minimum energy. Under this condition equations (11-8) and (11-9) become

Compton electron:

$$E_{max} = h\nu_0 \frac{2(h\nu_0/m_0 c^2)}{1 + 2(h\nu_0/m_0 c^2)} \qquad (11\text{-}10)$$

Photon energy:

$$h\nu_{1min} = h\nu_0 \frac{1}{1 + 2(h\nu_0/m_0 c^2)} \qquad (11\text{-}11)$$

Since the exchange of energy due to Compton interactions is involved with the free electrons, it is essentially independent of atomic number since most materials have about the same number of electrons per gram (i.e., approximately 3×10^{23}). Therefore, the absorption of radiation per gram by Compton scattering is nearly the same for all materials, such as tissue, bone, and muscle. Thus energy absorption per unit thickness is directly proportional to the density and thus the atomic number Z. If we let σ be the fraction of x-ray photons removed from the x-ray beam by Compton effect per unit thickness, it can be shown that

$$\sigma \propto \frac{Z}{E}$$

where Z = atomic number of absorbing material

$\quad\quad E$ = energy of incident photons

Thus the probability of a Compton interaction decreases with increasing quantum energy. The Compton effect predominates at intermediate energies from 600 keV to 2.5 MeV.

Example: Compton Scattering

Determine the maximum energy of a recoil electron and the minimum energy of the scattered x-ray photon when the incident photon has an energy of 6 MeV.

Solution. From equations (11-10) and (11-11) we have

$$E_{\max} = 6 \times \frac{2(6/0.511)}{1 + 2\,(6/0.511)} = 5.75 \text{ MeV}$$

$$h\nu_{1_{\min}} = 6 \times \frac{1}{1 + 2\,(6/0.511)} = 0.25 \text{ MeV}$$

Total energy = 5.75 + 0.25 = 6 MeV, as expected.

Pair Production

When the energy of an incident x-ray photon is greater than 1.02 MeV, and it passes close enough to the nucleus of an absorber atom to interact with its electric field, the x-ray photon may disappear, giving up all its energy to two new particles of energy 0.51 MeV each. These particles are the positron and the electron (see Figure 11-3). This process of pair production

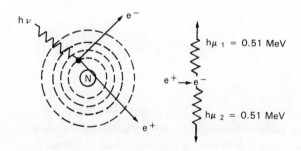

Figure 11-3 Radiation attenuation due to pair production.

represents the conversion of energy into matter. Energy in excess of the 1.02 MeV threshold required for pair production goes into the kinetic energy of each of the newly produced particles. In this process no net charge is produced since the charges on the positron and electron balance out electrically. Thus an energy balance equation for this method of energy absorption is

$$hv = 1.02 + E_+ + E_- \tag{11-12}$$

where E_+ = kinetic energy of the positron, MeV

E_- = kinetic energy of the electron, MeV

hv = energy of the incident x-ray photon, MeV

Both the electron and positron dissipate their kinetic energy via ionization, excitation, and Bremsstrahlung production in the same manner as any high-energy electron does. When the kinetic energy of the positron is about expended, it combines with a free electron which converts mass back to two 0.51-MeV photons of energy. These photons of energy may undergo Compton scattering or photoelectric absorption.

Example

A 10-MeV photon is passing through an absorber. List the probable interactions and energy loss for the photon, starting with pair production.

Solution. Consider a 10-MeV photon as it is passing through an absorber and undergoes a pair-production interaction, producing two particles of energy 4.5 MeV. Approximately 1 MeV is lost in producing the positron and electron. The pair lose their energy as mentioned previously until the positron interacts with a free electron, producing two photons of energy 0.51 MeV. Thus of the 10 MeV the incident photon possessed, 8.98 MeV is expended and 1.02 MeV is left for further interaction, such as Compton scattering, thus reducing the energy of the particle even more.

In summary, at lower energies photoelectric production is important and is the dominant mechanism of energy absorption. As the energy of the incident photon increases, Compton scattering becomes dominant. As the energy of the x-ray photons is increased still higher, pair production becomes the most important type of energy absorption.

THE LINEAR ATTENUATION COEFFICIENT

From the material presented previously in this chapter on photoelectric effect, Compton scattering, and pair production, it was seen that the mechanism of energy absorption for x-ray photons will depend on the energy of the incident photon and the atomic number of the absorbing material. The energy absorption of an x-ray photon is either total in its interaction with an absorber or it is unaffected. It is like an "all-or-nothing" interaction.

Because of this, x-radiation cannot be completely absorbed and can only be reduced in intensity. This is in contradistinction to atomic radioactivity producing alpha and beta particles, whose energies can be completely absorbed in finite lengths of absorbing material.

Consider the test configuration shown in Figure 11-4. Let the number of incident photons on the absorber of thickness Δx be N_0 and let the number of photons recorded on the other side be N. Because of the "all-or-nothing" concept of x-ray photon absorption, if there is any intensity attenuation, some photons ΔN must have been removed from the x-ray beam. The number ΔN is directly dependent on the number of incident photons impinging on the surface of the absorber. Thus if the number of incident photons is increased by a factor K, the probability of interaction is increased by the same factor K. This means that the number of photons removed will increase by the factor K. The same type of linear relationship holds true if the thickness of the absorber is also changed by a multiplying factor since the number of atoms in the path of the x-ray photon is increased. Thus ΔN changes as a function of the thickness Δx of the absorbing material.

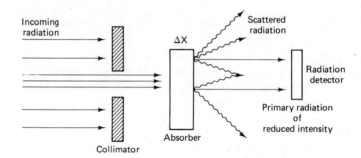

Figure 11-4 Test configuration for determination of the linear attenuation coefficient.

For a fixed incident x-ray beam containing N photons, the differential change in the beam due to a differential thickness of absorbing material can be written

$$\frac{dN}{dx} = -\mu N \tag{11-13}$$

where the negative sign indicates that the photon beam is being decremented and μ is a constant of proportionality which is a function of the atomic number Z of the absorber and the energy E of the incident photons and has a definite value for fixed values of Z and E.

To find the value N of the x-ray photon, we solve the first-order differential equation given in equation (11-13) by separation of variables and integration. Thus

$$\frac{dN}{N} = -\mu dx \qquad (11\text{-}14)$$

Integrating, we have

$$\int \frac{1}{N} dN = -\mu \int dx + c \qquad (11\text{-}15)$$

$$\ln N = -\mu x + c \qquad (11\text{-}16)$$

where c is a constant of integration. Exponentiating equation (11-16), we have

$$N = ce^{-\mu x} \qquad (11\text{-}17)$$

The boundary condition needed to solve for the constant c is that at $x = 0$, $N = N_0$, where N_0 is the initial number of x-ray photons at zero absorber thickness. Therefore,

$$N = N_0 e^{-\mu x} \qquad (11\text{-}18)$$

where μ is now referred to as the linear attenuation coefficient. If we solve equation (11-18) for μ, we have

$$\mu = -\frac{dN}{N} \frac{1}{dx} \qquad (11\text{-}19)$$

Multiplying the numerator and denominator by the photon energy, $h\nu$, we have

$$\mu = -\frac{dN \cdot h\nu}{N \cdot h\nu} \frac{1}{dx} \qquad (11\text{-}20)$$

But $dN \cdot h\nu$ is the energy decrement of the beam and $N \cdot h\nu$ is the energy of the beam. So equation (11-20) is the fraction of energy removed from the beam per unit length of absorber, which is μ, the linear attenuation coefficient. Tables of linear attenuation coefficients are available for a variety of absorbing materials.

Example

Compute (a) the thickness of lead and (b) the thickness of concrete required to reduce a 0.1-MeV x-ray beam by 90%. $\mu_{Pb} = 59.7/cm$, $\mu_{con} = 0.397/cm$.

Solution

(a) For concrete we have

$$\frac{1}{10} = e^{-0.397x}$$

$$\ln 10 = 0.397x$$

$$x = \frac{\ln 10}{0.397} = \frac{2.3}{0.397} = 5.79 \text{ cm of concrete}$$

(b) For lead we have

$$\frac{1}{10} = e^{-59.7x}$$

$$x = \frac{2.3}{59.7} = 0.0385 \text{ cm of lead}$$

or 0.385 mm of lead for the same attenuation. If we take the ratio of concrete to lead, we have

$$\frac{5.79}{0.0385} = 150.4$$

or that it takes about 150 times as much concrete as lead to attenuate the x-ray beam under these conditions. The linear attenuation coefficient is a quantitative measure of attenuation per centimeter of absorber. See Table 11-1 for linear attenuation coefficients for various materials.

Example

Calculate the amount of radiation in percent that is transmitted through a slab of material 20 cm thick which has a linear attenuation coefficient of 0.435 cm^{-1}.

Solution

$$\mu x = (0.435)(20) = 8.70$$

$$\frac{N}{N_0} \times 100 = 100e^{-8.70} = 0.02\%$$

Therefore, 0.02% is transmitted.

The concept of the linear attenuation coefficient as developed so far is for monochromatic radiation and is defined for a specific absorber at a fixed x-ray photon energy. As an example, consider aluminum as an x-ray absorber. For x-ray photons of an energy of 0.1 MeV, μ is equal to 0.435 cm^{-1}. At an energy of 0.5 MeV, μ is equal to 0.227. This is due to the fact that the mechanisms of energy absorption are energy dependent. Thus as the energy of the x-ray beam is increased, the number of x-rays that are attenuated decreases, as does the linear attenuation coefficient.

Since the number of photons at a given energy is proportional to the x-ray beam intensity, we may also write the attenuation in terms of intensity as

$$I = I_0 e^{-\mu x} \tag{11-21}$$

where I_0 is the initial intensity of the beam.

HALF-VALUE LAYER

Because not all x-radiation is absorbed by an absorber, we cannot refer to a thickness of material that will stop all the radiation incident on it. As a more convenient measure of a material's ability to attenuate x-radiation, the concept of half-value layer is used. The half-value layer of one absorbing

TABLE 11-1 Linear Attenuation Coefficients (cm^{-1})

	$\rho(g/cm^3)$	Quantum Energy (MeV)												
		0.1	0.15	0.2	0.3	0.5	0.8	1.0	1.5	2	3	5	8	10
C	2.25	0.335	0.301	0.274	0.238	0.196	0.159	0.143	0.117	0.100	0.080	0.061	0.048	0.044
Al	2.7	0.435	0.362	0.324	0.278	0.227	0.185	0.166	0.135	0.117	0.096	0.076	0.065	0.062
Fe	7.9	2.72	1.445	1.090	0.838	0.655	0.525	0.470	0.383	0.335	0.285	0.247	0.233	0.232
Cu	8.9	3.80	1.830	1.309	0.960	0.730	0.581	0.520	0.424	0.372	0.318	0.281	0.270	0.271
Pb	11.3	59.7	20.8	10.15	4.02	1.64	0.945	0.771	0.579	0.516	0.476	0.482	0.518	0.552
Air	1.29×10^{-3}	1.95×10^{-4}	1.73×10^{-4}	1.59×10^{-4}	1.37×10^{-4}	1.12×10^{-4}	9.12×10^{-5}	8.45×10^{-5}	6.67×10^{-5}	5.75×10^{-5}	4.6×10^{-5}	3.54×10^{-5}	2.84×10^{-5}	2.61×10^{-5}
H_2O	1	0.167	0.149	0.136	0.118	0.097	0.079	0.071	0.056	0.049	0.040	0.030	0.024	0.022
Concrete[a]	2.35	0.397	0.326	0.291	0.251	0.204	0.166	0.149	0.122	0.105	0.085	0.067	0.057	0.054

[a] Ordinary concrete of the following composition: 0.56% H, 49.56% O, 31.35% Si, 4.56% Al, 8.26% Ca, 1.22% Fe, 0.24% Mg, 1.71% Na, 1.92% K, 0.12% S.

Source: National Bureau of Standards Report No. 1003 (1952).

material is the thickness of the material that is required to reduce the incident radiation by half.

Let $X_{1/2}$ be the thickness of the half-value layer of absorbing material. Then

$$\tfrac{1}{2} N_0 = N_0 e^{-\mu X_{1/2}} \tag{11-22}$$

$$\tfrac{1}{2} = e^{-\mu X_{1/2}}$$

$$\mu X_{1/2} = \ln 2$$

$$= 0.693$$

$$X_{1/2} = \frac{0.693}{\mu} = \text{HVL} \tag{11-23}$$

From this we see the relationship between the half-value layer and the linear attenuation coefficient, μ. From this formulation of the half-value layer we can rewrite the relationship of the reduced number of photons as follows:

$$N = N_0 e^{-\mu x} = N_0 e^{-0.693x/X_{1/2}} = N_0 \cdot 2^{-x/X_{1/2}} \tag{11-24}$$

Example

Calculate the HVL in aluminum for 0.1-MeV photons for $\mu = 0.45$ cm^{-1}.

Solution

$$X_{1/2} = \frac{0.693}{\mu} = \frac{0.693}{0.45 \text{ cm}^{-1}} = 1.54 \text{ cm of aluminum}$$

Thus 15.4 mm of aluminum will reduce the intensity of the 0.1-MeV x-ray photons by one-half. It should be remembered that the number of photons of energy 0.1 MeV has been reduced by 50%, but those photons that do not interact with the absorber still possess 0.1 MeV of energy. This is true for monoenergetic photons.

MASS ATTENUATION COEFFICIENT

Since x-ray beam attenuation is a function of the interaction between the x-ray photons and the electrons in the absorber, attenuation is dependent on the electron density of the material, thus making the linear attenuation coefficient density dependent. A more fundamental unit of attenuation is obtained from the linear attenuation coefficient by dividing it by the density ρ represented by μ/ρ. Using dimensional analysis on the units, we have that the units of μ/ρ are

$$\frac{\mu}{\rho} = \frac{1/\text{cm}}{\text{g/cm}^3} = \text{cm}^2/\text{g}$$

The mass attenuation coefficient can be thought of as the fractional reduction in x-ray intensity produced by a layer of thickness 1 g/cm^2. The unit of the mass attenuation coefficient is "per g/cm^2," which is usually written cm^2/g.

Example

A beam of 1.5-MeV photons impinges on a carbon filter of thickness 4.0 cm. Find (a) the linear attenuation coefficient, (b) the percentage of photon that is transmitted through the filter, and (c) the half-value layer for this material given $\mu/\rho = 0.0517$ and $\rho = 2.25$ g/cm^3.

Solution

(a) For $\mu/\rho = 0.0517$ cm^2/g:

$$\mu = 0.0517\rho \frac{\text{cm}^2}{\text{g}} = (0.0517)(2.25) = 0.1163 \text{ cm}^{-1}$$

(b) $\dfrac{N}{N_0} \times 100 = 100e^{-0.0517(4)} = 81.32\%$

(c) HVL $= \dfrac{0.693}{0.1163} = 5.96$ cm

The energy absorption properties of an irradiated material are a function of the atomic number of the absorber and the energy of the x-ray photons. This can be seen in Figure 11-5 for carbon, aluminum, copper, and lead. In this figure we see that the curves fall rapidly with increased photon energy because of the increased degree of energy absorption by the photoelectric effect. In the range from about 0.1 MeV to about 5 MeV the rate of change of the curves for the mass attenuation coefficient is less due to the dominance of the Compton effect in this energy range. Beyond 5

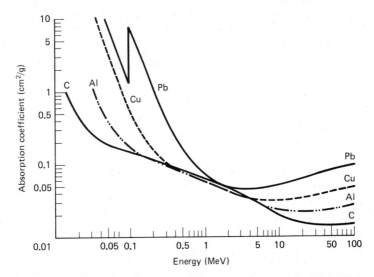

Figure 11-5 Curves illustrating the systematic variation of absorption coefficient with atomic number of absorber and with quantum energy. (Reprinted by permission. From *Introduction to Health Physics* by Herman Cember, copyright 1969 Pergamon Press, Inc., Elmsford, N.Y.)

MeV the curves begin to increase in value due to the dominance of pair production. In the energy range between 0.75 and 5 MeV most materials have about the same energy-absorbing properties when put on a mass basis.

MEASUREMENT OF RADIATION EXPOSURE

Exposure can be thought of as a property of the radiation field at a point in the field. It would be desirable to be able to determine the intensity of the x-ray beam at any point in space. In practice it is difficult to measure the energy of a beam of x-ray photons. Thus an indirect measure is used to define the energy content of the beam.

The Roentgen

The unit of exposure is defined in terms of the energy transferred to a volume of air by x-ray photons. As the x-ray photons move through the air, energy is absorbed from the beam and produces secondary electrons due to pair production, Compton electrons, and photoelectrons. The kinetic energy of these electrons is lost in the air by ionization. The electrical charge produced is used as a measure of the exposure. Exposure is defined as the relationship between the charge Q and the mass of air in the test volume.

The unit of exposure is the roentgen (R). An exposure equal to 1 R = 2.58×10^{-4} C/kg air. 2.58×10^{-4} C/kg air is equivalent to 1 electrostatic unit (esu) per cc of air at standard temperature and pressure. If the mass of 1 cc of air at STP is 1.293×10^{-3}, then

$$1 \text{ R} = \frac{Q}{M} = \frac{1 \text{ esu}}{1.293 \times 10^{-3}} = 2.58 \times 10^{-7} \text{ C/g air}$$

The energy transferred to 1 g of air is calculated as follows for an exposure of 1 R. $E_{\text{air}} = 33.7$ eV to produce one ionization in air. 1 R produces 1.61×10^{12} ionizations in 1 g of air. Therefore,

energy of 1 R = $(1.61 \times 10^{12})(33.7) = 5.43 \times 10^{-13}$ eV

1 eV = 1.6×10^{-12} erg

Therefore,

energy of 1 R = $(5.43 \times 10^{13})(1.6 \times 10^{-12}) = 86.88$ ergs/g air

Absorbed Dose—Rad

The damage produced by x-radiation depends on the amount of energy absorbed from the radiation field and is proportional to the concentration of absorbed energy in tissue. Thus the rad (radiation absorbed dose) is defined in terms of absorbed energy per unit mass of tissue in the following way: 1 rad is an absorbed dose of 100 ergs per gram. Thus the absorbed dose in air of 1 R is 0.8688 rad. If a sample of air is exposed to K roentgens, the absorbed dose is

$$D = 0.8688 \frac{\text{rad}}{\text{roentgen}} \times K = 0.869K \text{ rad}$$

and the relationship to exposure to dose *in air* is

$$0.869 = \frac{\text{rad}}{\text{roentgen}}$$

The Concept of Dose Equivalent

The concept of dose of radiation basically relates the amount of energy absorbed per unit mass for the exposure. It has been shown that equal amounts of energy absorbed do not always have the same biological effect. The biological effects depend on the rate of energy loss per unit distance. If we were dealing with radioisotopes, the distribution of the isotope would also be considered. The effect of each of these factors is multiplied by the dose in rads to produce a new unit called a rem (a unit of dose equivalence).

The rate of energy loss per unit distance is called the *linear energy transfer* (LET). The LET of the radiation is given by

$$\text{LET} = \frac{dE}{dl}$$

where dE = mean energy loss or energy given up of a particle

   dl = distance traversed in the loss of the energy dE

Generally, the higher the LET, the more effective it is in producing biological damage for a given absorbed dose. In order to relate the biological effects of radiation from different sources of radiation that produce the same biological effect, a normalizing factor called the *quality factor* (QF) is introduced. It is a factor that relates the effectiveness of different sources of radiation based on LET. Values of quality factor as a function of LET are assigned on the basis of animal experiments.

The concept of "dose equivalent" then puts all forms of radiation on a normalized scale based on biological effect. Dose equivalent is in units of rem and is given by the equation

dose equivalent (rem) = absorbed dose (rad) $\times$ QF   (11-25)

For x-rays, electrons, and positrons, QF = 1. Thus for x-rays the rad is numerically equal to the rem. Standards for radiation protection are given in terms of the rem unit by putting this in terms of maximum allowable radiation dose in rems or millirems.

Example

Consider that a person is exposed to 5 mrads of x-rays and 1 mrad of fast neutrons. What is the dose received by this person?

$$
\begin{aligned}
&\text{QF} \\
\text{x-ray}&1 \text{dose x-rays} = 5 \times 1 = 5 \text{ mrems} \\
\text{fast neutron}&10 \underline{\text{dose neutron} = 1 \times 10 = 10 \text{ mrems}} \\
&\text{total dose} = 15 \text{ mrems}
\end{aligned}
$$

The Design of Shielding for X-Ray Facilities

Since different organs and tissues have different sensitivities to radiation, the greatest health hazards are associated with the irradiation of particular organs and tissue. Thus dose limitations for individuals are based primarily on the doses that are acceptable for the most critical tissues. These are:

1. Red bone marrow
2. Lens of the eye
3. A fetus (for pregnant women)
4. Gonads

The amount of radiation a patient receives for a particular diagnostic or therapeutic procedure is determined by a health physicist or knowledgeable radiologists and the associated risk–health benefit assessment made. Thus, for the patient, x-radiation exposure is determined by the medical or dental procedure involved. For the worker who comes in contact with or works in proximity to x-ray equipment, engineering intervention in the form of facility design and safe work practices is used to reduce exposure to radiation to a "safe" level.

Exposure can be reduced from external sources of radiation by application of one or more of the following techniques:

1. Minimizing exposure time to the radiation
2. Maximizing the distance from the x-ray source
3. Shielding the x-ray source

Exposure time.   If the parameters of distance and shielding are held constant, the effect of time on exposure is a simple linear relationship of the form

total exposure (mR) = exposure rate (mR) × time of exposure (hr)

This equation holds for dose rate since the quality factor for x-rays is 1; then

total dose = (dose rate) × (exposure time)

This illustrates that reducing the working time is the simplest way of minimizing exposure.

Distance. If exposure time cannot be reduced, with the shielding held constant, then the distance becomes a factor. If the source of radiation is considered a point source, it can be shown that the relationship between radiation intensity or exposure and distance is an inverse-square relationship; that is, the radiation exposure decreases as the square of the distance as one moves away from the source of radiation. Thus if E_1 is the exposure rate at a distance d_1 and E_2 the exposure rate at a distance d_2, then

$$\frac{E_2}{E_1} = \frac{d_1^2}{d_2^2} \tag{11-26}$$

Example

If the leakage radiation from a therapeutic x-ray tube housing is 30,000 mR/hr at a point 5 cm from the surface when the tube is operated at its maximum rating for voltage and current, what is the exposure rate 1 m from the tube housing?

Solution

$$(\text{mR/hr})_2 = \frac{(0.05)^2}{1^2}\,(30{,}000 \text{ mR/hr}) = 75 \text{ mR/hr}$$

Shielding. The exponential relation for the attenuation of x-radiation as given in equation (11-18) holds for a monoenergetic x-ray beam. Thus it is a simple computation to determine the thickness of shielding material necessary to reduce the radiation intensity to the desired level. However, in the engineering design of an x-ray facility, the exponential relationship for monoenergetic photons does not hold. In practice, shields are seldom used under the ideal conditions used to develop equation (11-18).

Because some of the scattered radiation makes it through the absorbing material to the detector, it becomes part of the total radiation incident on a detector on a person which leads to an underestimate of required shielding thickness. Thus, to make accurate calculations, a factor must be included in the attenuation equation to include this increase in particles/cm^2/per second due to the scattered radiation. This factor is called a *buildup factor* and is denoted by B.

$$I = BI_0 e^{-\mu x} \tag{11-27}$$

Buildup is a complex function of four factors:

1. The energy of the radiation
2. The atomic number of the shield
3. The geometric shape of the shield
4. The geometric shape of the source

The buildup factor is always greater than 1. Charts of buildup factors are given in terms of relaxation lengths. One relaxation length is that thickness

of shield which will attenuate a narrow beam $1/e$ of its original intensity.
Thus

$$\frac{I}{I_0} = \frac{1}{e} \qquad (11\text{-}28)$$

Let X_R be the relaxation length. Then

$$\frac{1}{e} = e^{-\mu X_R} = \frac{1}{e^{\mu X_R}} \qquad (11\text{-}29)$$

so

$$\mu X_R = 1 \qquad (11\text{-}30)$$

or

$$X_R = \frac{1}{\mu} \qquad (11\text{-}31)$$

The relaxation length is numerically equal to the reciprocal of the absorption
coefficient of the absorber. The relaxation constant is to distance and
attenuation of radiation what the "time constant" is to a dynamic system.

X-RAY SHIELDING

Besides the concept of buildup which is taken into account in the engineering
design of a medical radiation facility, other factors for determining the
thickness of the protective shielding necessary to reduce the x-ray exposure
to a predetermined level are:

1. The quality of radiation production
2. The quantity of radiation produced in a given time period
3. The distance from the x-ray tube to the area of interest
4. The degree and nature of occupancy of the area
5. The type of area
6. The material to be used

Primary Protective Shielding

The primary protective barrier is shielding used to reduce radiation
exposure of the primary or useful beam to the required level. The maximum
permissible weekly dose rates of 0.1 R/week for a controlled area and 0.01
R/week for an uncontrolled area are the required levels in or near an x-
radiation facility.

The maximum dose rate at any occupied point at a distance d meters
from the target in the x-ray tube is given by

$$D_m = \frac{P}{T} \qquad \text{R/week} \qquad (11\text{-}32)$$

TABLE 11-2   Occupancy Factors for X-Ray Facilities

Full occupancy ($T = 1$):
  Control space, wards, workrooms, darkrooms, corridors large enough to hold desks,
  waiting rooms, rest rooms used by occupationally exposed personnel, children's play
  areas, living quarters, occupied space in adjacent buildings

Partial occupancy ($T = 1/4$):
  Corridors too narrow for desks, utility rooms, rest rooms not used routinely by occupationally exposed personnel, elevators using operators, and uncontrolled parking lots

Occasional occupancy ($T = 1/16$):
  Stairways, automatic elevators, outside areas used only for pedestrians or vehicular
  traffic, closets too small for future workrooms, toilets not used routinely by occupationally exposed personnel

---

where P = maximum permissible dose rate

  T = occupancy factor (1, 1/4, or 1/16, depending on condition); see
  Table 11-2 for values of T

Invoking the inverse-square law to bring the dose rate to the dose rate at
1 m from the x-ray target, we have

$$D = d^2 D_m = \frac{d^2 P}{T} \qquad \text{R/week at 1 m}$$

where d is the distance from the x-ray target to the shield. Let

W = work load, which is a measure of the amount of use of the x-ray
  machine in units of milliampere-minutes per week

U = use factor, which is the fraction of the work load during which the
  useful beam is pointed in the direction under consideration

For radiographic use the following values of U have been adopted:

  Floor: $U = 1$
  Wall: $U = 1/4$
  Ceiling: $U = 1/16$

The product of WU in mA-minutes/week is the work load responsible for
the dose rate. Define a ratio

$$K = \frac{D}{WU} = \frac{d^2 P}{WUT} \qquad \text{R/mA-min at 1 m}$$

Values of K for various thicknesses of concrete and lead at different kVp
values have been determined by experiment and are given in Figures 11-6
through 11-10.

Example

Find the primary protective barrier thickness necessary to protect a controlled area
32.8 ft from the target of an x-ray machine operating at 100 kVp. The work load

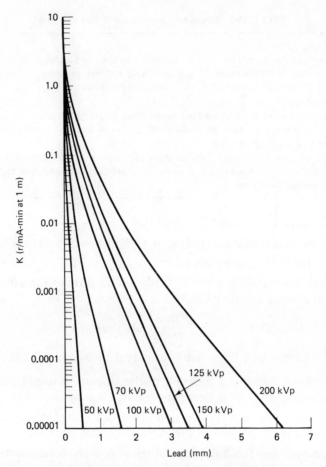

Figure 11-6 Attenuation in lead of x-rays produced by potentials of 50- to 200-kV peak. (From *Medical X-ray Protection Up to Three Million Volts*, National Bureau of Standards Handbook 76.)

is estimated at 1000 mA-min/week, and the occupancy factor is 1 for the area to be protected.

$$P = 0.1 \text{ R/week (maximum permitted dose rate)}$$

$$d = \frac{32.8}{3.28} = 10 \text{ m}$$

$$W = 1000$$

$$U = 1/4$$

$$T = 1$$

Then

$$K = \frac{0.1 \times 100}{1000 \times 1/4 \times 1} = 0.04$$

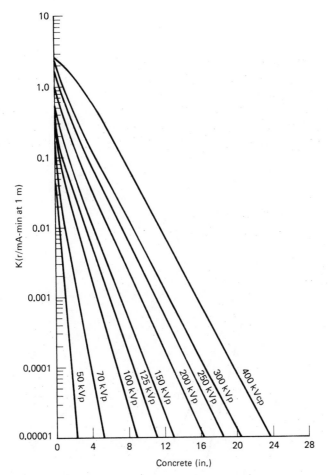

Figure 11-7 Attenuation in concrete of x-rays produced by potentials of 50 to 400 kV. (Density = 2.35 g/cm³.) (From *Medical X-ray Protection Up to Three Million Volts*, National Bureau of Standards Handbook 76.)

From the charts we get 0.4 mm of lead and 1½ in. of concrete. If construction materials other than concrete or lead are to be used for shielding purposes and no attenuation data are available, then as a first approximation to equivalency in attenuation, a ratio of densities of the elements can be used.

Example

The density of concrete is 2.35 g/cm³. Assume that 2.54 cm of sand plaster are used for shielding. What is it equivalent to in terms of concrete? The density of sand plaster is 1.54 g/cm³.

Solution

$$\text{concrete equivalence} = \frac{1.54}{2.35}(2.54) = 1.66 \text{ cm}$$

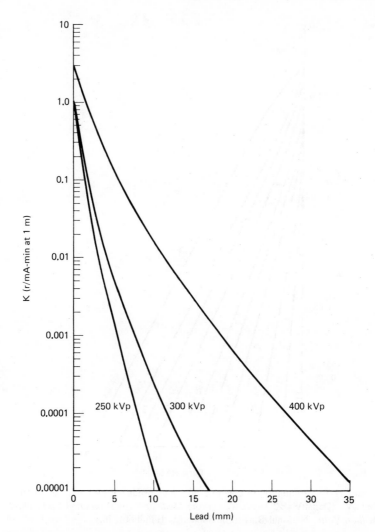

Figure 11-8 Attenuation in lead of x-rays produced by potentials of 250 to 400 kV. (From *Medical X-ray Protection Up to Three Million Volts*, National Bureau of Standards Handbook 76.)

Secondary Protective Shielding

A secondary barrier is one that shields against leakage and scattered radiation only. Since the quality of each of these forms of radiation is significantly different, their barrier requirements must be computed separately.

Scattered radiation. The amount and energy of scattered radiation depends on the following factors:

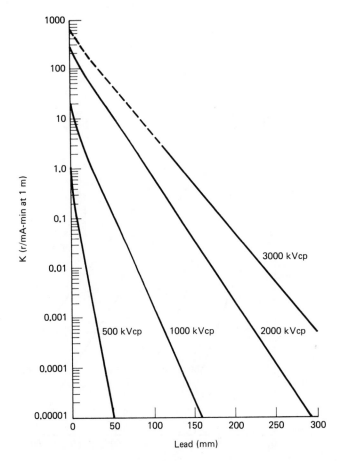

Figure 11-9 Attenuation in lead of x-rays produced by potentials of 500- to 3000-kV constant potential. (From *Medical X-ray Protection Up to Three Million Volts,* National Bureau of Standards Handbook 76.)

1. Incident exposure rate
2. Cross-sectional area of beam at the irradiated object
3. Absorption of the object
4. Angle of scattering
5. Operating anode voltage of the machine

For facilities design calculations, a number of simplifying assumptions can be made that still provide for good radiation protection. They are:

1. For engineering purposes, the 90° scattered radiation generated from a useful beam at a potential of 500 kV or less is, to a first approximation, assumed to have the same average energy as the useful beam. This

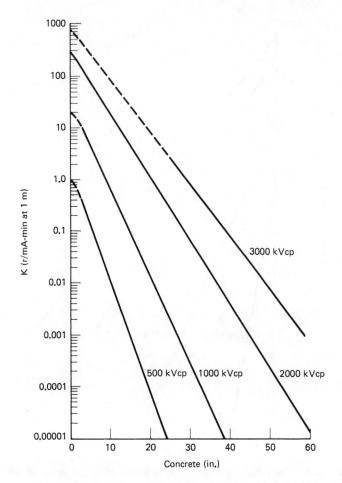

Figure 11-10 Attenuation in concrete of x-rays produced by potentials of 500- to 3000-kV constant potential. (Density = 2.35 g/cm³.) (From *Medical X-ray Protection Up to Three Million Volts,* National Bureau of Standards Handbook 76.)

is due to the fact that at anode voltages below 500 kV, Compton scattering does not greatly degrade the photon energy. The irradiated object also acts as an absorber of lower-energy photons.

2. The 90° scattered radiation produced by a useful beam at anode voltages greater than 500 kV is, to a first approximation, equivalent to the energy distribution of x-rays generated by a 500-kV anode potential regardless of the anode voltage of the useful beam.

3. The dose rate of radiation scattered through 90°, at a distance of 1 m from the scatterer, is 0.1% of the incident dose rate on the irradiated

object. The K value computed for scattered radiation may be 1000 times greater than that for the useful beam.

To account for the increasing x-ray output with increasing kV, the value of K is reduced by a factor f that depends on the applied kV. The values of the correction factor f corresponding to different levels of kV are as follows:

kV	f
500 or less	1
1000	20
2000	120
3000	300

For scattered radiation a value of 1 for the use factor U is used in the computation of the K value. Thus for scattered radiation the equation used for the calculation of K for use in determining the thickness of shielding is

$$K = \frac{1000 \, d^2 P}{fWT} \qquad (11\text{-}33)$$

Example: Shielding for Scattered Radiation

Determine the thickness of scatter-protective shielding required to protect a controlled area 2.1 m from the target for a 250-kVp x-ray therapy machine which has a weekly work load of 20,000 mA-min. The wall in question has an occupancy factor of 1 ($T = 1$). The machine has a continuous current rating of 20 mA at 250 kVp.

Solution

$$d = 2.1 \text{ m}$$
$$P = 0.1 \text{ R}$$
$$f = 1$$
$$T = 1$$
$$W = 20{,}000 \text{ mA-min/week}$$

Then

$$K = \frac{1000(2.1)^2(0.1)}{(1)(20{,}000)(1)} = 0.022$$

From the chart for lead at potentials 250 to 400 kV, we have

$$2.5 \text{ mm Pb}$$

From the chart of concrete at the same voltage, we have

$$6.5 \text{ in. concrete}$$

Leakage radiation from x-ray tube housing. Since the quality of radiation between scattered and leakage radiation may be considerably different, their shielding requirements must be computed separately. The number of half-value layers of secondary shielding needed for leakage radiation reduction depends on:

1. The type of tube house (diagnostic or therapeutic)
2. The operating potential of the tube
3. The weekly operating time of the tube
4. The distance from the tube to the occupied area
5. The nature and degree of occupancy
6. Permissible exposure per week

By design, x-ray therapy equipment operating at 500 kV or above has a maximum leakage radiation rate of 0.1% of the useful beam exposure rate at 1 m from the source. The weekly leakage exposure at a distance d from the source is given by

$$D = \frac{0.001\dot{D}t}{d^2} \tag{11-34}$$

where $\dot{D}$ = beam exposure rate, dD/dt

t = beam on-time, min/week

d = distance from the source

But $\dot{D}t$ is equal to W in R/week at 1 m. Thus

$$D = \frac{0.001WUT}{d^2}$$

Consider the transmission factor B_{LX} for x-ray shielding that will reduce the weekly exposure to P. Thus

$$P = B_{LX}(D) = \frac{B_{LX}(0.001WUT)}{d^2} \tag{11-35}$$

Solving for B_{LX}, we have

$$B_{LX} = \frac{1000Pd^2}{WT}$$

$U = 1$ for leakage radiation. For therapeutic-type x-ray tube housing operating at below 500 kV, the leakage radiation is limited to 1 R/hr at 1 m from the source with the tube operating at maximum rated tube current I. The weekly leakage exposure at a distance d from the source is given by

$$D = \frac{1}{d^2}\frac{t}{60} \tag{11-36}$$

Where $WT = It$,

$$D = \frac{1}{d^2} \frac{WT}{60I}$$
(11-37)

The weekly exposure P at a distance d is given by

$$P = B_{LX}(D) = \frac{B_{LX}WT}{d^2 60I}$$
(11-38)

Solving for the transmission factor, we have

$$B_{LX} = \frac{Pd^2 60I}{WT}$$
(11-39)

for therapy machines below 0.5 MV. For a diagnostic machine, the leakage rate at the housing is 0.1 R/hr at 1 m from the source. The transmission factor for this type of unit becomes

$$B_{LX} = \frac{Pd^2(600I)}{WT}$$
(11-40)

Calculation of secondary shielding requirement due to leakage. The thickness of the energy-absorbing shield, S_N, is then computed by $S_N = N(\text{HVL})$, where N is the number of half-value layers corresponding to a transmission factor of B_{LX}.

Example

Determine the thickness of leakage-protective shielding required to protect a controlled area 2.1 m from the target for a 250-kVp x-ray therapy machine which has a weekly work load of 20,000 mA-min. The wall in question adjoins an area with an occupancy factor of 1. The x-ray machine has a maximum continuous tube current rating of 20 mA at 250 kV. That is:

$$P = 0.1 \text{ R}$$
$$d = 2.1 \text{ m}$$
$$W = 20,000 \text{ mA-min}$$
$$T = 1$$
$$I = 20 \text{ mA}$$

Solution

$$B_{LX} = \frac{(0.1)(2.1)^2(60)(20)}{20,000 \ (I)} = 0.265$$

From the nomograph Figure 11-11:

$$\text{HV } S_N = 5.2$$
$$S_n = 5.2(.88) = 4.57 \text{ mm Pb}$$
$$S_n = 5.2(2.8) = 14.4 \text{ cm concrete}$$

Total secondary shielding requirements must take into account both the scattered and leakage radiation produced by the x-ray machine. If the

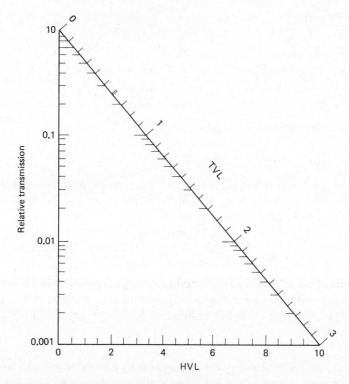

Figure 11-11 Relation between the transmission, B_{LX} or B_{LG}, and the number of half-value layers, N, or tenth-value layers, n. (From NCRP Report 49, Washington, D.C., by permission.)

shielding thicknesses for leakage and scattered radiations are found to be approximately the same, one HVL should be added to the larger one to obtain the required total secondary shield thickness. If the two differ by a large amount (i.e., difference of 3 HVLs), the thicker of the two will be adequate.

Total secondary shielding calculation.   For the various example problems on secondary shield calculation for scatter radiation, the thicknesses of shielding required were

$$2.5 \text{ mm lead}$$

$$6.5 \text{ in. concrete}$$

The thicknesses of shielding required for leakage radiation protection were

$$4.57 \text{ mm lead}$$

$$14.4 \text{ cm concrete} = 5.67 \text{ in. concrete}$$

At 250 kV the HVL of lead is 0.88 mm and 2.8 cm (1.1 in.) of concrete (Table 11-3). The difference in lead shielding is given by

$$\text{number of HVL} = \frac{4.57 - 2.5}{0.88} = 2.35 \text{ HVL of lead}$$

The difference in lead is 2.35 HVLs of lead, which is not quite 3 HVL. It would seem that 4.57 mm of lead should be adequate for scatter shielding purposes. Thus a secondary shield made of 4.57 mm of lead should be adequate. If one has a doubt, 1 HVL can be added or (4.57 + 0.88 =) 5.45 mm of lead can be used for shielding against scatter and leakage radiation.

TABLE 11-3 Half-Value and Tenth-Value Layers[a]

	Attenuation Material					
	Lead (mm)		Concrete (cm)		Iron (cm)	
Peak Voltage (kV)	HVL	TVL	HVL	TVL	HVL	TVL
50	0.06	0.17	0.43	1.5		
70	0.17	0.52	0.84	2.8		
100	0.27	0.88	1.6	5.3		
125	0.28	0.93	2.0	6.6		
150	0.30	0.99	2.24	7.4		
200	0.52	1.7	2.5	8.4		
250	0.88	2.9	2.8	9.4		
300	1.47	4.8	3.1	10.4		
400	2.5	8.3	3.3	10.9		
500	3.6	11.9	3.6	11.7		
1,000	7.9	26	4.4	14.7		
2,000	12.5	42	6.4	21		
3,000	14.5	48.5	7.4	24.5		
4,000	16	53	8.8	29.2	2.7	9.1
6,000	16.9	56	10.4	34.5	3.0	9.9
8,000	16.9	56	11.4	37.8	3.1	10.3
10,000	16.6	55	11.9	39.6	3.2	10.5
Cesium-137	6.5	21.6	4.8	15.7	1.6	5.3
Cobalt-60	12	40	6.2	20.6	2.1	6.9
Radium	16.6	55	6.9	23.4	2.2	7.4

[a] Approximate values obtained at high attenuation for the indicated peak voltage values under broad-beam conditions; with low attenuation these values will be significantly less.
Source: NCRP Report #49, Washington, D.C.; by permission.

PROBLEMS

1. What fraction of energy is lost by a 1-MeV x-ray photon if it is scattered through an angle of 90°?

2. Assume that a 1-MeV x-ray photon grazes an electron such that the Compton electron is ejected at an angle of $\theta = 90°$ and the scattered photon emerges at $\theta = 0°$. Calculate the energy of the Compton electron and the scattered photon.

3. A 10-MeV photon interacts in a pair-production process. Calculate the energy of the positron if the electron emerges from the interaction with an energy of 2.0 MeV.

4. A low-energy beam of photons with an energy of 10.22 keV suffers Compton collisions. Calculate the minimum energy of the scattered radiation and the maximum energy the recoil electron may acquire.

5. A beam of photons with an energy 5.11 MeV strikes a scattering medium. Find the energy of the radiation scattered at angles of 45°, 90°, and 180°.

6. Calculate the fractional decrease in beam intensity when a beam of 50-keV x-rays passes through a 1.0-cm aluminum absorber.

7. Compute the half-value layer for 100-keV x-rays in aluminum and copper.

8. The exposure rate ratio of 50-keV x-rays incident to and transmitted from an aluminum absorber is 1000:1. What is the thickness of the absorber?

9. Find the photon energy of a beam of monoenergetic x-rays if the half-value layer is 0.61 cm of lead.

10. List five factors that influence the thickness of a protective barrier required to reduce the exposure in an area to a given safe level.

11. When dealing with radiation and humans, the units used are the roentgen, the rad, and the rem. Define each of these terms, explain what each is a measure of, and show how they are related mathematically.

12. A 250-kVp therapy unit is operated 7.5 hr/week in the direction illustrated in Figure 11-12. Work is done with a beam current of 45 mA. Determine the shielding required for walls A and B.

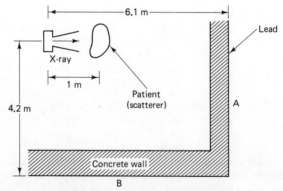

Figure 11-12   KVP therapy unit. Assume that $P = 0.1$, $T = 1$ for adjacent areas.

13. The plans for a busy radiographic facility are shown in Figure 11-13. If the x-ray machine is used for 7.5 hr/week at 300 mA and 150 kVp, determine the lead shielding for walls *A, B, C, D, E,* and *F*. Determine the thickness of concrete for the floor. The clinical engineering office is 3 m below the floor of the x-ray room.

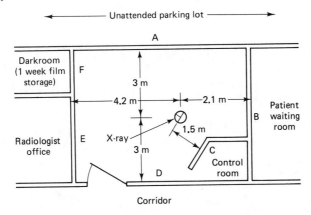

Figure 11-13   Plans for a radiographic facility.

REFERENCES

1. Cember, Herman, *Introduction to Health Physics* (Elmsford, N.Y.: Pergamon Press, 1976).

2. Christensen, Edward E., Thomas S. Curry III, and James Nunnally, *An Introduction to the Physics of Diagnostic Radiology* (Philadelphia: Lea & Febiger, 1972).

3. NCRP Report 49, *Structural Shielding Design and Evaluation for Medical Use of X-Rays and Gamma Rays of Energies Up to 10 MeV* (Washington, D.C.: National Council on Radiation Protection and Measurements, 1976).

4. Shapiro, Jacob, *Radiation Protection: A Guide for Scientists and Physicians* (Cambridge, Mass.: Harvard University Press, 1972).

5. U.S. Department of Commerce, *Medical X-Ray Protection Up to Three Million Volts,* Handbook 76 (Washington, D.C.: National Bureau of Standards, 1961).

6. Ziemer, Paul L., and Alan L. Orvis, "Hospital radiation safety," in *CRC Handbook of Hospital Safety,* Paul E. Stanley (ed.) (Boca Raton, Fla.: CRC Press, 1981).

12

ENGINEERING PRINCIPLES OF NUCLEAR DETECTION AND MEDICAL IMAGING SYSTEMS

Nuclear medicine began more than 50 years ago when radioactive tracers were used to measure human blood velocity. The significance of this work was that radioactive tracers could be used to monitor physiologic events. The use of radiotracer methods from a diagnostic point of view had to wait until the 1950s for engineering and technological advances to be made in the area of detection transducers, electronic circuitry, and short-lived radiopharmaceuticals. In particular, this included the development of the sodium-iodide (NaI) scintillation detector and the Anger scintillation camera. These developments have given rise to the area of nuclear imaging systems, which can be electronic or computer based. Nuclear imaging is playing an important role in the diagnosis of many clinical problems since it is useful in measuring the distribution of radioactivity in the body in either a dynamic state as used in cardiac studies or a static state as in the study of thyroid conditions.

INSTRUMENTATION FOR THE DETECTION AND MEASUREMENT OF NUCLEAR RADIATION FOR MEDICAL USAGE

Scintillation Counters

Essentially all of the principal interactions of nuclear radiation with matter result in the production of ion pairs. This process is the basis of operation or the transduction principle for almost all radiation detectors.

For beta and alpha particles, detectors based on gas ionization and subsequent increased conductivity are common, as with the Geiger–Mueller counter. For medical purposes gamma and x-radiation are commonly used. Since the density of a gas in a gas detector is low compared with a solid, the chances of interaction with a beam of gamma or x-radiation is small and the sensitivity of these detectors is low. Far more radiation is absorbed in a solid, making this type of radiation detector potentially more sensitive. This is particularly important when dealing with the detection of highly penetrating gamma radiation.

A number of different substances, inorganic and organic, will produce light or scintillate when under the influence of high-energy radiation. This property is the transduction property, which is used in scintillation counters or detectors. Scintillation is the emission of visible or ultraviolet radiation following an ionizing event. Usually, a gamma ray loses its energy in a scintillator through the photoelectric, Compton, or pair-production mechanism. Electrons generated in this manner lose their energy in turn through ionization and excitation of the scintillator molecules. In the decay from the excited states these molecules emit light.

Sodium iodide crystals containing traces of thallium, used as an activator, are used almost exclusively in nuclear medicine. The amount of light it produced per unit of absorbed energy in an NaI crystal is one of the highest of any scintillation material.

To be useful, a NaI(Tl) scintillation crystal must be used in conjunction with associated electronics. The system consists of the following components:

1. Photomultiplier tube
2. Preamplifier circuit
3. Linear amplifier
4. Pulse-height analyzer
5. Scaler and timer
6. Rate meter

Photomultiplier Tube

The basic concept behind the photomultiplier (PM) tube is the initial ejection of electrons from a cathode due to the "photoelectric effect" by incident radiation and the subsequent multiplication of the number of electrons through the mechanism of secondary emission from a sequence of dynodes (anode). Figure 12-1 is an illustration of the electron multiplication process. Usually there are nine or more dynodes in a PM tube. Each dynode is held at a voltage more positive than the preceding one in order to accelerate the electrons due to secondary emission to the next dynode. The input radiation of energy $h\nu$ is incident on a photocathode which ejects photoelectrons as a result of the photoelectric effect. The ejected electrons from

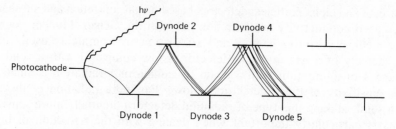

Figure 12-1 Schematic internal operation of a PM tube. The photon striking the cathode releases a single electron. The cascading effect through the dynodes is illustrated.

the photocathode are attracted to the first dynode since it is more positive than the photocathode. To avoid nonlinear problems due to gas ionization, the PM tube is evacuated. The gain of the PM tube is controlled by the design and placement of the dynodes. The overall electron multiplication over the original photoelectrons is a factor of between 10^5 and 10^7 by the time they reach the last dynode. Thus, by the process of sequential secondary emission events, a relatively large pulse of electrons arrives at the anode. The pulse duration may be only a few nanoseconds long. The supply voltage for the tube can vary from 600 to 3000 V. Figure 12-2 illustrates the power supply connection to the PM tube and the use of a resistor voltage-divider network to supply the increasing dynode voltage.

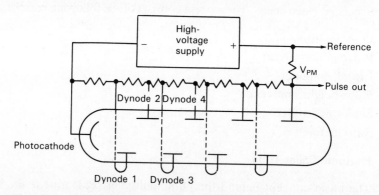

Figure 12-2 Schematic of power supply connection to a PM tube to maintain the dynodes at the appropriate voltages.

Figure 12-3 shows a set of characteristic curves for a PM tube giving the relationship between anode current as a function of PM voltage and light input. The current that flows in the PM tube when no light is striking the cathode is called the "dark current."

The effective gain, G, of a photomultiplier tube can be considered to be the number of electrons in each anode pulse per scintillation. If the

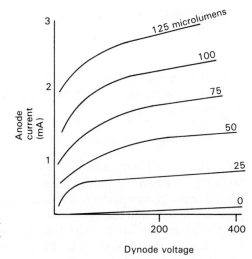

Figure 12-3 Photomultiplier anode current plotted as a function of dynode voltage for constant illumination.

charge on an electron is 1.6×10^{-19} C and assuming an average electron gain of 10^6, the average charge delivered to the anode for each photon reaching the first dynode is $(1.6 \times 10^{-19})(10^6) = 1.6 \times 10^{-13}$ C per anode pulse. Thus we can obtain an estimate of the charge delivered to the anode per event or per unit time interval using the average charge, $\overline{Q}$, as the estimator. This can be computed as

$$\overline{Q} = N_a \overline{G} q_e \qquad (12\text{-}1)$$

where $\overline{Q}$ = average at anode, coulombs/event

N_a = number of anode pulses/event

$\overline{G}$ = average electron gain, electrons/anode pulse

q_e = charge on an electron

If the total charge output from the anode due to an input flux of light photons produced by a scintillation from an NaI crystal is the measured quantity, the photomultiplier tube acts as a transducer with current output.

Preamplifier Circuit

The output of the PM tube is a voltage pulse. The function of the preamplifier circuit is to convert the low-amplitude pulse from the PM tube to a pulse having sufficient amplitude and pulse shape to drive the pulse-height analyzer. The circuits used in the preamplifier should be electronically stable and relatively insensitive to pulse amplitude and high count rates. Linearity is essential to preserve the correspondence between the energy absorbed by the scintillation crystal and the pulse output of the preamplifier.

The basic functions of the processing electronics into which the photomultiplier tube output is fed are:

1. Amplification
2. Discrimination
3. Counting

The preamplifier itself functions to couple the output of the detector to the processing electronics, which may be remote from the scintillation detectors and photomultiplier tube.

Pulse-Height Analyzer

The pulse-height analyzer (PHA) is an electronic device that allows counting of electrical pulses output from the photomultiplier tube which results from gamma radiation falling within a selected energy range. The principle involved in the PHA makes use of the fact that the pulse amplitude of the pulses leaving the PM tube is directly proportional to the energy absorbed by the crystal. The PHA can be set to accept pulses whose amplitudes fall within a preselected range. The lower level of acceptable energy is the window baseline, and the upper level of acceptable energy is the upper energy window setting. The range of energy ΔE that these limits represent is called the *energy window width*.

Basic Pulse Circuits: Need for Pulse Shaping

Since the output of the photomultiplier tube is a voltage pulse, an understanding of basic pulse circuits and associated electronics which are used in nuclear particle analysis systems will first be discussed.

The three main divisions of a nuclear counting system are:

1. Data production and selection (detector)
2. Data storage (scaler, PHA, or rate meter)
3. Data handling (calculation and display)

Detector outputs are usually voltage signals which constitute the input for an amplifier. The amount of charge produced and the time over which it is delivered to the amplifier depends on the detector. A detector can produce charges in the time range 10^{-15} to 10^{-10} nanosecond to several milliseconds.

An important aspect of nuclear detection and processing systems is their ability to resolve adjacent pulses. One wishes to distinguish between closely spaced pulses in time even when the mean rate of the nuclear events is low. Consider the detector output illustrated in Figure 12-4a. By applying the signal in Figure 12-4a to a signal shaper or processor (in this case a differentiation circuit), the detector pulses A through F are transformed into the set of pulses in Figure 12-4b. The signal processing converts the detector outputs to pulses with a zero baseline without loss of timing or amplitude information.

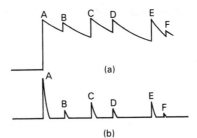

(a)

Figure 12-4 Differentiating of detector output: (a) detector output; (b) differentiation of (a).

(b)

RC low-pass filter.   Consider the characteristics of the RC network shown in Figure 12-5. The transfer function for the network is given by

$$\frac{V_o}{V_i} = \frac{1}{1 + j\omega RC} \tag{12-2}$$

For ω small, $V_o/V_i \simeq 1$ and the system acts as a low-pass filter. For $\omega \gg 1$, $V_o/V_i \to 0$. The amplitude response is down 3 dB at $f_L = \omega_c/2\pi = 1/2\pi RC$ and the rate of amplitude decrease is -6 dB/octave. For an input signal $V_i(t)$ to the circuit, the equation relating to the output voltage becomes

$$\frac{dV_o}{dt} = \frac{i(t)}{c} = \frac{V_i - V_o}{RC}$$

If $RC \gg 0$, $V_o \ll V_i$ and we have

$$\frac{dV_o(t)}{dt} = \frac{V_i(t)}{RC} \tag{12-3}$$

Solving for V_o, we have

$$V_o(t) = \frac{1}{RC} \int V_i(t)\, dt \tag{12-4}$$

which illustrates the fact that this circuit acts like an integrating circuit, although this is just an approximation.

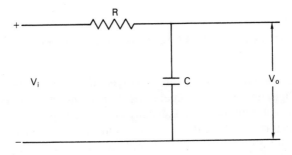

Figure 12-5 RC low-pass filter.

RC high-pass filter.   Consider an RC network of the form shown in Figure 12-6. The transfer function for this network is given by

$$\frac{V_o}{V_i} = \frac{j\omega RC}{1 + 1\ \omega RC} \tag{12-5}$$

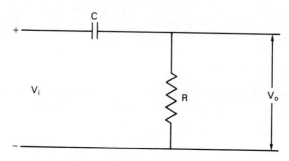

Figure 12-6 RC high-pass filter.

For ω small,

$$\left|\frac{V_o}{V_i}\right| \to 0$$

For $\omega \gg 1$,

$$\left|\frac{V_o}{V_i}\right| \to 1$$

For $\omega RC \ll 1$, then,

$$\frac{V_o}{V_i} \simeq \frac{j\omega RC}{1}$$

In the time domain this becomes

$$V_o(t) \simeq \frac{dV_i}{dt}$$

This circuit approximates a differentiation operation where the output voltage is the derivative of the input voltage.

CHARACTERISTICS OF ELECTRICAL SIGNALS FROM DETECTOR OUTPUT

Because of many factors that are involved in the conversion of gamma radiation to light and finally to an electrical pulse, no two events of detection produce electrical outputs that are exactly alike. For this reason, the resulting electrical signal is statistically limited in its integrity as a measure of gamma radiation. Normally, the detector output signal is a current pulse, the time integral of which (its total charge) is proportional to the energy of the gamma photon. The circuit shaping and amplifying this current pulse must exhibit the following characteristics:

1. The shaping must allow a precise energy analysis.
2. The signal-to-noise ratio must be high.
3. The pulse must be short in duration and must allow high pulse rates.

The output from a NaI(Tl) photomultiplier detector is given in Figure 12-7 together with a smoothed pulse. If an amplitude proportional to the energy of the gamma photon is desired, the current output from the detector must always be integrated, irrespective of the actual pulse shape. The integration can be accomplished in two different ways.

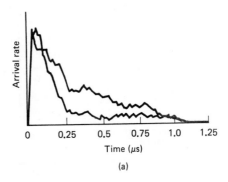

(a)

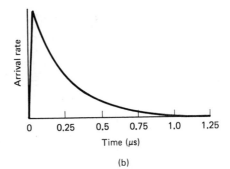

(b)

Figure 12-7 Arrival rate of electrons at anode of photomultiplier tube coupled to NaI(Tl) crystal, in response to detection of two identical-energy gamma photons: (a) original data; (b) smoothed data.

1. The current pulse is integrated by means of an *RC* circuit directly at the detector output and the network output further amplified. Figure 12-8 shows a schematic drawing of this method of primary pulse shaping and processing.
2. The integration is carried out using an operational amplifier integration circuit, shown in Figure 12-9a.

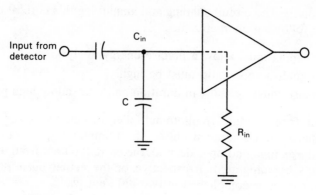

Figure 12-8 Schematic of a basic pulse-shaping circuit.

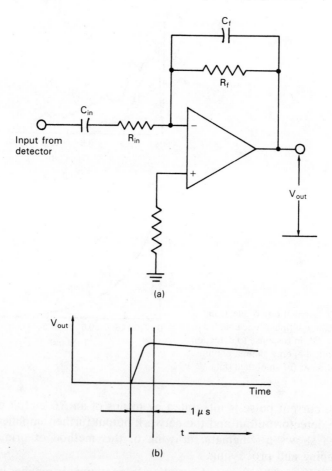

(a)

(b)

Figure 12-9 (a) Integration using an operational amplifier; (b) voltage output from integration amplifier.

The operational amplifier circuit of Figure 12-9a can be thought of as a charge-to-voltage converter. For this circuit, the change in output voltage is related to the charge by

$$\Delta V_o = -\frac{\Delta Q}{C_f} \tag{12-6}$$

The lower -3-dB frequency of this circuit is

$$f_L = \frac{1}{2\pi R_f C_f} \tag{12-7}$$

and the upper -3-dB frequency of this circuit is

$$f_U = \frac{1}{2\pi R_{in} C_{in}} \tag{12-8}$$

The purpose of filtering or smoothing the output pulses from the photomultiplier tube is to condition the output waveform for further signal processing. Resistor R_f in parallel with C_f is used to discharge or "bleed" off charge from C_{in} and C_f. If R_f were not present and the leakage paths for stored charge were too small, the charge buildup due to input pulses would soon drive the amplifier circuit into saturation. In Figure 12-9b, the voltage output of the integrator is given. It should be noted that when the voltage stops increasing, the integration process is complete.

Integration reduces noise in the system by the fact that if we consider the noise to be made up of a sum of a large number of sinusoids at frequencies $n\omega$, $n = 1, 2, 3, ...$, then integration essentially divides each of the sinusoidal components of the noise by $n\omega$, that is,

$$\int \sin n\,\omega t\,dt = \frac{-1}{n\omega}\cos n\,\omega t \tag{12-9}$$

Insufficient integration time, which is the same as premature termination of integration, causes a maximum voltage to be obtained before the entire population of electrons has arrived and increases statistical uncertainty, which in turn degrades pulse-height resolution. In scintillation cameras this is manifested as loss of image detail.

Differentiation Following Integration

Once the output voltage has reached a maximum value, its information content is maximum and it should be reduced to the baseline value to allow the amplifier to process the next signal. Termination of the integration process by differentiation is one way to accomplish this. This can be done with the simple RC differentiation circuit or use of an operational amplifier high-pass filter, as illustrated in Figure 12-10. The differential equation relating the output voltage, V_o, to the input voltage, V_i, is given by

$$V_o = -R_2 C_1 \frac{dV_i}{dt} \tag{12-10}$$

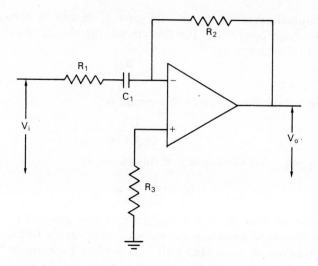

Figure 12-10   Operational amplifier differentiation circuit.

More Integration Needed

After first integration and differentiation, the basic pulse shaping is complete where maximum pulse height occurs near the end of electron arrival at the PM anode for a given event. Despite the signal processing so far, the signal may still contain significant noise. Additional integration will improve the pulse shape and create a better signal-to-noise ratio. This can be easily accomplished using an operational amplifier integrator.

ELECTRONIC PULSE-HEIGHT ANALYZER

After the initial output from the photomultiplier tube, the current pulse is amplified and shaped to minimize the effect of noise on pulse shape and the timing of the nuclear events. The objective of this form of signal processing is to produce a pulse that is linearly related to the detector pulse amplitude and thus to the energy of the detected particle. At this stage, the pulses must be separated and counted according to their related energy.

Once the linear pulses have been selected according to their height and the information content has been extracted, the pulses are of no further value. Therefore, the pulses are fed into a trigger circuit which produces a "logic" pulse of standardized height and duration. These logic pulses carry with them time information about the original nuclear events and so are distributed randomly in time. At this time these pulses are not yet in a form to be processed by a digital computer.

Requirements of Electronic Pulse-Height Discriminator

Integral pulse-height discriminators.   This type of discriminator passes a pulse whose height is above a given threshold. Electronically, this can be implemented using the biased diode concept. Consider an ideal diode whose characteristics are given in Figure 12-11. Thus a threshold

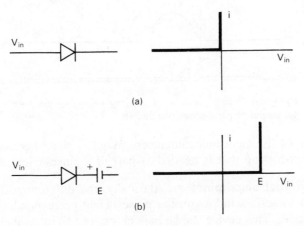

(a)

(b)

Figure 12-11   (a) Circuit symbol and voltage–current characteristic of the ideal diode; (b) biased diode with associated voltage–current characteristics.

circuit of the form shown in Figure 12-12 could be used. If a trigger circuit like a monostable multivibrator or "one-shot" is used at the output of the

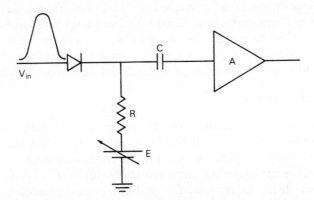

Figure 12-12   Threshold circuit for pulse-height discrimination.

amplifier, the output is as given in Figure 12-13c. This circuit is but one of many that can accomplish the same goal using a variety of devices ranging from tunnel diodes to Schmidt triggers and more. It is limited only by the

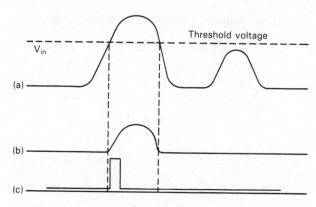

Figure 12-13 (a) Voltage input to circuit; (b) voltage output from threshold circuit; (c) output of pulse-generating circuit.

imagination of the electronics engineer. What is given here demonstrates the pulse processing that is needed as part of a pulse-height analyzer.

Differential discriminator. Basically, the differential discriminator is designed to set a window on the pulse height corresponding to present energy interval. This device can be built up out of two integral discriminators in parallel set to an upper and lower threshold and run through an anti-coincidence circuit. Consider the circuit given in Figure 12-14. In this configuration the bias on the input diode D_1 sets the lower threshold as in the previous circuit. Diode D_2 receives the truncated pulse of the second diode. The window width is then set by the bias on diode D_2. These two circuits illustrate the types of circuits used and how they must function in a pulse-height analyzer; they demonstrate well the principles involved. There are still problems due to the inherent problems brought on by high pulse rates. Designs of these circuits are many and varied; all make use of integrated-circuit digital design and will not be discussed here.

Counting Circuits

In the nuclear electronics area, the terms *counter* and *scaler* are used synonymously. "Scaler" is the older term and has been superseded by the term "counter." Since practically all counting circuits today utilize two-state or binary devices, there are many different ways in which the pulse outputs from the pulse-height analyzer can be counted and coded.

The most convenient and generally used binary coding is binary-coded decimal, BCD (see Figure 12-15). Using position notation in the base 2, we find that it requires 4 binary digits, "bits," to encode the number 10. Thus numbers between 0 and 10 can be encoded in the following way.

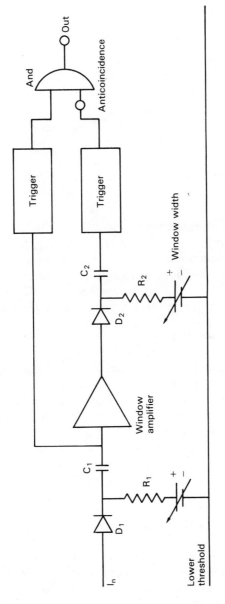

Figure 12-14 Schematic diagram for a differential discriminator.

389

Decimal	Binary
0	0 0 0 0
1	0 0 0 1
2	0 0 1 0
3	0 0 1 1
4	0 1 0 0
.	
.	
.	
9	1 0 0 1
10	0 0 0 0

Electronically zero to count in decades

Figure 12-15 Table of decimals and their binary equivalents.

Each "Logic 1" represents a power of 2 in the form

$$2^3 + 2^2 + 2^1 + 2^0$$

Construction of counting registers. A counting register is one in which the count information is stored in a coded digital form such as binary or BCD which will increment the stored number on command. After each increment command, the code that represents the number of counts received is revised to represent the next-higher number. The JK flip-flop is commonly used to construct a counting register for counting nuclear events.

Properties of the JK flip-flop. The JK flip-flop has three inputs, J, K, and T. The J and K inputs are used for binary input and T is the clock, toggle, or trigger input. There are two complementary outputs, Q and $\overline{Q}$. There is also a direct set input and a direct clear input. A logical "0" at either of these inputs immediately sets or clears the inputs and overrides all other functions. The operation of the JK flip-flop is controlled by signals at the J and K inputs. Figure 12-16 is a summary of the input/output relationships of the JK flip-flop. When J and K are both HI, binary 1, the flip-flop changes state for each pulse input to the T input part. Thus in this condition it behaves as a triggered flip-flop. The truth table for this device is given in Figure 12-17.

Ripple counter. This is the most elementary counting circuit. This circuit is made up of a sequence of JK flip-flops which can store information about the accumulated number of input pulses in a binary-coded form. Figure 12-18 illustrates the arrangement of the JK flip-flops into a simple counter. The operation of the counter requires that Q_0 change state at the falling edge of each pulse and that all the other Q's make a transition when and only when the output of the preceding flip-flop changes from 1 to 0.

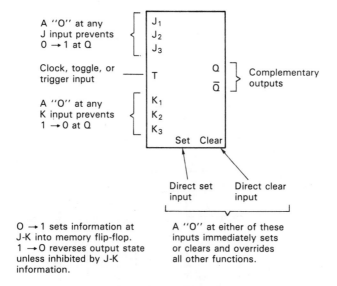

A "0" at any
J input prevents
0 → 1 at Q

J_1
J_2
J_3

Clock, toggle, or
trigger input

T

A "0" at any
K input prevents
1 → 0 at Q

K_1
K_2
K_3

Q
$\overline{Q}$
} Complementary
outputs

Set   Clear

Direct set   Direct clear
input   input

0 → 1 sets information at
J-K into memory flip-flop.
1 → 0 reverses output state
unless inhibited by J-K
information.

A "0" at either of these
inputs immediately sets
or clears and overrides
all other functions.

Figure 12-16 JK flip-flop.

J	K	Q	$\overline{Q}$	
LO	LO	No change		
HI	LO	HI	LO	with first clock or trigger pulse only
LO	HI	LO	HI	
HI	HI	switch at each trigger input		

Figure 12-17 Truth table for JK flip-flop.

This negative transition "ripples" through the counter from the least significant bit to the most significant bit. Figure 12-19 is a waveform diagram for the four-state ripple counter.

For the ripple counter, the only limitation on pulse rate input is set by the transition or slew rate of the first flip-flop. Although the ripple-through effect may have no significance in a simple counter, in different applications it may cause misinterpretation of the counter contents. This could be the case when the counter is sampled by other circuitry when counting is in progress. This problem can be avoided with the use of a synchronous counter in which all the outputs move to their new state simultaneously. Figure 12-20 illustrates just such a synchronous binary counter implemented from JK flip-flops.

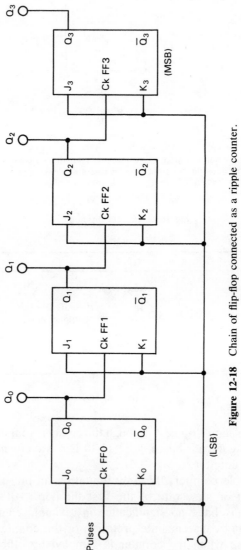

Figure 12-18 Chain of flip-flop connected as a ripple counter.

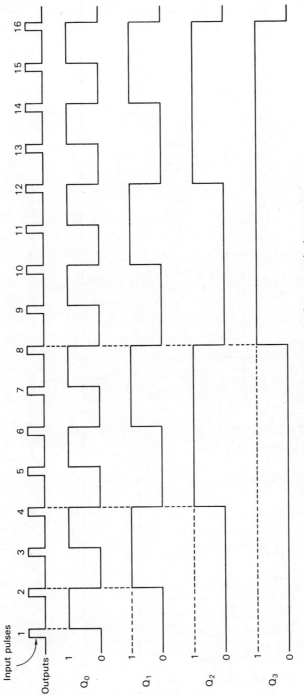

Figure 12-19 Waveform chart for the four-stage ripple counter.

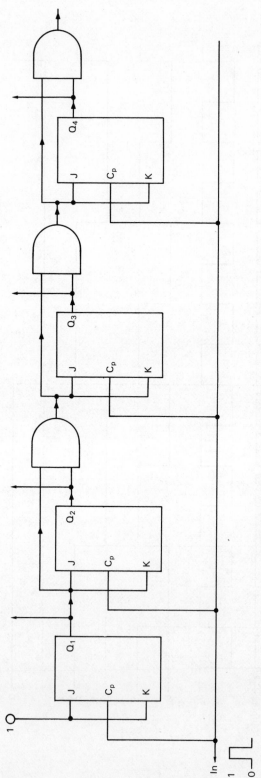

Figure 12-20 Synchronous binary counter assembled from JK bistables.

THE ANGER SCINTILLATION CAMERA

The Anger camera, named after its inventor, Hal O. Anger of Berkeley, California, is a device that permits imaging of static and dynamic radionuclides distribution in the body. The basic components of the Anger camera are:

1. A collimator
2. A position-sensitive gamma ray detector
3. A coordinate calculating electronics
4. A pulse-height analyzer
5. A display system

The Anger camera produces a picture on a cathode ray tube of the distribution of an administered radionuclide within the target organ of a patient. Basically, use is made of the emitted gamma rays to produce light flashes in a disk-shaped NaI crystal. The system determines the location of each scintillation and produces a spot of light on the face of a cathode ray tube in an analogous position.

The collimator normally consists of a large piece of lead with many small holes through it. The holes are of equal constant cross section. Only gamma rays emitted from patients in a direction perpendicular to the position-sensitive detector are allowed to enter the detector. The channel diameter and length of the collimator strongly affect the spatial resolution and sensitivity of the system.

A conventional Anger camera uses 37 two-inch-diameter photomultiplier tubes arranged in a matrix. Figure 12-21 illustrates the basic component of a nuclear imaging system. The gamma-ray flux is partitioned by the collimator, and intensity data are transferred from the detector to the display. Position data are transferred by electronic means in the gamma camera, which will be discussed later in the chapter.

Collimator Design for Anger Camera

Collimators provide an interface between the detector and the organ structure being imaged. They are designed with an array of cylinders arranged so as to permit only those photons which enter the collimator and travel on an axis parallel to the axis of the cylinder or at an angle such that the following relationship holds. Let t be the depth of the collimator and d the diameter of the cylinder (Figure 12-22).

$$\text{maximum entry angle } \phi = 90° + \theta = 90° + \tan^{-1}\frac{d}{t} \qquad (12\text{-}11)$$

We can think of a collimator as a filter for photons whose filtering action is a function of the direction of travel of the incident photon. The parallel hole collimator is used in the Anger camera. It consists of a thick plate

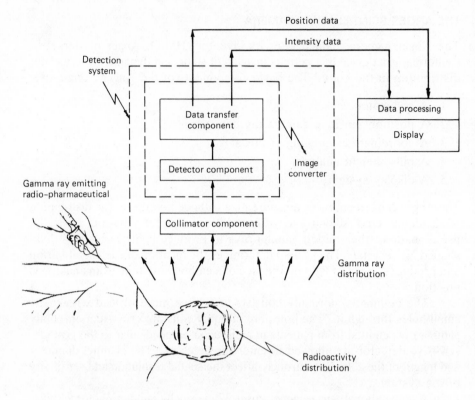

Figure 12-21 Basic components in a nuclear imaging system.

having thousands of parallel channels distributed in a uniform fashion over the surface of the plate. Generally, the collimator is made of lead, although other materials have been used. The usable detection area is defined by the size of the crystal and the hole geometry. Crystal diameter size varies from 10.5 to 16.5 in. Since the collimator selectively passes some photons while rejecting others, it is acting in a manner similar to an x-ray grid in x-ray imaging systems. In general, approximately 0.01% of the gamma rays emitted by a labeled organ are detected and used for image production.

Immediately following the collimator is the NaI(Tl) crystal, which is optically coupled to the photomultiplier tubes via a crystal window interface. The photomultiplier tube now produces electrical pulses in response to the gamma-ray-induced scintillation. The output pulses of the PM tube are

Figure 12-22 Collimator cylinder of depth t and diameter d. ϕ is the maximum entry angle for an unattended photon.

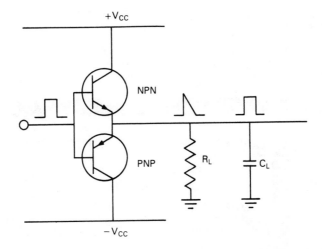

Figure 12-23 Pulse amplifier output stage for pulse transmission down a coaxial cable. R_L, load resistance; C_L, capacitance of coaxial cable.

further amplified. These pulses must be sent from the camera head to the console. To do this without distorting the pulses because of the distributed capacitance of the coaxial cable, a complementary-symmetry emitter follower, shown in Figure 12-23, is used for pulses that are both positive and negative and is used as a line driver amplifier. A block diagram of a commercial radioisotope camera is shown in Figure 12-24.

Delay-Line Amplifiers

When the signals from the preamplifier and line drivers reach the console, they are applied to a delay-line amplifier (Figure 12-24). Here the triangular 100-μs pulse is reduced to a negative 2-μs pulse of a more rectangular shape (Figure 12-25). This is done to reduce camera dead time, which means that pulses separated by more than 2 μs do not have an effect on one another.

Baseline Restoration (Diode Clamp Circuit)

Because of capacitive coupling of the photomultiplier tube output and subsequent amplifier stages, the voltage level of the pulse baseline varies depending on the count rate. Thus the next step in signal processing is the restoration of the dc baseline, as this helps to make the system less dependent on count rate. A baseline restoring circuit is usually placed immediately before the pulse-height analyzer to which it will be dc coupled. Consider the clamping circuit given in Figure 12-26. During the negative portion of the cycle the diode D conducts and charges up capacitor C to the voltage V_B, at which time the diode conduction stops. Then during the positive

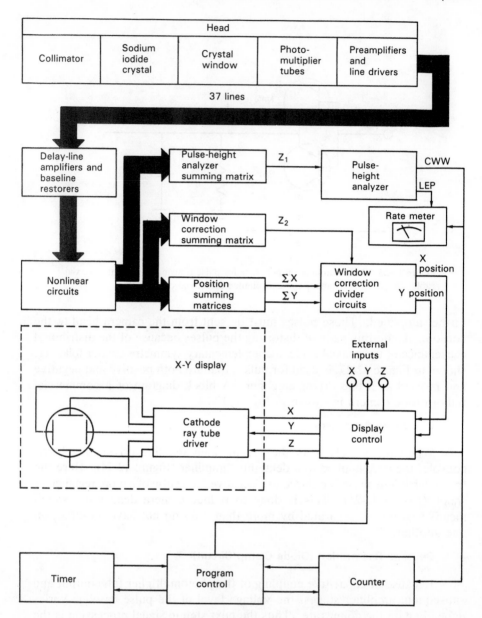

Figure 12-24 Block diagram of a radioisotope camera.

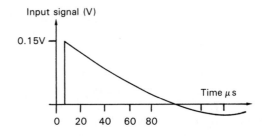

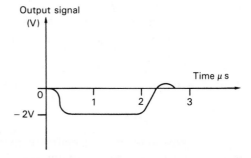

Figure 12-25 Reduction of a 100-μs triangular pulse to a negative 2-μs pulse.

portion of the cycle, the voltage V_B is added to the input voltage to restore
the baseline. The essential problem with this simple circuit is that it cannot
perform well if the pulses are of differing heights, have overshoots, or there
is noise present. To handle these problems, other, more involved electronic
circuits have been developed to overcome the difficulties of the simple
baseline restoration circuit.

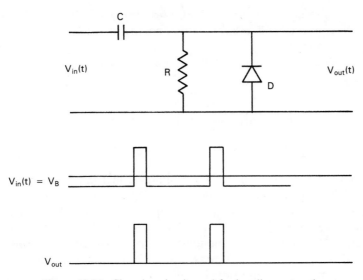

Figure 12-26 Clamping circuit used for baseline restoration.

Nonlinear Circuits and Summing Network

The first of these circuits is called the *energy circuit*. This circuit makes use of a type of nonlinear pulse compression that attenuates larger signals more than smaller ones in order to compensate for the nonlinearities of the optics of the detector.

When the output from each detector transmission line reaches the summing matrix, the signals are combined in a weighted manner (Figure 12-27). A weight-summation network is needed to allow for the fact that the total amount of light received by the PM tubes varies depending on where in the crystal a photon is absorbed. The resistors are adjustable and set at the factory so that the absorption of a given amount of gamma radiation energy anywhere in the crystal's usable volume will produce the same amplitude of output signal from the network. This is intended to produce uniform response over the camera aperture.

Position Circuits

The position-summing matrix consists of four circuit configurations on the order of the energy summation network. Each network sums the input from all PM tubes corresponding to $X > 0$, $X < 0$, $Y > 0$, $Y < 0$ given the notation $X+$, $X-$, $Y+$, $Y-$ in that order. The X-axis positioning signal, required by the cathode ray tube display, is proportional to the difference between $(X+)$ and $(X-)$. The difference is obtained by feeding the two signals into a differential amplifier, where it is scaled. The result is given by

$$\sum X = K[(X+) - (X-)] \text{ proportional to } X \qquad (12\text{-}12)$$

This result is normalized as a function of energy by using the voltage output from the summing resistor matrix of the PM tubes, which provides a voltage z proportioned to the energy of the gamma-ray photon.

$$\sum X = \frac{K}{z}[(X+) - (X-)] \qquad (12\text{-}13)$$

A similar operation is performed for the Y axis:

$$\sum Y = \frac{K}{z}[(Y+) - (Y-)] \text{ proportional to } Y \qquad (12\text{-}14)$$

The variable resistors in the position-summing matrix are adjusted so that the signal X is linearly proportional to the X coordinate of the scintillation within the NaI crystal, and that Y is proportional to the Y coordinate. In order to develop a signal that corresponds to the X, Y coordinates of a scintillation, the voltages corresponding to X coordinates must be proportional to their distances from the Y axis, and voltages corresponding to the Y coordinates must be proportional to distances from the X axis.

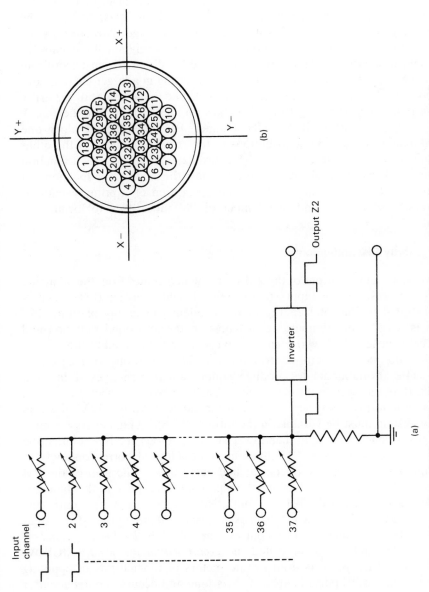

Figure 12-27 (a) Weighted summing network input to signal inverter; (b) photomultiplier tube locations in crystal detector head.

Since the input into the amplifier is a voltage appearing across a resistor, the node is a summation of currents from the position matrix of resistors. Thus the current output is proportional to the reciprocal of the resistance (mhos). A larger current output from a particular PM tube resistor pair corresponds to greater distance from a coordinate axis, so PM tubes close to a given coordinate axis will have a relatively large matrix of resistors associated with them and will produce correspondingly small currents for the summation process. Thus PM tubes located at a distance from a coordinate will have proportionally smaller resistors, resulting in a larger current contribution to the summation process. The PM tube in the center of the array at the point corresponding to $(0, 0)$ makes no contribution to either the X or Y composite signal. In a similar manner, PM tubes that lie along one coordinate axis make no contributions to the composite signal of the other coordinate. The accuracy of the X-Y coordinates of an event is a function of the total energy absorbed by the crystal, the number of locations of Compton interaction that occur for the absorption of a primary photon, conversion efficiency, and the number of PM tubes used to localize an event.

Energy Discrimination

The weighted output of the PM tubes, which comes from the summing matrix, represents the inputs to the pulse-height analyzer (PHA) and is proportional to the total energy of the incident gamma-ray photon. The PHA is used to discriminate against noise and background radiation and to accept those pulses whose energy levels can be attributed to the nuclide distribution and are useful contributions to the system output image.

Different radionuclides produce gamma radiation energies of different levels; thus the system must be able to respond over a range of energies. Since gamma rays possess energies in the range 44 to 662 keV, the gain of the system must be variable in the ratio 15:1. Since high-energy gamma photons produce more light than lower-energy photons in a NaI scintillation detector, the gain of the PM tubes must be adjusted correspondingly to accommodate the energy of the nuclide being used. The gain of the PM tubes can be adjusted by changing the high voltage supplied to the dynodes. The window width of the PHA is controlled by a potentiometer controlling the voltage to the threshold electronics, as discussed previously. The output of the PHA is a sequence of two pulses, one called the *leading edge pickoff* (LEP) *pulse* and the other called the *count within the window* (CWW) *pulse*. The LEP pulse is used in rate meters or other forms of counting circuit. The CWW pulse is about 0.2 μs long and occurs about 1 μs after the LEP pulse. The 1-μs delay puts the CWW pulse about in the center of the X-Y deflection voltage pulses, which allows the CRT beam to be

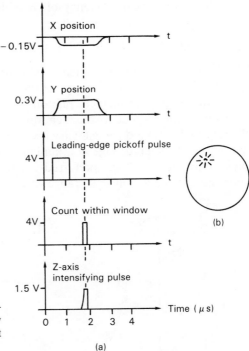

Figure 12-28 (a) Camera voltage waveforms; (b) cathode ray tube display showing dot produced by adjacent waveform.

positioned before it is turned on by the Z-axis intensifying pulse. Typical waveforms are given in Figure 12-28.

OPERATIONAL CHARACTERISTICS OF SCINTILLATION CAMERAS

The performance of scintillation cameras is generally evaluated in terms of the following parameters:

1. Sensitivity
2. Scintillation resolving time
3. Field uniformity
4. Picture spatial resolution

Sensitivity

Anger camera sensitivity is defined as the number of counts per second that the system obtains for each unit of activity being viewed. It is dependent on:

1. The geometrical efficiency of the collimator
2. The efficiency of the crystal, function of geometry, and atomic structure
3. The pulse-height-analyzer window width

The scintillation crystal efficiency is a measure of the crystal's ability to scintillate as a function of energy of the incident photon. As seen in Figure 12-29, the percent of incident photons that are completely absorbed varies inversely with respect to their energy. For a desired output count rate, the required patient dose will vary inversely as the crystal efficiency. Since one wishes to keep dosage at a minimum, radionuclides producing gamma rays below 300 keV are preferred because gamma cameras have higher counting efficiencies in their range of energies.

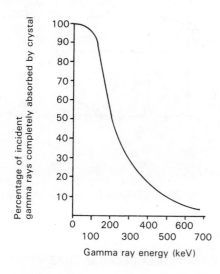

Figure 12-29 Scintillation camera crystal efficiency.

Scintillation Resolving Time

The rate at which a radioisotope camera can process incoming radiation information is dependent on the electronic processing time and anticoincident time of the system. The processing time interval can be as long as 10 μs. The shorter the interval, the greater the counter rate the camera system can handle. We find that for an Anger system that as activity increases, the count rate increases in some instances to $90K$ counts per second. If activity increases still further, the count rate begins to decrease. The reasons for the decreases are the increased probability that two pulses will arrive at the same time and be rejected by the anticoincidence circuit. If the pulses are close together but not close enough to be considered coincident pulses, the overlapping flashes in the crystal, which produce an additive light intensity, will be interpreted as a single high-energy photon and be rejected by the pulse-height analyzer. High count rates also produce baseline shift in systems without a baseline restoration circuit. This shift rejects some pulses by putting them outside the PHA window. The total effect is to produce a falling count rate with increasing activity.

Field Uniformity

In the ideal case, a scintillation camera should have a uniform response across its field of view. This means that absorption and position determination accuracy should be independent of where in the crystal the photon interacts. Unfortunately, this is not the case in practice. It is possible for the system to vary as much as $\pm 15\%$ over the entire crystal. Even though this seems like a lot of variation, it is found in practice that the corresponding variations across the display are usually so gradual that performance is not significantly degraded.

Spatial Resolution

Spatial resolution is a measure of the ability of the imaging system to resolve small details of the object. The most commonly used measure of spatial resolution is obtained from the camera's response to a line source of radiation. This produces what is known as a *line spread function* and the measure of resolution becomes what is known as the *full width at half maximum* (FWHM) (see Figure 12-30). This concept is not unlike that of bandwidth, which is used in assessing filters or electronic amplifiers.

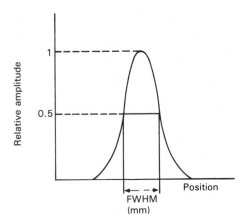

Figure 12-30 Relative signal amplitude as a function of spatial resolution. Line source produces the line spread function shown, illustrating the concept of full width at half maximum.

If a filter is a narrow-band notch filter, it can be assessed on the basis of its bandwidth. For a gamma camera, its spatial resolution is measured by its "bandwidth" as given by FWHM. All imaging systems have an all-over system spatial resolution R_o which can be broken down into two components: R_g, the geometric spatial resolution, which is solely a function of the geometric configuration of holes in the collimator, and R_s, the resolution due to scatter in the medium. Each collimator design has a radius of view R_v that defines a circle in a given plane from which photons must originate if they are to interact with the detector. Another way to look at this is that the collimator defines a cone of acceptance where a gamma photon must originate in order to interact with the crystal detector (see Figure 12-31).

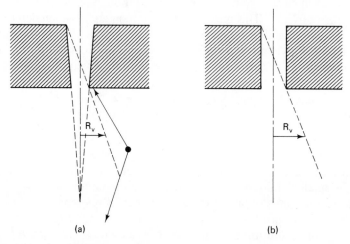

(a)   (b)

Figure 12-31 Schematic of radius of view R_v for (a) focusing collimator and (b) parallel hole collimator.

Because of scatter within the medium, it is possible for photons outside the "cone of acceptance" or circle of view defined by the collimator geometry to be scattered into the detector (see Figure 12-32). This produces an apparent increase in the circle of view which then degrades the overall resolution. The effect of a scatter medium on the system resolution is a function of:

1. Source configuration
2. Depth of medium
3. PHA setting

but is the same for all collimator types. Thus scatter is treated as a second component of overall resolution, as given by R_s. For rectilinear scanner, detection of an event and the determination of the coordinates of the event are independent, so the overall spatial resolution becomes

$$R_o = \sqrt{R_g^2 + R_s^2} \qquad (12\text{-}15)$$

Anger cameras have an additional resolution component known as intrinsic resolution, R_i. R_i is a measure of how well an imaging device can

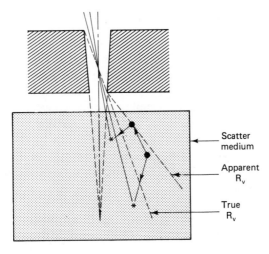

Figure 12-32 Diagram showing how scatter medium can cause apparent increase in radius of view of given collimator.

localize an event. Most cameras today have an intrinsic resolution of 5 to 8 mm (FWHM) for 14-keV gamma rays. For the Anger camera, the overall system spatial resolution can be given as

$$R_o = \sqrt{R_g^2 + R_s^2 + R_i^2} \qquad (12\text{-}16)$$

From Table 12-1 we can see that changing the collimator from high sensitivity to high resolution produces an improvement of 17.67% in total system resolution. Thus the biggest gain in spatial resolution performance can be obtained by proper selection of the collimator for the camera.

TABLE 12-1 Comparison of Overall Spatial Resolution Performance[a]

Collimator Type	R_c Collimator Resolution FWHM (mm)	R Intrinsic Resolution FWHM (mm)	R Scatter Component (mm)	R_o Overall Resolution FWHM (mm)
High resolution	7.82	7	10	14.50
	7.82	5	10	13.64
No purpose	9.35	7	10	15.38
	9.35	5	10	14.58
High sensitivity	12.62	7	10	17.56
	12.62	5	10	16.86

[a] Technetium-99m imaging in scatter medium with 20% energy window. The scatter component is the resolution equivalent measured within a water scatter medium 10 cm from the face of the collimator.

Modulation Transfer Function

Limitation of indices of performance such as FWHM to measure spatial resolution performance in scatter media has led to the use of better descriptors, such as the modulation transfer function (MTF), to indicate imaging system performance. The MTF was first used to analyze the performance of electronic imaging systems, and the concept is now being used to describe the spatial resolution performance of imaging systems in radiology and nuclear medicine. The basic concept relates to the theory that all images may be resolved into a spectrum of spatial frequencies by means of Fourier transformer. The MTF predicts the response of the detection system to these spatial frequencies.

Consider an object having a sinusoidal distribution of activity with a spatial frequency f cycles per centimeter. If the maximum activity is T_O and the minimum activity is NT_O, we define a function called the input object contrast C_O as

$$C_O(f) = \frac{T_o - NT_o}{T_o + NT_o} \tag{12-17}$$

If the imaging device had perfect spatial resolution, the image of the spatially distributed activity would be the same as that of the object. Since imaging devices are not perfect due to line spread, there will be a loss in contrast of the resulting image. The image contrast is then

$$C_I(f) = \frac{T_I - NI_I}{T_I + NT_I} \tag{12-18}$$

The modulation transfer function for a given spatial frequency is defined as the ratio of image contrast to the object contrast (see Figure 12-33), that is,

$$\text{MTF}(f) = \frac{C_I(f)}{C_O(f)} \tag{12-19}$$

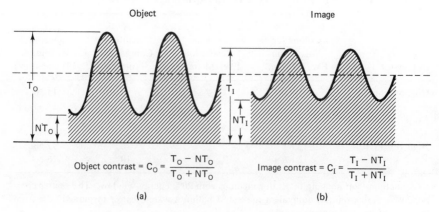

$$\text{Object contrast} = C_O = \frac{T_O - NT_O}{T_O + NT_O} \qquad \text{Image contrast} = C_I = \frac{T_I - NT_I}{T_I + NT_I}$$

(a)   (b)

Figure 12-33 Object contrast and image contrast.

When spatial resolution is perfect, the image and object contrasts are the same and MTF is unity; that is, when

$$C_I(f) = C_O(f) \qquad \text{then MTF}(f) = 1$$

If $C_I(f) = 0$, then $\text{MTF}(f) = 0$ $(I_I = NT_I)$; thus

$$0 \leq \text{MTF} \leq 1$$

In practice, imaging systems are not perfect and thus do not have perfect spatial resolution. Scanning a line source produces a normally distributed count rate about the center (see Figure 12-34). Let the line spread function be defined as $A(\epsilon)$, which defines the count rate distribution. Next, consider a one-dimensional sinusoidal distribution of activity that has a concentration $\rho(x)$, where $\rho(x)$ varies sinusoidally as a function of x about a mean concentration ρ_0. Then

$$\epsilon = \text{displacement of live source from central}$$
$$\text{axis of collimator}$$
$$f = \text{spatial frequency, cycles/cm}$$
$$\rho(x) = \rho_0 + \rho_0' \cos 2\pi f x$$
$$A(\epsilon) \quad \text{is experimentally determined} \qquad (12\text{-}20)$$

For a sheet of activity the concentration $\rho(x)$ at any point P in the image of the sinusoidal distribution is the convolution of the measured line spread function and the concentration:

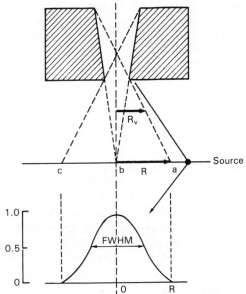

Figure 12-34 Normally distributed amplitude produced by a line source.

$$\rho(x) = \int_{-\alpha}^{\alpha} A(\epsilon) \left[\rho_0 + \rho_0' \cos 2\pi f(x - \epsilon)\right] d\epsilon \tag{12-21}$$

$$\rho(x) = \rho_0 \int_{-\alpha}^{\alpha} A(\epsilon) \, d\epsilon + \rho_0' \int_{-\alpha}^{\alpha} A(\epsilon) \cos 2\pi f(x - \epsilon) \, d\epsilon \tag{12-22}$$

Normalizing with respect to the area under the normal distribution,

$$\rho(x) = \rho_0 + \rho_0' \frac{\displaystyle\int_{-\alpha}^{\alpha} A(\epsilon) \cos 2\pi f(x - \epsilon) \, d\epsilon}{\displaystyle\int_{-\alpha}^{\alpha} A(\epsilon) \, d\epsilon} \tag{12-23}$$

Expanding the cos term as a double angle and simplifying, we obtain

$$\rho(x) = \rho_0 + \rho_0' \, \text{MTF}(F) \cos \left[2\pi f x - \psi(f)\right] \tag{12-24}$$

where $\psi(f)$ is a phase distortion term and $\psi(f) = 0$ for a nuclear problem. The MTF can now be expressed as

$$\text{MTF} = \frac{\displaystyle\int_{-\alpha}^{\alpha} A(\epsilon) \cos (2\pi f \epsilon) \, d\epsilon}{\displaystyle\int_{-\alpha}^{\alpha} A(\epsilon) \, d\epsilon} \tag{12-25}$$

Thus the MTF of a system can be calculated for all values of spatial frequencies if the line spread function is known.

Although indices such as FWHM of the line spread function can be useful, the MTF provides a more versatile characterization of spatial resolution. The MTF decreases as spatial frequency increases. If the spatial resolution performances of several imaging devices are compared, the imaging device having an MTF close to unity for the greatest range of spatial frequencies has the best resolution.

PROBLEMS

1. What are the six components of a gamma scintillation counter? Explain in detail the function of each component. Draw a block diagram of the scintillation counter.

2. Discuss the need for pulse-shaping circuits in a pulse-height analyzer. Illustrate typical circuits used for this purpose. Show mathematically and with diagrams what these circuits do to the pulses produced by the detector.

3. Consider a photomultiplier with nine dynodes, with dynode resistors of 210 kΩ. If the anode current in the tube is 60 μA and the high-voltage power supply is 1100 V, what is the approximate change in gain produced by the anode current in the last dynode stage if the overall tube gain is 1×10^6?

4. List the component parts of an Anger camera and describe the function of each component.

5. Describe, using any appropriate equations, the method by which the occurrence of a scintillation event in an NaI(Tl) crystal is used in an Anger camera.

6. In defining the spatial resolution of an imaging system, the line spread function and the concept of full width at half maximum (FWHM) are used. Define the latter term and explain how it is used with respect to an Anger camera.

7. The concept of FWHM can also be used as a measure of energy resolution for a scintillation detection system. The energy resolution or discriminating capability of the detector system is expressed as

$$\text{percent energy resolution} = \frac{\text{FWHM} \times 100}{\text{peak voltage}}$$

If FWHM $= 0.5$ V and the peak voltage is 5 V, what is the energy resolution of the detector? Explain the meaning of the concept of percent energy resolution.

8. Explain the concept of the modulation transfer function and its use in imaging systems.

REFERENCES

1. Chandra, Ramesh, *Introductory Physics of Nuclear Medicine* (Philadelphia: Lea & Febiger, 1976).

2. Jackson, Herbert L., *Mathematics of Radiology and Nuclear Medicine* (St. Louis, Mo.: Warren H. Green, 1971).

3. Parker, Roy P., Peter H. S. Smith, and David M. Taylor, *Basic Science of Nuclear Medicine* (Edinburgh: Churchill Livingstone, 1978).

4. Quimby, Edith, Sergie Feitelberg, and William Gross, *Radioactive Nuclides in Medicine and Biology: Basic Physics and Instrumentation,* 3rd ed. (Philadelphia: Lea & Febiger, 1970).

13

DEFIBRILLATION AND CARDIOVERSION

Ventricular fibrillation is an emergency condition characterized by an electrical disturbance in the ventricles, resulting in uncoordinated, unsynchronized ventricular contractions. The contracting or polarized areas of the ventricles are interspersed with relaxing or depolarizing areas. This interweaving of polarized and depolarized cells in their refractory periods results in random contractions of the ventricular tissue. Because there is no unified ventricular action, there is an immediate loss of cardiac output and a drastic reduction in blood pressure. The result of this is a rapid deterioration of life functions due to tissue hypoxia, acidosis, and a buildup of metabolic waste products in the cells of the body. Needless to say, if the condition is not reversed quickly, the person will die. Diagnosis of ventricular fibrillation is made on the basis of the absence of an arterial pulse and replacement of the R and T waves of the ECG by a random rapidly fluctuating waveform called a *fibrillation wave*.

Ventricular fibrillation generally will not spontaneously revert back to a normal cardiac rhythm. The most effective way to achieve ventricular defibrillation known at this time is to deliver a pulse of electrical current through the heart. This has the effect of depolarizing a critical mass of cells that have repolarized and prolongs the refractory period of those already in a depolarized state.

Ventricular fibrillation was first described by Ludwig and Hoffa in 1850 and was given its present name by Vulpian in 1874. Shortly afterward, Prevost and Batelli described a series of experiments on dogs in which they produced ventricular fibrillation with alternating current, direct current, and transient current from a discharging capacitor. In 1899 they also discovered that ventricular fibrillation could be terminated with a current of just the right magnitude and duration. This was the first reported successful defibrillation attempt.

AC DEFIBRILLATORS

The first defibrillator that was successful both clinically and commercially was the ac defibrillator by Zoll in 1956. This defibrillator was capable of producing 60-Hz currents of approximately 15 A for 150 ms at 1-s intervals. Figure 13-1 is a simplified schematic of the Zoll ac defibrillator. The ac defibrillator consisted of a step-up transformer with a tapped secondary and a relay timer. This instrument would deliver a maximum of 750 V into a typical 50-Ω patient load. This resulted in a 15-A current flow that was timed for 150 ms. The current burst could be repeated once every second. The various taps on the secondary of the transformer provided for an adjustable voltage output that could be changed to meet the needs of the situation.

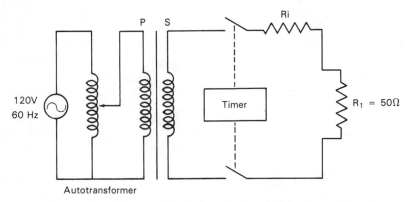

Figure 13-1   Circuit diagram of a simple defibrillator. The relay is a high-voltage relay. R_i, internal resistance of the defibrillator; R_1, equivalent tissue resistance.

The electrodes, or paddles as they are sometimes called, used for external defibrillation consist of flat, round metal disks about 7.5 cm in diameter that are axially mounted on insulated handles if they are to be used as anterior/anterior electrodes. A variation of these two electrodes is the anterior/posterior electrode set. Here one electrode is similar to the anterior/anterior electrode but the second electrode looks more like a ping-pong paddle with a wire in the handle. A third type of electrode pair is the internal electrode. These electrodes are edge mounted instead of axially mounted and the disk portion of the electrode is slightly concave, so the electrode pair can be used to cup the heart when a low-level defibrillation

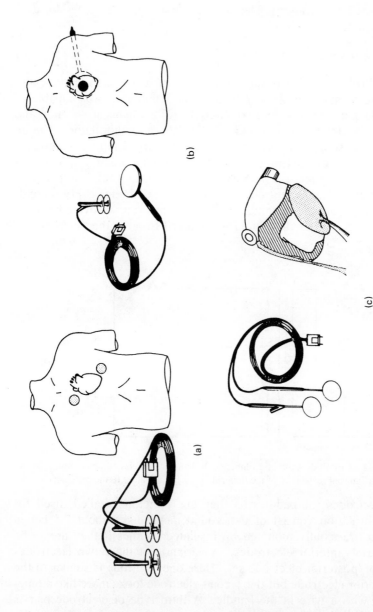

Figure 13-2 Positioning of electrode paddles: (a) anterior/anterior electrode placement; (b) anterior/posterior placement; (c) internal electrodes and model heart. (From Harry E. Thomas, *Handbook of Biomedical Instrumentation and Measurement*, by permission of Reston Publishing Co., a Prentice-Hall Company, Reston, Va., © 1974.)

shock is applied to the heart during open-chest surgery. Figure 13-2 illustrates these different types of defibrillation electrodes.

In use, external electrodes would be placed on the chest either in the anterior/anterior locations, which is across the wall of the chest, centered and in line with the heart median position, so that the electrical energy from the defibrillator is channeled directly through the heart. The anterior/posterior electrodes are placed with the larger paddle-shaped electrode under the back of the patient below the heart, and the anterior electrode is placed over the heart, near the sternum.

Since the surfaces of the external defibrillation electrodes are usually flat, it is hard to make uniform contact with the surface of the skin. Uniform contact is necessary to prevent a high current density from forming under the electrode at the point of contact with the patient. The problem is localized joule heating under the electrode, which can cause a burn. Since the number of watts of heat generated is related to the product I^2R in any one location, it is important to keep the skin resistance as low as possible and the current density uniform over the surface of the electrode.

For these reasons, the defibrillation electrodes must be covered with a conductive paste or saline pads to reduce the skin resistance of the patient and to promote a more uniform current distribution under the electrode. It is important that the clinical engineer, biomedical equipment technician, or medical personnel see to it that a tube of electrode jelly is always kept with the defibrillator.

One inherent limitation of the ac defibrillator was that it could not be used in the treatment of atrial fibrillation. Because of the large transformer needed in this device and the combined weight of the rest of the device, one could hardly call this system portable. Another practical limitation of this device was the amount of current it drew from the power mains. At maximum output, the peak current in the primary of the transformer was over 90 A. It could, and would, raise havoc with other devices connected to the same power line. In spite of these limitations, the ac defibrillator was an outstanding success. Almost instantaneously it changed the technique of defibrillation and made it a routine emergency procedure.

DC DEFIBRILLATORS

The capacitive discharge defibrillator was developed by Lown [5] and Peleska [8] and is commonly called a dc defibrillator. Fourier analysis of the output waveform of this device demonstrates that there is a dc component to the wave but the defibrillator output waveform is not dc. Perhaps *pulse defibrillator* would be a more appropriate name for this device. Figure 13-3 illustrates a typical pulse defibrillator. In the capacitive-discharge pulse defibrillator, a capacitor is charged to a voltage of 7 kV or less depending

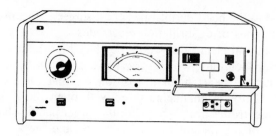

Figure 13-3 Pictorial view of a typical defibrillator.

on what level of electrical energy is to be delivered to the patient. The energy stored on the defibrillator capacitor is given by the formula

$$\text{energy} = \frac{1}{2}CV^2 \qquad (13\text{-}1)$$

where C is the capacitance in farads and V is the voltage on the capacitor. A simplified schematic of this type of defibrillator is given in Figure 13-4. A high-voltage step-up transformer with an autotransformer in the primary is used for voltage adjustment. The output of the transformer goes through the diode used as a half-wave rectifier. This, in turn, charges the capacitor. When medical personnel want to discharge the stored energy through the patient, the button or buttons on the paddle electrode handle is/are pushed and the high-voltage relay disconnects from the charging circuit and connects the capacitor in series with an inductor and the patient load, usually considered to be in the neighborhood of 50 Ω.

Because this discharge circuit is an *RLC* second-order circuit that is slightly underdamped, the output waveform will look typically like the one shown in Figure 13-5. This waveform has a rise time of less than 500 μs and a peak amplitude of about 2.68 kV. The time to the first zero crossing is about 3.65 ms. The function of the series inductor is to limit the initial current delivered to the patient without dissipating much of the delivered energy from the capacitor. If the inductor were not in the circuit, then upon discharge of the defibrillator the full capacitor voltage would appear across the patient. The equivalent circuit for this circuit configuration is a series *RC* circuit as shown in Figure 13-6. This is a first-order system with an exponential decaying current. At the moment the high-voltage relay closes, the full initial voltage of the capacitor, which could be as high as 7 kV, appears across the patient resistance of 50 Ω. By Ohm's law, the initial current through the patient would be

$$\frac{7000}{50} = 140 \text{ A} \qquad (13\text{-}2)$$

The power dissipated at the electrodes is a function of the square of the current (I^2R) and thus there would be a great amount of heat generated at

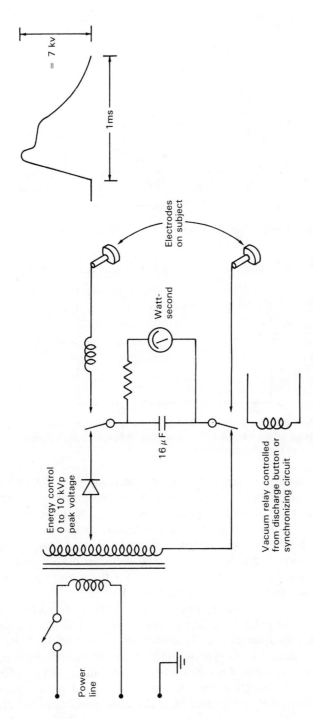

Figure 13-4 Schematic diagram of a capacitive discharge defibrillator.

= 7 kv

1 ms

Electrodes
on subject

Watt-
second

16 µF

Energy control
0 to 10 kVp
peak voltage

Vacuum relay controlled
from discharge button or
synchronizing circuit

Power
line

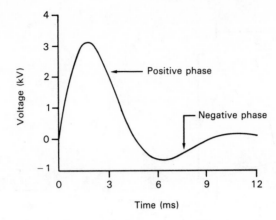

Figure 13-5 Typical voltage waveform of a capacitive discharge defibrillator discharged across a 50 Ω load.

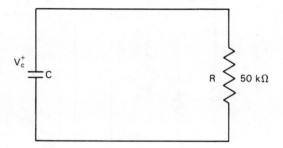

Figure 13-6 Equivalent circuit for defibrillator with no inductance. V_c^+ is the initial voltage of the capacitor after charging. R is the equivalent resistance of body tissue at high current levels.

the electrode surface, causing burns at the electrode site. Thus the inductor serves to limit the initial current and prolong the pulse since the current through an inductor cannot change instantaneously. With the series inductor the current waveform through the patient load is as shown in Figure 13-5.

Most defibrillators are designed to store up to 400 watt-seconds or joules of energy. There have been some defibrillators designed to store 500 W-s or more, but the real clinical value for the use of such high-power defibrillators has not yet been established.

Bourland et al. [2] have done studies on the effects of strength and duration of a trapezoidal waveform on the ability to defibrillate a fibrillating heart. These experiments were carried out using "Joule," the world's largest defibrillator. In this type of defibrillator, the trapezoid is generated by truncating an exponential waveform from the RC decay of the defibrillator's capacitor. When enough energy has been delivered to the patient, the capacitor is shorted by a shunting SCR that then terminates the discharge.

There are a number of advantages to this type of defibrillator, which are that there is less peak current since the voltage is controlled on the capacitor, no heavy inductor is required, and the need for a high-voltage relay is eliminated. A circuit described by Tacker [10] for a trapezoidal defibrillator is given in Figure 13-7. Any advantages, in terms of effectiveness, of the trapezoidal waveform defibrillator over the *RLC* types for clinical use have not been demonstrated. Both systems have been used with success.

Even though 50 Ω is used as the nominal impedance for the body under defibrillation conditions, chest impedance can vary depending on the current delivered [3]. Because of the way in which tissue reacts to the heavy flow of current through it, impedance decreases with successive defibrillation attempts on the same subject using the same energy settings [4]. Pantridge et al. [7] have shown that successful defibrillation can be accomplished with sequential low-level shocks of approximately 200 J or less. One possible mechanism that operates here is that if the first shock is not successful, it will have lowered the thoracic resistance enough so that the second shock at the same voltage level will cause more current to flow through the heart due to the reduced resistance of the chest, and it is the current that is important rather than the voltage.

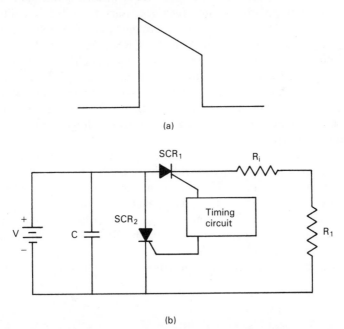

(a)

(b)

Figure 13-7 Trapezoidal waveform (a) and circuit (b). SCR_1 controls the discharge of the defibrillator and SCR_2 is used to terminate the discharge of the capacitor *C*. R_i is the internal resistance of the device. R_1 is the tissue resistance of the patient between the defibrillator paddler. *V* is the voltage produced by a high-voltage power supply.

Tacker et al. [11] have worked to establish a dose concept for ventricular defibrillation of humans. Specific doses have been reported for animals and are expressed in units of joules per kilogram of body weight. However, exact doses have not been agreed upon for humans. Still, Tacker has suggested that for the damped sinusoidal waveform delivered to the closed chest, 1 J per pound be used for children and 2 to 3 J per pound for adults.

SYNCHRONIZED OPERATION: CARDIOVERSION

Atrial fibrillation, as its name implies, is fibrillation of the atria only. This type of arrythmia causes reduced cardiac output and can be a precursor to ventricular fibrillation. During atrial fibrillation, the AV node is inundated by impulses which are irregular in rate and rapid in frequency. The ventricles do not respond to all of the atrial impulses, but ventricular rate is often rapid, which results in a decreased filling time and in decreased blood pressure and cardiac output.

Atrial fibrillation is not as critical an arrythmia as ventricular fibrillation, and the use of a defibrillator to restore a normal cardiac rhythm is an elective procedure. The use of a defibrillator in this manner is called *cardioversion*. During cardioversion it is necessary to deliver a defibrillating pulse at a point in the cardiac cycle that avoids the vulnerable period. This is easily accomplished by detecting the QRS complex and delaying the energy delivery by about 30 ms, so the pulse is delivered at a time that corresponds to the falling edge of the R wave (see Figure 13-8). Cardioversion is performed with minimal energy and is often performed without discomfort on patients who are not comatose or anesthetized. In some cases this may even be essential because anesthetizing a critically ill patient may represent an unacceptable risk.

In essence, a cardioverter is nothing more than a defibrillator, used at a lower energy level, with a synchronization circuit that inhibits the firing of the unit until the next R wave occurs after the discharge button or buttons are pushed. An ECG signal is picked up from the body using at least two leads to an ECG amplifier. Some defibrillators have the ability to use the paddle electrodes to pick up the ECG signal, eliminating the need for a set of regular ECG leads.

The discharge criteria in the cardioverter must include the ability to

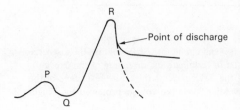

Figure 13-8   ECG waveform showing the point of discharge of the defibrillator in the cardioversion mode of operation.

decide whether any particular input feature is a legitimate R wave for the triggering of the defibrillator. Two properties of the QRS complex are used to distinguish an R wave from a large-amplitude T wave, for instance. These properties are amplitude and rise time of the wave. Threshold detection of the highest-amplitude part of the ECG wave is necessary but not sufficient for R-wave detection. The implication of the fast rise of the QRS complex is that it contains higher frequencies than other portions of the ECG. Thus the firing criteria use both threshold detection and a high-pass filter to detect the presence of an R wave.

PERFORMANCE TESTING OF DEFIBRILLATORS

Since the defibrillator is a lifesaving device, it must be fully functional at all times, as there is no room for device failure here. Adequate maintenance and performance testing of this device is essential. Since the defibrillator is such an important device in the hospital, the frequency at which performance tests are run should be a function of how often the defibrillator is used. In many hospitals, the defibrillators in ICU and CCU areas are sometimes checked for function only by the new shift of nurses at each shift. Unfortunately, the test sometimes used is to short the paddle electrodes together and fire the defibrillator. If the high-voltage relay is activated and the system discharges, two things happen: (1) the nurse or physician who has just performed this "test" is lulled into believing that the device is performing correctly, and (2) they helped contribute to the ultimate failure of the defibrillator.

With the defibrillator electrodes shorted together, the resistance in the discharge path is now only a few ohms instead of the design load of 50 Ω. This will cause larger than normal currents to flow in the device and can lead to failure of the electrolytic capacitor. This is no way to test a defibrillator! A 50-Ω test load should be located at each defibrillator station for use in testing to see if the device is operational. As we will see in following sections, the fact that the defibrillator discharges does not ensure that it is performing properly but, unfortunately, is taken for a sign of complete functionality.

The easiest way, but by all means not the only way, to determine the performance of a defibrillator is to use one of the defibrillator analyzers that are currently on the market. The two important functions that this piece of test equipment will perform are to calculate the energy delivered to the 50-Ω load in the tester in watt-seconds, and to provide a means of making a record of the defibrillator waveform.

The watt-second meter will help to determine how well the energy delivered to the load corresponds to the energy meter reading on the defibrillator. Also, if there is a break in the cables, the applied voltage to the electrodes is high enough to arc across the break in the cable, but energy is lost in the process. If the delivered energy is significantly less than the energy indicated by the watt-second meter on the defibrillator, the cables should be checked for continuity with an ohmmeter. Readings should be taken at 100, 200, 275, 350 W-s and the maximum setting of the defibrillator. These values should be recorded and compared to previous values from the same defibrillator.

It is good practice to choose one of these values, say 200 W-s, to get a recording of the waveform delivered to the load. The waveform can be compared to that supplied by the manufacturer to see if it conforms. Also, if there are any problems in the circuits or the high-voltage relay, they will show up in the output waveform long before they are capable of producing other symptoms, such as reduced energy output. Thus subtle changes in the defibrillator waveform may indicate component degradation and warn of impending component failure.

If the defibrillator has a synchronized mode of operation, it should be tested with the defibrillator analyzer. Some of the newer test instruments have the ability to obtain the time interval between the onset of a simulated QRS complex and discharge of the defibrillator. It is important that the delay between QRS detection and discharge of the defibrillator is not prolonged or the discharge may occur during the vulnerable period of the cardiac cycle.

At this point, all the performance tests necessary for a defibrillator have been done. What is left to do is to inspect the defibrillator visually to determine the physical condition of the device and to perform an electrical safety test. Electrical safety testing is discussed in Chapter 16. A sample form that could be used to keep records for a defibrillator is given in Figure 13-9.

Some defibrillators have incorporated as part of the instrument an ECG monitoring scope and strip-chart recorder. Some of the newer defibrillator analyzers have a programmed sequence of test input signals designed to test for the frequency response and linearity of the monitor and strip-chart recorder. This is a convenient feature since the defibrillator analyzer cannot be used to test the entire device performance for both ECG and defibrillator. Figure 13-10 is representative of the defibrillator analyzers that are on the market today.

If a defibrillator analyzer is not available, the foregoing test procedures can be performed using a 50-Ω load and a 1000:1 voltage divider with oscilloscope jacks, a storage oscilloscope, a signal generator, and a frequency meter with timing capability. These are standard items that the clinical engineer should have in his or her electronics test area. Using the test load and voltage divider, the defibrillator waveform can be observed on the storage oscilloscope. The waveform can either be photographed or traced off the screen of the scope. The energy can be obtained from the waveform by integrating the square of the voltage over time. This can be done using

DC DEFIBRILLATOR INSPECTION FORM

Location:	Hospital inventory number:
Model:	
Serial number:	Date of inspection:
Manufacturer:	Inspection time intervals (in wks):

I. Visual Inspection

	OK	Comments
1. Plug, line cord and strain reliefs _____		
2. Paddles, cables, connectors, and switch _____		
3. Fuse, (spare fuse _____ yes _____ no) _____		
4. Condition of control, indicators and meter _____		
5. General condition _____		

II. Operation

6. Output energy (watt-seconds):

Stored Energy Setting	Delivered Energy Setting	Delivered Energy	Previous Value	± Change	OK	Comments
100						
200						
275						
350						
Max.						

7. Charging time to maximum energy setting:

_____ sec to _____ W-sec   Previous value:

_____ sec to _____ W-sec

(a)

Figure 13-9 Typical inspection form illustrating the performance tests to be performed on a defibrillator. (From *Health Devices*, Emergency Care Research Institute, Plymouth Meeting, Pa.)

DC Defibrillator Inspection Form, Side 2

8. Leakage current to chassis:

	OFF	ON	OK	Comments
Properly grounded	_____ μA	_____ μA		
Ungrounded, normal polarity	_____ μA	_____ μA		
Ungrounded, reversed polarity	_____ μA	_____ μA		

9. Leakage current to paddles:

Properly grounded	_____ μA	_____ μA		
Ungrounded, normal polarity	_____ μA	_____ μA		
Ungrounded, reversed polarity	_____ μA	_____ μA		

10. Internal discharge of stored energy:

( ) Slow   ( ) Acceptable   ( ) Fast

11. Synchronizer operation _____		
12. Waveform analysis _____		
13. Other features _____		

14. Energy delivered after 1 minute, at maximum setting with energy control returned to zero:_____ W-sec

15. Output of tenth repeated discharge at maximum setting:_____ W-sec

16. ( ) Previous conversion tag acceptable
   ( ) No tag, new tag added
   ( ) New tag added due to recalibration or update

17. ( ) Photograph of waveform at this control setting
   ( ) Examination of waveform at this control setting

Inspected by: _____

(b)

Figure 13-9   (*cont.*)

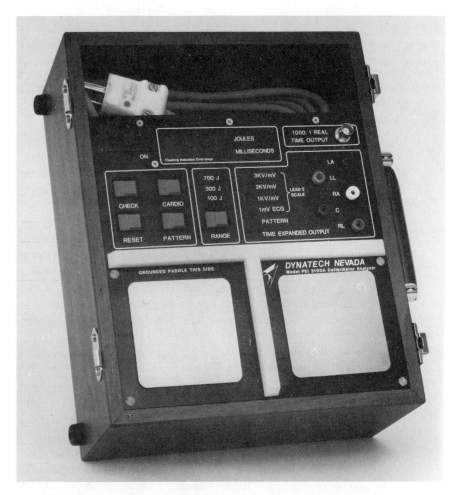

Figure 13-10 Automated defibrillator analyzer designed to test all performance aspects of a defibrillator and provide both real-time and expanded-time outputs of the defibrillator waveform for analysis and record-keeping purposes. (Courtesy of Dynatech Nevada, Inc., Carson City, Nevada.)

basic numerical approximations. The formula for the energy is

$$\text{energy} = \frac{1}{R} \int V^2(t)\, dt \tag{13-3}$$

The synchronized mode of operation can be tested using the timing mode of operation that comes in some frequency counters. The start timing signal would come from the simulated R-wave input to the defibrillator, and the stop timing signal would come from the leading edge of the defibrillator waveform. In this way, the same characteristics of the defibrillator are tested as with defibrillator analyzers, but not as rapidly or conveniently.

Defibrillator Safety Features

Features that should be incorporated in a defibrillator system that are essential for safe operation are:

1. Reliable grounding of the instrument chassis via the power cord. This requires a three-wire line cord terminated in a heavy-duty or "hospital-grade" plug.
2. Ungrounded output is essential so that the output current flow is from electrode to electrode and does not find an alternative path to ground. If this did happen, the patient could be burned at the contact point where the current exited to ground. Even more important, the current density through the heart may be reduced below the threshold necessary to stop ventricular fibrillation.
3. The capacitor should automatically be discharged into an internal resistor when the power to the unit is turned off. This will prevent an apparently dead machine from accidentally shocking someone.
4. During a cardiac emergency the main concern is the patient. A defibrillator inadvertently left in the "synchronized only" mode may not be noticed and will not function without a triggering QRS complex input. Automatic reversion to a defibrillate-only mode of operation is essential to allow the defibrillator to discharge when it is needed; otherwise, the unit will wait for a QRS wave that is not forthcoming.
5. The operator must have complete control of when the energy is delivered to the patient. Generally, energy can be delivered only if both paddle switches are pushed simultaneously. This prevents another person from accidentally delivering energy before the operator is ready.
6. There should be insulating guards on the paddles to prevent shocking the operator in case his or her hands slip on the paddles or excess electrode paste comes around the electrode. These guards should be in good condition, without cracks or missing pieces.

PROBLEMS

1. Explain the difference between a synchronous and an asynchronous defibrillator.
    (a) Draw a block diagram of each type of defibrillator.
    (b) When would each type be used on a patient?
    (c) What is the maximum allowable delay between QRS detection and firing of the defibrillator for safe use of the device? Explain how you arrive at your answer.
2. In the defibrillator of Figure 13-4, the 16-μF capacitor is charged to 7000 V and allowed to discharge through an inductor of 58-mH inductance and 13-Ω resistance. The patient represents a 50-Ω load to the defibrillator.
    (a) Draw the equivalent circuit for this situation.

(b) Derive the equation for the output voltage across the patient between the paddles of the defibrillator.

(c) What is the peak current passing between the electrodes and through the patient?

(d) Discuss the possible complications that might take place if the electrodes are not in complete contact with the surface of the patient's skin when the defibrillator is discharged.

3. Depending on the design of the dc defibrillator, there can be a difference between the stored energy on the capacitor and that which is actually delivered to the patient during discharge of the device. List and explain what you believe to be reasons for the difference and what can be done to reduce it.

4. Because of the life-and-death importance of the defibrillator as a clinical device, it is necessary for the clinical engineer to have a performance testing program on all cardiac defibrillators in the hospital. Discuss what you believe to be the important parameters that should be measured on this device, and explain why each test is performed. Discuss how each of these tests would be performed and what instrumentation would be required to perform the tests.

5. Some persons who work with defibrillators in hospitals feel that they should be tested daily or even at the beginning of every shift in areas where these devices are used. Sometimes the defibrillator is tested by touching the two electrodes together and discharging the device. Explain what is wrong with this procedure, using circuit diagrams and equations to illustrate your point. Is there any value in testing a defibrillator at such frequent intervals?

REFERENCES

1. Biloon, Frank, *Medical Equipment Service Manual* (Englewood Cliffs, N.J.: Prentice-Hall, 1978).

2. Bourland, J. D., W. A. Tacker, and L. A. Geddes, "Strength-duration curves for trapezoidal waveforms of various tilts for transchest defibrillation in animals," *Med. Instrum.,* 1978, pp. 38–41.

3. Dahl, Charles F., "Transthoracic impedance to direct current discharge: effect of repeated countershocks," *Med. Instrum.,* Vol. 10, No. 3, May–June 1976.

4. Geddes, L. A., "Electrical ventricular defibrillation," in *IEE Medical Electronics Monograph,* D. W. Hill and B. W. Watson (eds.) (London: Peter Peregrinus, 1976).

5. Lown, B., "Cardioversion of arrhythmias," *Mod. Concepts Cardiovasc. Dis.,* Vol. 33, 1964, pp. 863–868.

6. O'Dowd, W. J., "Defibrillation design and development—a review," *J. Med. Eng. Technol.,* Vol. 7, No. 1, Jan.–Feb. 1983, pp. 5–15.

7. Pantridge, J. F., et al., "Electrical requirements for ventrical defibrillation," *Br. Med. J.,* May 1975, pp. 313–315.

8. Peleska, B., "Cardiac arrhythmias following condenser discharges led through an inductance," *Circ. Res.,* 1965, pp. 11–18.

9. Simmons, David A., "Defibrillators," in *Hospital Instrumentation Care and*

Servicing for Critical Care Units, Robert B. Spooner (ed.) (Pittsburgh, Pa.: Instrument Society of America, 1977).

10. Tacker, W. A., "Defibrillators," in *CRC Handbook of Clinical Engineering,* Vol. I, Barry N. Feinberg (ed.) (Boca Raton, Fla.: CRC Press, 1980).

11. Tacker, W. A., F. Galitoto, E. Guiliani, L. A. Geddes, and D. G. McNamara, "Energy dosage for human trans-chest ventricular defibrillation," *N. Engl. J. Med.,* 1974, pp. 214–215.

14

ELECTROSURGICAL SYSTEMS

Electrosurgery involves the intelligent use of radio-frequency energy for surgical procedures where cutting and/or controlled bleeding is desired. Electrosurgery for many years has provided a means for accomplishing this with less patient trauma than by using other means. Often, a person using a tool with great skill may not understand its total function. It is like asking an auto mechanic about the combustion characteristics of the engine that he or she is working on or about the composition of the steel used in the tools. The electrosurgical unit is used as a tool by the surgeon without his or her understanding of the actions of high-frequency radio energy or how it is produced. It is this lack of understanding that leads to the misuse of the term "electrocautery" instead of the term "electrosurgery." There is a clear distinction to be made if we are to keep the meaning of the words clear.

Electrocautery is the application of heat to tissue for the control of bleeding using a wire heated by the passage of current through it. In *electrosurgery*, high-frequency bursts of RF energy conducted through the tissue produce the heat within the tissue to control bleeding and obtain the result the surgeon needs.

An electrosurgical unit is nothing more than a radio-frequency generator whose output is not matched into an antenna for radiating RF power into the air. Instead, it is conducted by wires to a narrow bladelike electrode embedded in an insulating handle. A second electrode is used as part of the return path that the RF energy takes back to the generator. This return electrode is sometimes called a *dispersive electrode* because it has a surface area that is very large compared to the cutting electrode in order to reduce the current density at the exit point. Originally, and for many years, this electrode was a large sheet of relatively rigid stainless steel called a "patient plate." Today, smaller, flexible, disposable electrodes have come into common use. Thus the term "dispersive electrode" has become the more common name for this return electrode. The patient is interposed between the cutting electrode and the dispersive electrode and is the conducting pathway for the RF current from one electrode to the other. Figure 14-1 shows a typical electrosurgical configuration.

The exact mechanism by which the RF current can be made to cut tissue and reduce bleeding by coagulation is not fully understood. Current theory puts forth the concept that the RF energy concentrated at the tip of the "cutting" electrode produces two forms of heating at the tissue interface. Since current is flowing from the electrode to the tissue by means of a conducting plasma arc, there is the expected I^2R loss in the tissue, where R in this case is the tissue resistance. The second source of heat is the RF heating of the water molecules in the cells of the tissue in the immediate vicinity of the cutting electrode. This type of heating is similar to that produced in a microwave oven, where the RF energy produces heat in water molecules due to the agitation produced by the interaction of the RF electric field and the bipolar water molecule.

All of this heating takes place very quickly relative to the transfer of this heat to the surrounding tissue. This creates a rapid buildup of heat in the tissue close to the electrode causing the water in the cells to boil and turn to steam, rupturing the cell walls and the tissue, producing cell separation or cutting. A standard scalpel produces tissue separation mechanically.

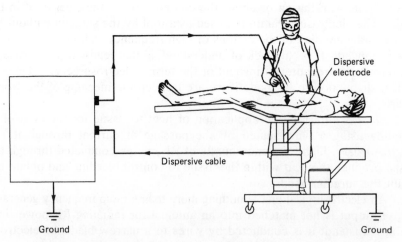

Figure 14-1 Arrows indicate the correct flow of RF current from electrosurgery through the patient and back to the generator via the dispersive electrode.

Because the edge of the scalpel blade is so thin, its contact area is very small. When the blade is applied to the skin and pressure applied, the force per unit area at the bottom of the blade is so high that it physically ruptures the cell membranes, producing tissue separation and cutting. The RF waveform that produces pure cutting generated by the electrosurgical unit is a continuous sine wave. Other waveforms have a different effect on the surgical field and will be discussed subsequently.

Surgical cutting using RF energy requires high power levels of RF energy. If low-power RF energy is used, desiccation of the tissue takes place. This is due to the fact that the rate of energy input is lower and thus water evaporates from the tissue due to the slower rise in temperature. Thus the water molecules evaporate from the tissue slowly rather than forming in a more explosive manner as steam.

If instead of a pure sine wave, a decaying oscillation or gated sine wave is used, coagulation or hemostasis is produced. This is one, if not the most important advantage of using electrosurgical units. This means that at the touch of a button or the pushing of a foot pedal, the surgeon can coagulate a bleeding vessel or with a combination or blended wave can have both cutting and coagulation at the same time. This is a great advantage for the surgeon. The waveforms used in the various functional modes of operation of the electrosurgical units are shown in Figure 14-2.

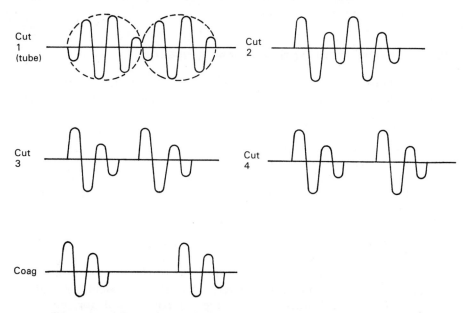

Figure 14-2   RF waveforms produced by a spark gap electrosurgical generator. Cut 1 is a pure cut waveform produced by the vacuum-tube oscillator. Cuts 2 to 4 are blends of cut and coagulation waveforms. COAG is a pure coagulation waveform.

HISTORY OF ELECTROSURGERY

According to Silva et al. [9], the use of radio-frequency current for applying
controllable heat to living tissue dates back to 1893, when d'Arsonval passed
500-kHz current through human subjects. These subjects sensed a feeling
of warmth but were otherwise unaffected by the current passing through
their bodies. The first report of the use of radio-frequency energy in surgery
was by Clark in 1911. He used a static electricity generator in conjunction
with Leyden jar capacitors, an inductor, and a spark gap to produce high-
frequency oscillations.

In 1925, Harvey Cushing became aware of the work of W. T. Bovie,
a physicist with the Harvard Cancer Commission, with the use of radio
frequencies on tissue. This was the beginning of the use of electrosurgical
systems in hospitals for surgical procedures. The electrosurgical unit that
Bovie developed for Cushing for use in neurosurgery still bears Bovie's
name. In some cases the term "Bovie" is synonymous with electrosurgical
units. Although Bovie's name is still carried on the electrosurgical units
manufactured by Liebel Flarsheim, a division of the Ritter Company, he
never received the level of monetary rewards for his invention that the
device brought the company. Unfortunately, this same problem exists today
for many creative engineers and scientists who provide most of today's
modern medical systems. Figure 14-3 is a block diagram of a spark gap
electrosurgical unit.

THEORY OF OPERATION: ACTIVE-DEVICE OSCILLATORS

There are two waveforms that are used in any electrosurgical generator:
the pure continuous sine wave used for cutting and the noncontinuous sine
wave (gated sine wave) or the damped sinusoid. These waveforms were
shown in Figure 14-2. There are two separate sections of the device that
are used to produce the waveform required by the surgeon.

The first section is the RF power oscillator used to generate the cutting
waveform. Just about any oscillator circuit could be used for this function,
but the Hartley vacuum-tube oscillator and the push-pull vacuum-tube oscillator are used the most. Schematic diagrams of these two types of
oscillators are shown in Figure 14-4. As mentioned earlier in this book,
there are still some forms of medical electronic equipment that use vacuum
tubes. The Bovie electrosurgical unit is a typical example of a current
piece of medical electronic equipment that is still manufactured with vacuum
tubes. As we shall see just a little later, a "spark gap" oscillator is also
used in electrosurgical units, and this type of power oscillator predates
even the use of vacuum. In fact, the spark gap oscillator was used in the
first radio transmitters at the turn of the century, as this was the only
practical means of generating high-power radio frequencies at that time.

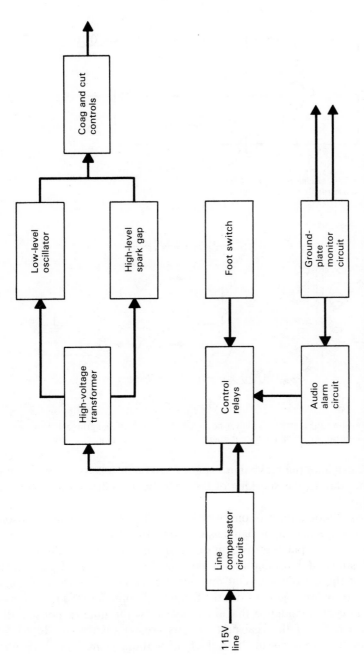

Figure 14-3 Simplified block diagram of a spark gap electrosurgical unit.

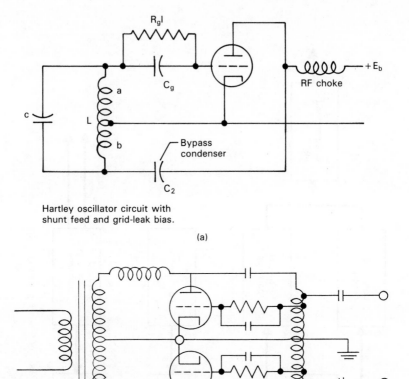

Hartley oscillator circuit with
shunt feed and grid-leak bias.

(a)

Figure 14-4 (a) Hartley oscillator circuit with shunt feed and grid-leak bias; (b) push-pull power amplifier output stage of electrosurgical unit.

This is the origin of the nickname "sparks" given the classic radio operators of the time, due to the sparking of the spark gap oscillators used for radio transmission.

An oscillator circuit is one whose output is a controllable sinusoidal signal with no applied input signal as would be the case in a classic oscillator circuit. From a Laplace transform point of view, for the condition of self-sustained sinusoidal oscillation to take place, the poles of the transfer function for the oscillator circuit must lie on the $j\omega$ axis of the s plane. In the inverse transform, going from the frequency domain to the time domain, there is the correspondence that for a pole at s_1 the time response will be of the form $e^{-s_1 t}$. If the real part of s_1 is greater than zero, $\text{Re}(s_1) > 0$, the pole of the circuit is in the right half of the s plane and corresponds to a growing exponential oscillation with time. On the other hand, if the real part of s_1 is less than zero, $\text{Re}(s_1) < 0$, the corresponding time function is a decreasing oscillation that will die out with time. Neither of these conditions exist with the oscillator circuits of interest to us. For sustained oscillation the real part of s_1 must equal zero, $\text{Re}(s_1) = 0$. This means that the pole of the transfer function lies on the $j\omega$ axis. Figure 14-5 illustrates these concepts.

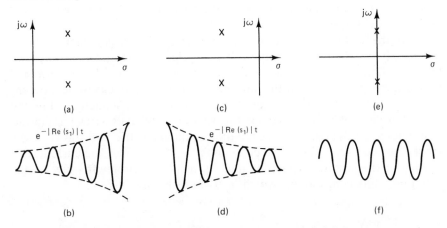

Figure 14-5 Effect of the real part of a natural frequency on the oscillation envelope: (a) pole–zero diagram, $\text{Re}(s_1) > 0$; (b) response for (a); (c) pole–zero diagram, $\text{Re}(s_1) < 0$; (d) response for (c); (e) pole–zero diagram, $\text{Re}(s_1) = 0$; (f) response for (e).

THE SPARK GAP OSCILLATOR

Figure 14-6 represents a schematic of a spark gap oscillator. This form of oscillator does not require an active device to produce oscillation, such as a vacuum tube, transistor, or operational amplifier. Instead, the properties of inductors, capacitors, and dielectric breakdown are exploited to produce the oscillations at the required power levels desired.

   A typical spark gap oscillator makes use of a series inductor and

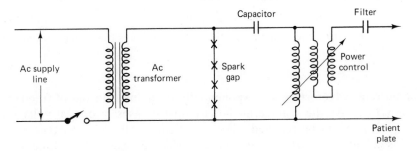

Figure 14-6 Schematic diagram of a spark gap oscillator used in electrosurgery.

capacitor to form what is commonly called a "tank" circuit. The transformer provides an alternating 60-Hz source of energy to the capacitor C in the circuit. During the portion of the cycle from 0 to 90°, the spark gap is an open circuit and the capacitor C is charged up through the inductor L. As the capacitor C charges up, the voltage on the capacitor is the same voltage that appears across the electrodes of the spark gap since the spark gap is in parallel with the capacitor.

As the voltage at the electrodes of the spark gap gets higher during the positive half of the applied voltage cycle, the electric field intensity between them increases. At some point in the cycle, which is determined by the applied voltage and the spacing between the gap electrodes, the air dielectric will "break down" or ionize and become conducting. For the sake of circuit analysis, the conducting spark gap behaves as a short circuit. The voltage on the capacitor at the time of the dielectric breakdown in the spark gap represents the initial conditions for the circuit. The current, $i(t)$, flowing in the circuit can be obtained by invoking Kirchhoff's law on the resulting equivalent circuit, which is given in Figure 14-7.

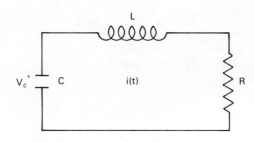

Figure 14-7 *RLC* equivalent circuit of the spark gap oscillator curing the time of conduction in the gaps. V_c is the voltage on the capacitor at the time when conduction begins in the spark gaps.

The equivalent circuit for this portion of the voltage cycle is a simple series *RLC* circuit. The second-order differential equation describing this circuit is

$$L \frac{d^2 i}{dt^2} + R \frac{di}{dt} + \frac{i}{C} = 0 \qquad (14\text{-}1)$$

The solution to this differential equation is of the form

$$e^{-at} \sin \omega t \qquad (14\text{-}2)$$

where a is the RC time constant of the circuit and ω is the radian frequency:

$$\omega = \frac{1}{\sqrt{LC}} \qquad (14\text{-}3)$$

The current waveform is an exponentially decaying sinusoid of frequency $1/(2\pi\sqrt{LC})$. As the voltage in the circuit decreases, a critical voltage at the spark gap will be reached when conduction will no longer take place. This is called the *extinguishing voltage*. This voltage is also a function of the spacing of the electrodes of the spark gap. When this voltage is reached, the air in the gap is no longer ionized and current flow stops. Thus the spark gap spacing will determine the amplitude and duty cycle of the current generated by this circuit.

During the negative portion of the transformer voltage, the capacitor again charges up through the inductor and at the dielectric breakdown voltage a second damped sinusoidal oscillation takes place. The frequency of occurrence of the damped sinusoids is twice the power-line frequency (i.e., 120 per second for 60-Hz power).

The advantages of using electrosurgical techniques are the time savings realized using this device, the assurance of asepsis, the absence of bleeding, and the elimination of cross infection from diseased to normal tissue. Also, wound healing is about the same as with a scalpel.

It turns out that the two modalities of operation, cut and coagulate, are such that they are not interchangeable; that is, the pure cut waveform does not perform coagulation well, and vice versa. Therefore manufacturers have provided a means of electronic mixing of the two waveforms so that the benefits of each could be utilized together. This is sometimes referred to as a *blended waveform.*

HAZARDS ASSOCIATED WITH ELECTROSURGERY

The most serious aspect of using an electrosurgical unit (ESU) is the possibility of burns. Burns occur when too much high-frequency electrical energy flows through a return path that does not distribute the electrical current over a large enough area. This causes the skin temperature to rise above the level where irreversible tissue damage takes place. Tissue can stand a temperature of 60°C for a few seconds, but 45°C is the upper limit of skin temperature that can be tolerated without burns appearing after prolonged periods of time.

Electrosurgical burns are debilitating and in some cases are devastating to the patient. Many cases of electrosurgical burns can be avoided if the operating room personnel understand their equipment well. It is important that the clinical engineer see to it that all personnel who make use of electrosurgical systems understand their equipment, especially the grounding or return path of the RF energy.

The majority of electrosurgical burns happen because the RF return path is faulty in some way and the RF current seeks an alternative return path to the electrosurgical generator. In doing this the current exits the body over a smaller area than intended, increasing the current density at the exit site. This increased current density produces burns at the exit site and is the source of the majority of problems associated with electrosurgical systems.

For most general surgical procedures, the dispersive electrode can be placed under the thighs or the lower portion of the buttocks of the patient.

Bony areas of the body, such as the sacral, tibial, pelvic, and scapular regions, should be avoided to prevent burns due to uneven distribution of the current over the electrode. Generally, the dispersive electrode should be placed as close as practical to the surgical site. This will help in reducing burns due to alternative return paths of the radio-frequency energy.

The ideal situation is to have the entire surface of the return (dispersive) electrode in contact with the patient. This will avoid localization of the current or "hot spots." If a plate or foil electrode is used, one should make sure that the surface is smooth before application, as bent corners or edges, creases, or dents can result in uneven contact. Since it is impossible to achieve the ideal state, it is useful to suggest to operating room personnel that conductive gel be used on metal surface return electrodes. The gel will fill in some of the voids between the patient and the metal surface, increasing the effective contact area between the patient and the return electrode and reducing skin resistance.

One of the insidious aspects of electrosurgical machines is that some of them will continue to operate even if there is a break in the return cable or if the return electrode is not applied to the patient at all. This is unlike most electrical devices, which usually will not function if one of the cables is broken or not there. In the case of the electrosurgical unit, the RF current can find alternative paths to ground through ECG electrodes or other grounded contact areas to the patient. Thus, if the return electrode is left off or not connected to the electrosurgical unit, the ESU will still produce cutting current, although often at a lower power level. It is possible that the operating room personnel may not know this and the surgeon may ask for more power to cut with. This should be a tipoff that there is something wrong with the ESU. If the condition goes undetected, the patient will be burned. Without an adequate return connection to the ESU, RF currents will flow to ground through points of low impedance, causing an increasing of current density and possibly producing a burn at the point of contact. At radio frequencies, capacitive coupling becomes important in determining the path of the RF current. A good solid connection of the patient dispersive electrode and good intimate contact of the electrodes with the patient, with the electrode as close to the incision site as possible, makes for the best insurance against RF patient burns that can be obtained.

OUTPUT CIRCUIT DESIGN

There are three basic output circuit design classifications in use today for electrosurgery units (ESUs): the grounded output system, referred-to-ground output system, and floating output system. In the *grounded output system* one side of the output transformer is grounded to the chassis as well as the return lead from the patient dispersive electrode. This is illustrated in Figure 14-8. In this type of system, RF current can return back to the

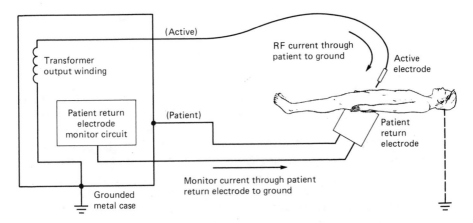

Figure 14-8 Diagram showing a typical grounded electrosurgical generator with a patient-return electrode monitor circuit.

ESU via any ground path. Thus it is extremely important that the patient electrode be connected and be put in place making good contact with the patient, or else any grounded object making contact with the patient may become a return site for the RF current.

In the *referred-to-ground output system,* shown in Figure 14-9, one side of the output transformer is grounded but the return patient cable is connected to the transformer through a return coupling capacitor. This capacitor acts as a high impedance to 60-Hz current and thus reduces any power-line leakage current that might appear at the dispersive electrode.

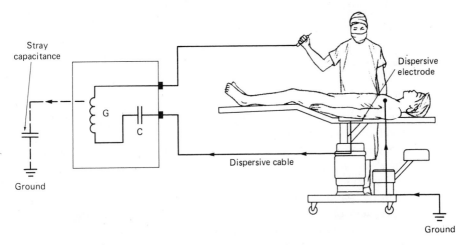

Figure 14-9 Referred-to-ground-output electrosurgical system. The capacitor *C* couples the RF energy back to the transformer while acting as a high impedance to 60-Hz current.

It can also protect the patient from failure of other 60-Hz equipment that might be used, as the capacitor will limit the current that will flow and will have an isolating effect relative to 60-Hz equipment. The RF characteristics of this type of output circuitry are basically the same as those for the grounded output system.

The *floating output system* is shown in Figure 14-10. This type of output system does not intentionally ground one side of the output transformer winding. The floating or isolated output assumes that the RF current will follow a current path from the active electrode through the patient to the dispersive electrode and return to the output transformer. If the return lead breaks or is not connected to the ESU or patient electrode, very little current will emanate from the active electrode. This is because the distributed "stray" capacitance represents a high-impedance return path to the ESU, thus limiting the flow of RF current in the circuit. This circuit configuration also offers the same 60-Hz isolation to the patient that the referred-to-ground system does against leakage current or component failure of a 60-Hz device.

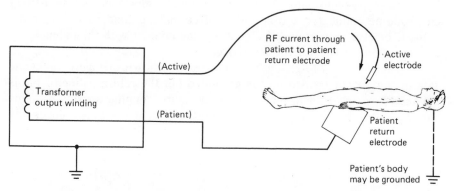

Figure 14-10 Diagram showing an isolated electrosurgical generator. If the patient-return electrode lead breaks, no RF current can flow even though the patient's body may be grounded. For this reason, the monitor circuit is not needed.

If, for any reason, the active electrode is grounded, the system reverts to the grounded output configuration. Under this circumstance the dispersive electrode becomes the active electrode, as it will be at a higher potential than the now-grounded cutting electrode. Current from the now-"active" dispersive electrode will seek a path to ground which may be an alternative path through the patient, causing an exit burn at the grounded patient site. Also, the problem of 60-Hz leakage or failure current flowing through the patient is the same as it was for the grounded output system.

From what has been said about electrosurgical units, it is clear that the most important factor relating to patient safety when using these devices is providing for a proper RF current return from the patient to the electrosurgical generator by means of a dispersive electrode. Any deviation from the conditions put forth by the manufacturer of the device increases the hazards for the patient. Every effort must be made by the clinical engineer to make personnel aware of the potential hazards of the ESU and other electronic equipment interactions with this device.

Most problems associated with the electrosurgery system can be minimized if the following procedures in the safe use of the ESU are adhered to.

1. Always operate the ESU at the lowest acceptable power level that will do the job.
2. If, for any reason, a higher than normal power setting is required for a procedure, or if the duration of activation is excessive, a fault probably exists in the system. All circuits should be checked immediately to see if they are connected or if there is continuity in all the leads. This should be done especially for the dispersive electrode and its connecting cables. The procedure should not proceed until the reason for the higher than normal power level is found. The well-being of the patient is at stake.
3. Never place any device that is connected to the patient closer to the active electrode than the dispersive electrode. This will minimize the probability that it will become an alternative RF current path.
4. Uniform and firm contact of the patient dispersive electrode with the patient's skin must be maintained throughout the surgical procedure. If the patient is moved, the return circuit for the ESU should be reinspected in case a cable has come loose or the electrode is no longer in good contact with the patient.

Electrosurgical units can also present an explosion hazard in the operating room. There is a live spark generated at the point of application as well as in the spark gap of the ESU. If the wound is cleaned with an explosive agent such as ether, then if the agent is not totally evaporated before the ESU is used, a fire might result. Also, a fire hazard is present when saline solution, water, or other conductive liquids are placed on top of the ESU. This device should not be used as a table! If any of these liquids spill and get into the internal compartment, the possibility of an electrical fire exists. To prevent this, no fluids of any sort should be stored on an electrosurgical unit.

PERFORMANCE AND SAFETY TESTING
OF ELECTROSURGICAL UNITS

The safety testing of an electrosurgical unit begins with a visual inspection. This inspection should include the following items where applicable.

1. Doors, drawers, panels, shelves, catches, latches, hinges, handles, knobs, and casters should operate smoothly, be mounted correctly, and be properly adjusted.

2. Control knobs, mechanical locks, and levers should be securely attached to the drive element and properly labeled.

3. Foot switches, hand pieces, outlet jacks, cabling, conductors, and accessories should be clean, free of cracks or splices, and fully functional.

4. Nuts, bolts, screws, and other hardware should be tight and in good condition.

5. Electrical connectors (jacks, receptacles, and plugs) should be free of cracks, breaks, and properly attached to the chassis, line cord, or cabling.

6. Cables and cords should be free of splices, frayed, or cracked insulation.

7. Cables, clips, studs, and terminals should be free of dirt, dust, and corrosion and should not be worn or pitted.

8. The grounding system (both external and internal) should be of an approved type, properly installed and functional.

9. Switches, circuit breakers, relay points, and selectors should be free of dirt, dust, and corrosion and should not be worn or pitted.

10. All cooling vents should be open and allow free passage of air.

11. All cooling fans should be securely mounted and functioning properly.

12. All electrical components should be properly mounted and operate without overheating.

The performance test for electrosurgical devices combines tests for functional operation and safety. Specifications for each instrument must be based on the manufacturer's published specifications for that particular model and serial number. There are no standard specifications for these devices.

The performance testing of electrosurgical systems should include, but is not limited to, the following performance factors.

1. Indicators such as pilot lights, meters, or dial faces should operate correctly as specified by the manufacturer.

2. Appropriate, specific responses should occur when placing controls in specific positions as specified by the manufacturer.

3. Cabinet or circuit interlocks must operate properly in accordance with the intention of the manufacturer. No interlock or alarm on the device should be defeated for any reason.

4. The ground resistance from the chassis to the power connector ground should not exceed 0.1 Ω.

5. The ground impedance from the patient electrode to the chassis for

Electrosurgical Unit
Inspection Form

Action
Not needed ☐ Needed ☐ Taken ☐
Removed from service ☐
Control no.
Date of inspection:
Next inspection due:

Type of unit: ☐ Isolated output   ☐ Grounded output

Location _____ Serial no. _____

Model _____ Manufacturer _____

	OK	Action needed	Action taken (date and initials)
1. Attachment plug			
2. Line cord and strain reliefs			
3. Foot-switch cable strain reliefs, and connectors			
4. Return electrode cable, strain reliefs, and connectors			
5. Return electrode			
6. Return electrode gel			
7. Storage provisions			
8. General condition of unit and cart			
9. Fuse or circuit breaker			
10. Antistatic provisions: _____ Ohms			
11. Indicator lights and audible tones			
12. Chassis grounding resistance: _____ Ohms			
13. Footswitch grounding resistance: _____ Ohms			
14. Return electrode grounding resistance: _____ Ohms			
15. Return cable sentry			
16. Active lead tester			

17. Leakage current: Output controls at zero settings.
Circle unacceptable values.
Operating mode: _____

Properly grounded	Ac off	Ac on (standby)	Output activated
Chassis	_____ μA	_____ μA	_____ μA
Active electrode	_____	_____	NA
Return electrode	_____	_____	*

*NA for isolated-output units.

(a)

Figure 14-11 General version of an inspection form used for electrosurgical system. (From *Health Devices*, Emergency Care Research Institute, Plymouth Meeting, Pa.)

Control No._____

17. Leakage current, Cont.

	Ac off	Ac on (standby)	Output activated	OK	Action needed	Action taken (date and initials)
Ungrounded, correct polarity						
Chassis	_____ μA	_____ μA	_____ μA			
Active electrode	_____	_____	NA			
Return electrode	_____	_____	_____ *			
Ungrounded, reverse polarity						
Chassis	_____	_____				
Active electrode	_____	_____	NA			
Return electrode	_____	_____	_____ *			

*NA for isolated-output units.

18. Output: ☐ Current ☐ Power   Load:____ Ohms

Settings	Operating mode			
Cut/Coag				Coagulate
/				
/				
/				
/				
/				
/				

19. Output isolation: Current/power to ground. For isolated-output units only.

Control settings: _____

Mode				Coagulate
Output				

20. Waveform analysis (describe)_____

Other tests (describe) _____

Comments and description of deficiencies (refer to item numbers) _____

_____

_____

_____

☐ Inspection tag affixed

☐ Equipment service request form completed

Inspected by _____

(b)

Figure 14-11 (*cont.*)

a grounded output system should not exceed 0.1 Ω. For an isolated output system the return to reference should not exceed 0.1 Ω at the operating frequency for the unit.

6. The unit should develop the proper output energy when tested across a load impedance of approximately 500 Ω. The 500 Ω is used to simulate tissue impedance. The waveforms generated at the output of the device should match those supplied by the manufacturer for every setting of the cut and coagulation modes of operation.

7. Sixty-hertz leakage current should be measured with the power off, power on, and with the unit activated in both the cutting and coagulation modes of operation since different circuits are used to generate each of these waveforms. Each of these modes of operation should be tested with the line cord properly grounded, ungrounded—normal polarity, and ungrounded—reversed polarity. (The rationale for this form of electrical safety testing is discussed in detail in Chapter 15.)

Figure 14-11 is a typical inspection form that covers the areas of performance and electrical safety that a clinical engineer or his or her designate should inspect and test.

PROBLEMS

1. Draw a simple block diagram of a typical electrosurgical unit, either vacuum-tube or solid-state type, and explain the function of each of the blocks.
   (a) Explain briefly the function of the spark gap in the Bovie electrosurgical unit.
   (b) What are some of the hazards of using an electrosurgical unit in the hospital?
   (c) In general, why doesn't an electrosurgical unit produce ventricular fibrillation in a patient even though it is producing currents as high as 2A through the patient? Use any graphs, equations, or other means to illustrate your answer.

2. Explain the operation of the spark gap and solid-state types of electrosurgical units.
   (a) How do the cut and coagulation waveforms differ in these units?
   (b) Draw typical waveforms illustrating each mode of operation for each type of electrosurgical unit.

3. Explain the theory of the spark gap oscillator in terms of the electric field across the spark gap, dielectric breakdown in the gap, plasma discharge, and "healing" voltage when the discharge is extinguished. How does the gap spacing determine the duty cycle of the coagulation waveform in a spark gap electrosurgical unit?

4. (a) For the coagulation waveform from the spark gap electrosurgical unit shown in Figure 14-12, calculate the frequency spectrum that this waveform produces. The second harmonic is 20 dB below the fundamental frequency in amplitude. The higher-order harmonics can be ignored.
   (b) Calculate the frequency spectrum for a gated (on/off) radio-frequency sine wave of the same fundamental frequency and duty cycle as the coagulation

Vertical: 10 mV/div.   COAG waveform

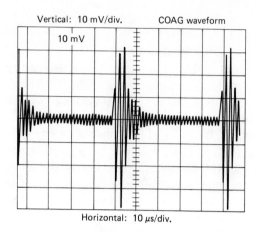

10 mV

Horizontal: 10 μs/div.

Figure 14-12

waveform in Figure 14-12. What difference, if any, do you think there would be using one waveform or the other on a patient in surgery?

REFERENCES

1. Battig, C. G., "Electrosurgical burn injuries and their prevention," *JAMA*, Vol. 204, June 1968, pp. 1025–1029.

2. Dobbie, A. K., "The electrical aspects of surgical diathermy," *Bio-Med. Eng.*, Vol. 4, May 1969, pp. 43–80.

3. *Health Devices*, Vol. 2, Nos. 8–9, June–July 1973, pp. 194–197.

4. *Health Devices*, Vol. 8, 1979, pp. 43–65.

5. Honing, W. M., "The mechanism of cutting in electrosurgery," *IEEE Trans. Biomed. Eng.*, Vol. BME-22, Jan. 1975, pp. 58–62.

6. Mitchell, J. P., and G. N. Lumb, "The principles of surgical diathermy and its limitations," *Br. J. Surg.*, Vol. 50, 1962, pp. 314–320.

7. Pearce, J. A., "A proposed method for quantitative performance evaluation of electrosurgical dispersive electrodes," *Med. Instrum.*, Vol. 13, Feb. 1979, pp. 52–54.

8. Schellhammer, P. F., "Electrosurgery—principles, hazards, and precautions," *Urology*, Vol. 3, No. 3, Mar. 1974, pp. 261–268.

9. Silva, L. F., et al., "Temperature distributions under electrosurgical dispersive electrodes," *Proc. AAMI 12th Annu. Meet.*, 1977, p. 160.

10. Watson, A. B., and J. Laughman, "The surgical diathermy: principles of operation and safe use," *Anesth. Intens. Care*, Vol. 6, No. 4, Nov. 1978, pp. 310–321.

15

HOSPITAL ELECTRICAL SUPPLY AND POWER SYSTEMS

Over the last two decades, the increased use of line-operated electrical and electronic equipment in the diagnosis and treatment of disease has made the hospital's electrical supply one of the more important components of the modern hospital. Because of this, it is important to good health care delivery that the electrical supply be correctly designed, installed, and maintained. How much the clinical or biomedical engineer is involved with the hospital power systems will vary from institution to institution, but he or she should have some understanding of the basics of the hospital electrical power system so as to make intelligent and proper decisions in matters concerning the installation and distribution of electrical power.

HOSPITAL ELECTRICAL SYSTEMS

The factors to be considered in the design of an electrical power system include safety, reliability, adequacy, distribution, grounding, and quality. The quality of electrical power is concerned with voltage stability, frequency stability, and waveform. Each of these factors is important but with different priorities depending on where in the hospital you are. Reliability and quality become equally important in surgery so that cardiopulmonary bypass machines function continuously and at constant speed.

Designation of the uses to which various areas of the hospital are to be put becomes an important parameter in the design of the total electrical supply system. The primary source for standards in wiring is the *National*

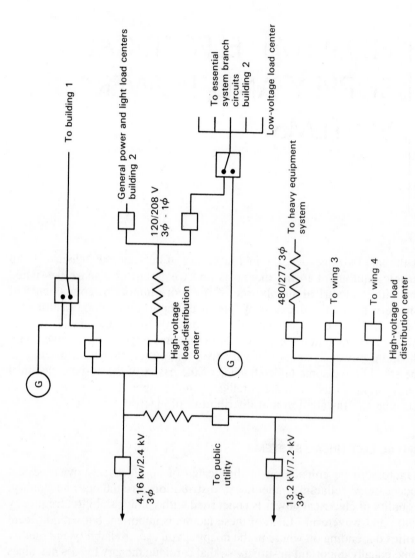

Figure 15-1 Hospital wiring diagram showing the distribution of electrical power.

To building 1

General power and light load centers
building 2

120/208 V
3φ - 1φ

High-voltage
load-distribution
center

To essential
system branch
circuits
building 2

Low-voltage load center

To heavy equipment
system

480/277 3φ

To wing 3

To wing 4

High-voltage load
distribution center

4.16 kv/2.4 kV
3φ

To public
utility

13.2 kV/7.2 kV
3φ

G

G

Electrical Code® (NEC), which is updated every two years, and supplemental standards published by the National Fire Protection Association (NFPA). These are complemented by various federal, state, and local codes as well as IEEE publications in the area.

Figure 15-1 shows a typical hospital wiring system diagram. When possible, the hospital should have two independent sources of power fed by separate distribution networks and substations. This is done to increase the reliability of the power to the hospital. Each source must have the ability to handle the entire load of the hospital. There should be an automatic transfer from one source of power to the other in the event of the failure of the primary source of power.

Even though there is a second source of power to the hospital, there must also be an emergency generator on the premises, independent from the other sources of power to provide uninterrupted partial service to "essential" areas of the hospital. The emergency generator should have the ability to restore power to essential areas within 10 s after the interruption of the primary power source. Extensive descriptions and definitions of the subsystems of auxiliary power supplies are given in the *National Electrical Code*® (NFPA 70).

GENERAL POWER AND LIGHTING SYSTEMS

The majority of the hospital load is supplied by the general power and lighting system. This system must be installed in such a way that a fault (short circuit line to line or to ground) or failure in this system will not interfere with the functioning of the other electrical systems in the hospital. What follows is a brief description of the other electrical systems in the hospital.

Essential electrical system.   The *essential electrical system* is the general term given to the auxiliary power supply and its associate equipment, such as transfer switches, feeders, and so on. This system is used during a disruption of the normal power supply. This system feeds both the *emergency system* and the *equipment system*.

Equipment system.   The equipment system is the division of the essential electrical system supplying the heating system, selected elevators, and other devices necessary for hospital function. Some delay in connecting this system to the emergency generator may be required to avoid a surge in the load on the generator. Primarily, this serves three-phase power equipment.

Emergency system.   The emergency system must be in full operation within 10 s after a fault in the general power supply. This system serves the *life safety branch*, the *critical branch*, and the *life support branch*.

Life safety branch.   The life safety branch supplies power to the equipment necessary for patient and personnel safety. Some of the equipment served includes hallway and stairway lighting, exit signs and directional signs, some hospital communication equipment, smoke detectors, alarm and alerting systems, and lighting in the vicinity of the emergency generator. This branch must be connected to alternative power sources by one or more transfer switches.

Critical branch.   The critical branch serves the patient care areas as well as areas related to patient care, such as nursing stations and pharmacy, providing power for task illumination. It also supplies power to isolation transformers (discussed later in this chapter) in anesthetizing locations.

Life support branch.   The life support branch, as its name implies, must be supplied with virtually uninterrupted power since it provides power to those areas where electrical power is essential for patient survival. These are areas of the hospital utilizing equipment such as respirators and heart-lung machines.

The general power and lighting system is the normal source of power in the hospital. National codes require that the components of the general system be physically separate from the essential system, such that a malfunction in the general system does not impair operation of the essential system.

HOSPITAL WIRING SYSTEMS

The following basic definitions apply to hospital power and wiring systems.

Branch circuit.   The branch circuit is the part of the electrical system extending from the last overcurrent protection device (fuse) to a receptacle or permanently connected apparatus. It carries a current load of 15 to 20 A. In a general care area, two or more branch circuits are used for each bed. Three or more branch circuits are needed in critical care areas, such as in the intensive care unit (ICU), cardiac care unit (CCU), and the emergency room (ER). All branch circuits to one bed location must come from the same distribution panel. In critical care areas, these distribution panels must be connected to the essential system. Each branch to a bed location is terminated in either one or two hospital-grade receptacles.

Low-voltage load distribution center.   This center receives three-phase 120-V power and distributes it to the distribution panels through overcurrent protection devices.

Distribution panel.   A distribution panel is fed by three-phase 120-V power and contains overcurrent protection devices and has branch circuits coming out of it.

Feeders.   These are lines leading from the final overcurrent devices at the start of the branch circuits to the service equipment (circuit breakers, switch gears, lightning arresters, etc.) at the entrance of the facility. Feeder circuits may include a step-down transformer.

Incoming supply.   This is the power supplied to the hospital from the public utility. It is usually a three-phase four-wire system, supplying the power from a generating substation to the hospital at 13.2/7.2 kV.

Intermediate voltages.   These are the voltages that are actually in the buildings. Typical values start at 4160/2400 V (three-phase, four-wire) and are later transformed down to 480/277 V and 208/120 V (again three-phase, four-wire) through a Y-type step-down transformer. These step-down transformers are arranged so as to keep the low-voltage lines as short as feasible, minimizing I^2R loss in the lines.

Grounding.   The purpose of grounding is to minimize the potential differences between all exposed metal surfaces in the patient vicinity and to avoid the possibility of any fault current flowing through the patient or an attendant. It is usually composed of a wire encased in a green insulating material, possibly with a yellow stripe, plus the conduit carrying the lines. Ground wires are connected to the metal frame of the building or to a metal water pipe with a minimum underground portion of 10 ft.

Overcurrent protection device.   This is a device with a fusible link or a resettable circuit breaker that is intended to open a circuit when excessive current flows in the circuit.

Ground-fault circuit interrupter.   This device detects a rise in the leakage current in the ground circuit above some predetermined value, usually 5 or 6 mA. The standard used by Underwriters' Laboratory is 6 mA. When the leakage current is greater than this value, the circuit is interrupted. The device is designed to detect the difference between the incoming (hot) and outgoing (neutral) current through a device. The difference is the loss due to leakage current or current going through a patient or medical personnel.

CODES: HOSPITAL CONSTRUCTION AND OPERATION

Both the construction and operation of hospitals are subject to the codes adopted by state and local governments. This includes, but is not limited to, the *National Electrical Code*® (NEC) as adopted by state and local governments plus added codes of their own for covering their jurisdiction. The NEC applies wherever the Occupational Safety and Health Act (OSHA) applies. The NEC is a general document, but section 517 of this code pertains specifically to hospitals. There are also other NFPA documents that pertain to some aspects of hospital safety that are not covered in the NEC. An example of this is NFPA 56A—Inhalation Anesthetics.

NFPA 76B divides the hospital electrically into three patient care classifications. What follows is a brief description of these three areas. For more detail, the reader should consult the document NFPA 76B.

1. *Class G:* General care areas where the patient contacts electrical instruments transcutaneously only. In this area the voltage between any two metal surfaces must not exceed 20 mV for new construction. The resistance of the ground wire between any two receptacles on the same branch circuit must not exceed 0.1 Ω, 0.5 Ω on separate branches, and 1 Ω on separate distribution panels.

2. *Class H:* Heavily instrumented patient care areas where the conductors enter the body, such as in the intensive care unit, cardiac care unit, and catheterization laboratory. The voltage between any two metal surfaces in the patient's vicinity must not exceed 20 mV for new construction. The resistance between receptacles should not exceed those specified for class G areas.

3. *Class W:* "Wet" areas such as hydrotherapy or dialysis. The voltage between any two metal surfaces should not exceed 40 mV for new construction. The resistances between receptacles should not exceed those values specified for class G areas. Ground-fault circuit interrupters which trip out at 6 mA or an isolated power system (if power interruption is not acceptable) are specified for these areas.

ELECTRICAL SAFETY IN THE HOSPITAL

It is a basic and fundamental concept of hospitals, attributed to Florence Nightingale, that "the hospital should do the sick no harm." Unfortunately, this is not always the case. It is one of the most important functions of the clinical engineer to see to it that the electrical environment of the patient and other medical personnel is as safe as it possibly can be. This means that the clinical engineer must take an active role in three main areas: (1) the electrical system in the hospital, (2) the electrical safety and performance of all equipment that comes in direct contact with the patient or is used on the patient by medical personnel, and (3) in-service education of medical personnel with respect to electricity, electrical safety, and the performance and proper utilization of medical devices. Since electrical shock is an unseen menace that can produce injury and death, it can only be combatted with an understanding of its particular mode of operation. It is the clinical engineer's job to educate hospital personnel so that they can perform their jobs safely relative to electrical systems used on or near the patient. Knowledge is the first requirement of electrical safety. Engineers, nurses, physicians, and technicians have an obligation to reduce the risk of electric shock by recognizing potentially hazardous conditions and properly handling power-line-operated devices on or near the patient.

Physiological Effects of Electrical Current

We will consider only the voltages 120 to 220 V, rather than higher voltages in the power system. Also, we will consider other forms of dangers due to extraneous current being conducted through the body that are common to the hospital environment. Electric shock comes about when medical personnel, patients, or others become part of an electric circuit where the body or parts of the body complete a current path. The parts of the body that become part of the circuit form an inhomogeneous volume conductor in which the distribution of current is determined by local tissue conductivity. The three major problems with current conduction through the body are:

1. Power dissipated as heat (I^2R loss) and resulting in tissue burns
2. Transmission of current through sensory and motor neurons evoking action potentials (produces the tingling sensation felt by the person in the circuit)
3. Producing ventricular fibrillation

Stimulation of motor neurons produces muscle contractions. If the current density is high enough, it produces total muscle contraction or "tetanus." One organ that is susceptible to electrical current passing through it is the heart. A tetanizing stimulation of the heart results in complete myocardial contraction, inhibiting pumping action and producing a life-threatening situation. When the tetanizing current is removed, there is a spontaneous resumption of heartbeat. This is the very principle on which the ac defibrillator operates.

Interestingly enough, lower current densities through the heart that excite regional areas of the myocardium can be more dangerous than higher current levels. When stimulation is during the vulnerable period (the latter portion of the T wave), ventricular fibrillation may occur due to local capture of a portion of the myocardium. Ventricular fibrillation is normally not spontaneously reversible; thus death will follow if a tetanizing current is not used to induce a regular heart rhythm. Ventricular fibrillation is the most frequent cause of death in fatal electrical accidents. Respiratory paralysis can also occur if the muscles of the thorax are tetanized by current flowing in the chest or the respiratory center of the brain.

So far, electrical intensity has been described in terms of electric current. Voltages necessary to cause these currents depend on the resistivity of the body. A large portion of body resistance is due to the epithelium (outermost layer of skin). Dry skin has a high resistance in the neighborhood of 300 to 500 KΩ. This resistance is substantially reduced when the epithelium becomes permeated with a conductive fluid such as water or electrode paste. This can reduce tissue resistance to 500 to 1500 Ω.

The hospital has its own special electrical shock problems that differ from those of other locations, such as one's home. This difference is due to the fact that in a hospital there are catheters and pacing leads that invade the body and are connected to the heart directly. In this case the natural protective barrier of the skin is lost, making electrical hazards to the patient greater. A distinction is made between the electrical shock produced by high-level currents introduced transcutaneously, called *macroshock,* and the electrical shock produced by low-level currents via devices that have invaded the body, called *microshock.*

Under macroshock conditions, the body responds differently to different levels of current. Four factors influence how the body will react to current flow through it: (1) the intensity of the current, (2) the frequency of the current, (3) the duration of the current, and (4) the resistance of the body tissue.

If body resistance and current duration are held constant, the effects of a macroshock current become more severe as current magnitude increases. At low levels, 100 μA to 1 mA, a mild tingling sensation is experienced due to the stimulation of neurons. As current is increased above this range, from about 1 to 10 mA, the result is violent muscular contractions, which are often painful. In some cases, these contractions result in the victim "jumping" free from the electrical contact, which could save his or her life.

In the range of 10 to 100 mA, temporary paralysis of muscles results. The individual's hand becomes clamped onto the conductor. Although there is no permanent damage associated with this current level, termed the "let-go" current, it is painful and potentially lethal since the chest muscles associated with respiration become paralyzed. If the current is maintained for a sufficient length of time, the person may become asphyxiated. Any current greater than 100 mA may cause ventricular fibrillation. Death is certain for currents greater than 1 A. Other causes of serious injury include direct respiratory arrest due to inhibition or destruction of the respiratory control centers of the central nervous system, severe internal and external burns, and the destroying of nervous, muscular, and visceral tissues.

Dalziel and Lee [1], in a classic paper on "Lethal electric currents," showed that both the level of perception and the let-go currents were functions of frequency. They showed that as the frequency increased, it required more current to produce the same levels of perception and let-go effect. As frequency increases, the body becomes less sensitive to the current. It is unfortunate that electrical power systems operate between 50 and 60 Hz, as this is in the frequency range of maximum sensitivity of humans to electrical currents. Figures 15-2 and 15-3 show the frequency dependence of perception and let-go currents. It should be recognized that the dangers of current flow through the body are a function of the current pathway and that it is the current pathway through the heart and the associated danger of ventricular fibrillation that is of greatest concern. One theory as to why this happens is that at higher frequencies, the current tends to stay more toward the surface of the body rather than going through it. The minimum value of fibrillating current in humans increases as frequency increases. It is for this reason that electrosurgical systems can be used on a patient without causing ventricular fibrillation.

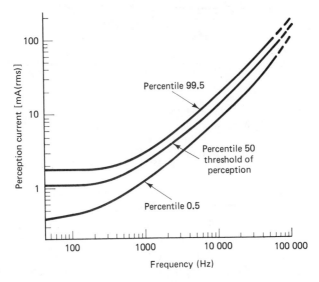

Figure 15-2 Effect of frequency on perception current with hand holding a small copper wire. Above commerical power frequencies, perception requires higher current. (From C. F. Dalziel, "Electric shock hazard," *IEEE Spectrum*, Feb. 1972, by permission.)

Microshock. Certain medical procedures require that conductive catheters be placed in the body of the patient. Catheters are used in cardiac pacing as well as in some diagnostic procedures, such as angiography. In these instances, the patient is deprived of the electrical protection of the higher-resistance epithelial layers of skin. Starmer and Whalen [14] showed

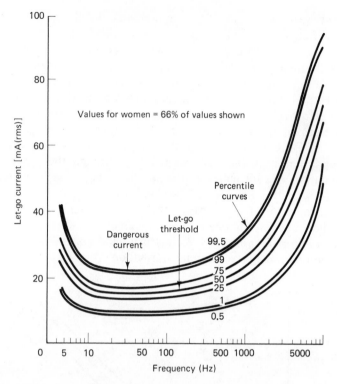

Figure 15-3 Let-go currents plotted against frequency. Currents become dangerous progressively to an increasing number of persons, as indicated by percentile values. $I_{\text{let}-\text{go}}$ at 500 Hz is over three times that at 50 Hz. (From C. F. Dalziel, "Electric shock hazard," *IEEE Spectrum*, Feb. 1972, by permission.)

that the current necessary to produce ventricular fibrillation can be as low as 180 μA when the current is applied to the heart directly. The greatest danger to a catheterized patient is ventricular fibrillation, which can be produced by very low level currents. A patient in this electrically compromised condition is termed an *electrically sensitive patient*.

The magnitude of the fibrillation current varies with catheter position as well as with current density. The minimum value of current causing fibrillation will occur for a catheter touching the left ventricular myocardium; the maximum current value occurs with the catheter touching the right atrium. Further, the minimum fibrillating current decreases as the current density increases.

Since ventricular fibrillation is the main cause of death in electrical shock cases, fibrillation produced by these small very low level currents is called "microshock." As will be studied later, it is possible for attending medical personnel to be an unwitting part of an electrical circuit causing current to flow to an electrically sensitive patient. Since the currents involved are below the level of perception of the attendant, that person will not be aware of any current flowing through his or her own body to the patient.

Electrical Hazards in the Hospital

As discussed in detail in Chapter 10, the electrical supply entering the hospital is in the form of a balanced three-phase power source. This is in contrast to a typical home's single-phase power. Figures 15-4 and 15-5 are diagrams of the hospital three-phase distribution and a home single-phase distribution. Because of the power considerations of some of the heavy equipment used in hospitals, such as x-ray and air-conditioning equipment, some of this three-phase power is used directly from the hospital's three-phase distribution system. At this point the phases are split into single-phase lines and fed into the wall junction boxes of the individual rooms. Receptacles are then wired from the wall junction box. Figure 15-6 illustrates how single-phase power is obtained from the three-phase electrical system. It should be noted that the "neutral" and "ground" lines are joined or "tied" together at one point in the power system. This should not be taken to mean that they are at the same potential, as they generally are not. The reason for this is that the normal power is not intended to flow in the ground wire but in the neutral. Because of the IR drop in neutral, it will be at a higher potential than the ground wire except at the point where they are tied together.

From Figure 15-6 we see that a separate ground wire is used in hospital wiring systems. Local codes for home wiring do not require this separate ground when metal conduit may serve as the ground return path. Hospitals require the use of a separate ground wire to keep ground resistance as low as possible and to prevent breaks in the ground path due to corrosion, which can happen with conduit.

The first defense against macroshock is to be sure that all outlets have a "good" ground, meaning that the ground wire is intact and is a low-resistance path for the flow of ground current. Second, the grounding on all electrical equipment is functional and is not circumvented by the use of two-wire extension cords or three-prong to two-prong adapters. In the event of a short to the case of an electrical device, no harmful voltage will appear on the chassis or case because the ground wire will provide a low-resistance path for the current to flow and will not allow the chassis voltage to rise to a dangerous level. At the same time, the current flowing will be large due to short circuit. This will cause the circuit breaker to trip out, disconnecting that branch from the power.

A good grounding system will not prevent macroshock if someone comes in direct contact with an electrically hot wire. This could be the case if a person touched an exposed metal conductor where the insulation was worn or broken away. Good visual inspection of the physical condition is the only way to prevent this type of electrical accident.

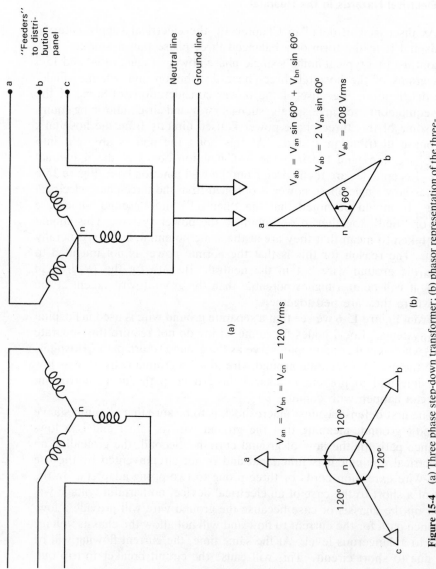

(a)

$V_{an} = V_{bn} = V_{cn} = 120 \text{ Vrms}$

$V_{ab} = V_{an} \sin 60^\circ + V_{bn} \sin 60^\circ$

$V_{ab} = 2 V_{an} \sin 60^\circ$

$V_{ab} = 208 \text{ Vrms}$

(b)

Figure 15-4 (a) Three phase step-down transformer; (b) phasor representation of the three-phase power system.

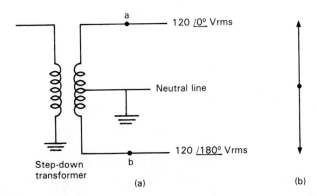

Figure 15-5 (a) Single-phase home power system with a center-tapped secondary providing 120 V ac between one side and neutral and 240 ac between each side of the transformer; (b) phasor diagram.

Microshock Hazards

The electrically susceptible patient is vulnerable to very low levels of current. The sources of this low-level current are not as apparent as the much larger currents that are responsible for macroshock. In the case of microshock, it is "leakage current" that can be the problem. Leakage current is an inherent flow of current from the live electrical parts of a device to the accessible metal case or part. "Leakage current" is an unfortunate name for this phenomenon, as it implies that something is faulty in the device. This is not the case, as leakage current exists in all power-line-operated equipment and is usually due to capacitive coupling and some resistive coupling to the chassis. Under normal circumstances, this current is conducted away by the ground wire.

Capacitive leakage current is developed any time two conductors that carry current are separated by a dielectric. In this case, the dielectric is the insulation on the wire. When an alternating voltage is applied between the conductors, a current will flow that is given by the familiar equation for current in a capacitor:

$$i(t) = \frac{C \, dV(t)}{dt} \tag{15-1}$$

When the frequency is held constant, the capacitive or leakage current is directly proportional to the power-side capacitance of device. The source of this capacitance in a medical electrical or electronic device is the sum of the capacitance due to distributed capacitance in the power cord and the primary windings and core of the power transformer. Also, if the device has RF filters, the capacitance to ground of the filter adds to the total capacitance to ground of the device. The greater the total capacitance, C, of the device, the greater the 60-Hz leakage current that will be present.

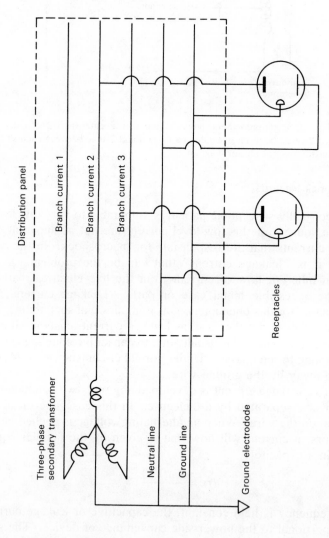

Figure 15-6 (b) Single-phase power obtained from three-phase power distribution panel.

The resistive component of leakage current comes from the fact that there is no such thing as a perfect insulator. Insulation today has such a high resistance that currents due to the Ohm's law relations $I = E/R$ are small enough to be ignored. Therefore, it is the inherent capacitance of the electrical device that poses the problem of microshock. As we will see later in this chapter, it is this capacitance to ground that each electrical device has that plays an important role in the level of isolation an isolation transformer can offer. We will make use of this concept of capacitance to ground later, in the section on isolation transformers.

For grounded equipment, leakage current is conducted to ground. If, for example, the leakage current were 100 μA and the ground resistance were 1 Ω, the chassis voltage would be 100 μV. This situation is shown in Figure 15-7. If a catheterized patient, who is also grounded, comes in

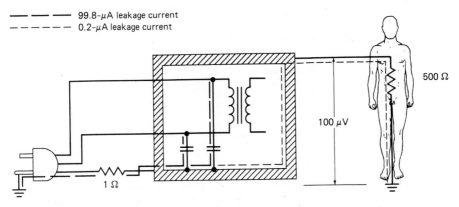

Figure 15-7 Medical device with wiring capacitance shown. Leakage current of 100 μA produces 100-μV chassis voltage.

contact with the chassis of the medical device, the chassis-to-ground voltage will be applied across the patient. Figure 15-8a is an illustration of this event and Figure 15-8b is an equivalent circuit of the same sequence of events. If we use the nominal impedance of 500 Ω for the catheterized patient, the current through the heart of the patient is, from Ohm's law, 100 μV/500 or 0.2 μA. This level of current flow through the heart is not known to produce ventricular fibrillation.

If the ground wire were broken, the situation changes. Figure 15-9 is an equivalent circuit of this situation. Now the chassis of the device has become a common node point connecting the device capacitance in series with the 500-Ω impedance of the patient. Under this condition, the current through the patient's heart is much larger than it was in the previous case with the ground wire intact and shunting current around the patient. This illustration demonstrates how important a grounded system is in pro-

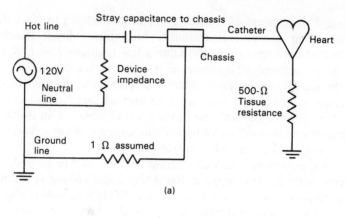

(a)

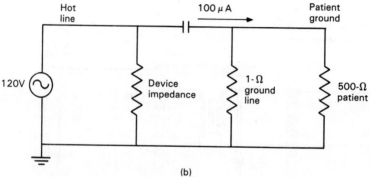

(b)

Figure 15-8 (a) Schematic of grounded catheterized patient who comes in contact with medical device chassis; (b) equivalent-circuit diagram for the grounded catheterized patient.

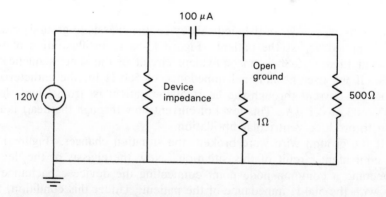

Figure 15-9 Circuit diagram illustrating the effect of an open ground wire in a medical device when touched by a grounded catheterized patient.

tecting the patient from both macroshock under a fault condition or from microshock due to inherent capacitance in the electrical device.

Hospital personnel or visitors to a patient may be an unwitting partner in causing death if this person touches the chassis of an ungrounded device and at the same time touches the pacing lead on the patient pacing catheter. Assume the impedance of the patient's attendant to be 10^5 Ω, the device capacitance to be 2500 pF (approximately 10^6 Ω at 60 Hz), and the patient grounded through a low-impedance monitor. Figure 15-10 is the equivalent circuit for this situation. The current flowing through the patient is calculated as

$$I = \frac{120\sqrt{2}}{\sqrt{(10^6)^2 + (10^5)^2}} > 100\ \mu A$$

Although this magnitude of current is dangerous to the catheterized patient, it is below the level of perception of the attendant. The only indication of a problem taking place is the possibility of interference on the monitor. Once again, this example is designed to illustrate the importance of good grounding of electromedical equipment for the safety of the patient. Also, pacing leads should be insulated from the casual inspection of the curious. If a pacing lead needs to be adjusted, it should be done without touching any part of the bed frame or the chassis of any other electrically operated device.

Since a two-wire device would act in the same way as an ungrounded three-wire device, all that has preceded would apply to two-wire devices. Thus no two-wire electrical device operating on 120 V ac should be used in any patient area. This is reason enough to prohibit the use of patient personal articles such as radios and television sets in the hospital. The only possible exception to this would be the use of low-voltage devices with a transformer at the wall plug. It is important that the clinical engineer

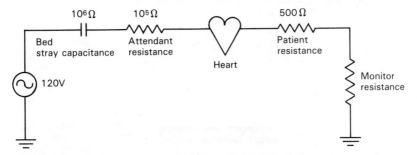

Figure 15-10 Schematic illustrating how hospital personnel or a visitor becomes part of a circuit with the patient, which can produce a hazardous current flow through a patient.

have a written policy on the use of patient electrical equipment that is accepted by the hospital.

With all that has been said here about grounds and grounded equipment, it was not meant to imply that all ground points are at the same potential. This is the ideal case and if no current were flowing in the ground lines, then indeed, they would all be at the same potential. The ground point of an electrical outlet is at the voltage $I_g R_g$, where I_g and R_g are the current in the ground at a point in time and the resistance of the ground line. Thus if two electrical devices are used on the same patient and plugged into different outlets with different ground line circuits, a difference of potential between the chassis of the two devices can exist depending on the currents in each ground circuit. Once again the electrically susceptible patient is subjected to hazardous conditions.

The solution to this problem is the *equipotential ground system*. This means that all the ground wires in a patient area are tied together electrically on a common bus bar, forcing each ground point in a patient area to be at the same potential as all the other ground points in the area. In this way all chassis voltages will be the same and prevent a difference of potential from developing between any two devices in the environment of the patient. Figure 15-11 is a diagram of an equipotential grounding system. All wires

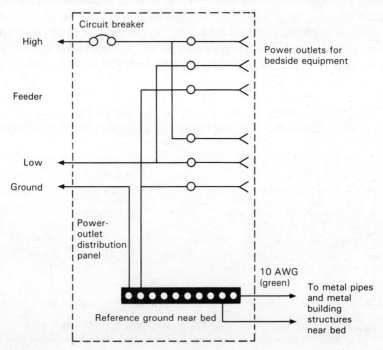

Figure 15-11 Diagram of a wiring installation illustrating an equipotential grounding system.

to the reference ground should be kept as short as possible and should be of the proper gauge.

Once again, the education of the user of hospital medical equipment is perhaps the most important part of an electrical safety program. The majority of physicians and nurses are trained to administer to the medical needs of the patient and are not educated in some of the concepts of electrical engineering. The medical staff must be made aware that plugging a device into a wall outlet, connecting it to a patient, and turning it on does not end their responsibility. The patient, in many cases, is unable to tell the attending personnel of shocks or burns because he or she may be unconscious or anesthetized. It is the responsibility of the physician or nurse to prevent a patient from being placed in a hazardous situation. An incentive for their learning is that many times it is not just the patient who is injured but a nurse, physician, or technician who receives a painful shock and is thrown to the floor. Awareness of the problems and adhering to simple precautions will do much in mitigating electrical hazards involving medical personnel.

ISOLATED POWER SYSTEMS

Because of the potential of an electrical spark igniting flammable anesthesia, the National Fire Protection Association (NFPA) recommended in their document NFPA 56A [9] that ungrounded ac (isolated) electrical distribution systems be used in anesthetizing locations in the hospital, the rationale being to reduce the hazards of electric shock and arcs in the event of insulation failure or accidental fault in electrical devices used in the area.

If an ideal isolated power system could exist, there would be no interaction of the system with its surroundings. Because of the distributed capacitance and inductance that exists in all electrical devices, perfect isolation cannot be achieved. A reasonable approximation to the ideal isolated power system is achieved using an isolation transformer.

An *isolation transformer* is nothing more than a transformer with a typical turns ratio of 1:1, designed to minimize the interwinding distributed capacitance. Figure 15-12 is an illustration of the ideal isolation transformer. In the isolated system the ground wire is connected to the neutral of the primary, not the secondary. In the event of a line-to-chassis or ground fault, as shown in the figure, there is no complete current path from line A to line B. In this ideal case, no current will flow in the circuit because there is an infinite impedance between the ground and the ungrounded side of the isolation transformer. This is precisely the advantage of using an isolation transformer. There is no heavy current flow or inductive sparking due to the infinite impedance in the circuit limiting the flow of fault current. If the fault were a person between one line of the secondary and ground, that person would be "protected" from large currents flowing through him

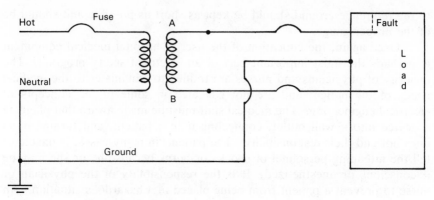

Figure 15-12 Diagram for an ideal isolation transformer.

or her. Any devices being supplied by the transformer will continue to operate in a normal fashion.

Figure 15-13 is a more accurate model of an isolated power system. Distributed capacitance in the transformer is now included in the model. Sometimes this distributed capacitance is referred to as "leakage capacitance." Capacitive coupling exists between the windings of the primary and secondary circuits and between the two sides of the transformer and the ground line. In the design of the isolation transformer, an electrostatic shield is used between the primary and secondary windings. This shield effectively reduces the transformer winding capacitance to the point where it can be ignored for the purposes of circuit analysis.

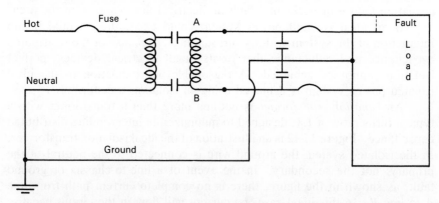

Figure 15-13 Nonideal isolation transformer, illustrating the effect of distributed capacitance on the device.

Consider Figure 15-14 as a new circuit model. Here the equivalent capacitive impedance from each line of the transformer to ground is 500 kΩ. If there were a short to ground in this circuit, the equivalent circuit becomes as shown in Figure 15-15. From Ohm's law, we have that the

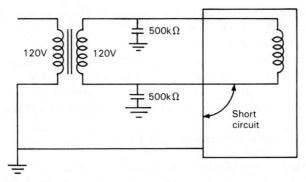

Figure 15-14 Circuit model of isolation transformer with a balanced 500-Ω distributed wiring capacitance to ground and a short to ground.

current flowing through the fault is $120/500,000 = 240\ \mu A$. This level of current is not enough to produce a spark due to the fault. If a patient or medical staff completed the circuit instead of there being a short, then, if the patient impedance were 1500 Ω, we would have the condition shown in Figure 15-16. This reduces the current even more so that a macroshock hazard does not exist for the staff member or patient.

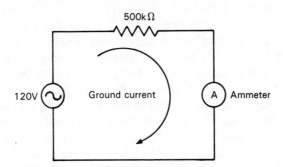

Figure 15-15 Equivalent circuit for isolation transformer having a short to ground in the secondary.

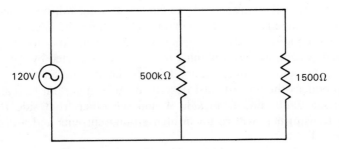

Figure 15-16 Equivalent circuit for isolation transformer with a patient or medical staff acting as a fault to ground on the secondary of the transformer.

These examples illustrate how an isolated power system can prevent a spark due to a system fault or prevent macroshock in someone who becomes part of a circuit to ground. We see that the maximum current that can flow under a one-fault condition is limited by the high impedance to ground of the isolation transformer. Another advantage of the isolated power system is that a single line-to-chassis fault does not render the equipment inoperative, as the current flowing to ground will not be enough to cause a circuit breaker to trip out.

The advantages of using an isolated power system are:

1. No spark is generated in a first fault to ground.
2. If someone becomes part of a fault to ground, there will not be a macroshock hazard to the person.
3. A first fault does not affect the operation of the equipment producing the fault.

LINE ISOLATION MONITOR

As we have seen from the preceding section, an isolated power system allows a fault to ground in a piece of equipment without the equipment showing any signs of a problem. Two problems result from the use of an isolated power system. The first is that it is possible for the fault in the electrical device to go undetected since it will appear to function normally under the fault condition. If the device is moved and operated in a nonisolated power area, it will trip a circuit breaker or blow an internal device fuse.

The second and most serious problem is that after a single line-to-chassis or ground failure, the isolated system reverts to a nonisolated system. This is the result of the fact that both the primary and secondary windings of the isolation transformer now have one line tied to ground. If there is a second fault on the system, heavy ground current will flow, tripping a circuit breaker and possibly causing a spark that could be very serious if flammable anesthetics are being used. If the second fault is produced by a person in contact between one line and ground, the full line voltage will be applied across the person and the current flowing will be a function of the electrical impedance of the person at the time the fault occurs. A macroshock condition exists here but will not be obvious to those using the isolated system, as they will think that they are protected by the system's isolation. One should study Figure 15-12 to satisfy oneself that a fault to ground produces the hazards just described. It is important to realize that a double line fault, one from side *A* and the other from side *B* of the isolation transformer, will result in high ground currents and a disruption of power.

To circumvent the problem of line faults going undetected, a device known as a *line isolation monitor* (LIM) is used as part of an isolated power system. The purpose of this device is to trip an audible alarm and a signal light if the impedance of either side of the isolated power to ground falls below a predetermined level. The LIM indicates the degree of isolation from ground of energized conductors in the isolated portion of the power system.

By monitoring the impedance to ground of the isolated power side of the electrical system, the LIM can be calibrated to indicate what current would flow to ground in the system if a fault were to happen. So, instead of being calibrated in terms of impedance, the LIMs have a meter that reads in current (mA). The current that is indicated on the meter of the LIM is the current that would flow to ground if a low-impedance fault were to connect one side of the power system to ground.

It must be emphasized here that the current value that is indicated by the LIM current meter is *not* the current that is flowing to ground at the present moment; it is indicating the current that will flow to ground if there is a low-impedance fault on the system. The LIM only predicts the flow of current in the event of a low-impedance fault. Most LIMs are designed to trigger at an impedance that will produce 2 mA when a low-impedance fault comes on the system. The LIM does not measure fault current directly!

The need for the LIM to be an impedance-measuring device is not apparent from low impedance or "shorts" to ground. It would seem that a simple "go/no go" type of detector should be sufficient. As mentioned earlier in this chapter, every electrical device has associated with it unwanted, stray, or leakage capacitance. This capacitance represents a 60-Hz impedance to ground for each device connected to the isolated power system. Figure 15-17 shows an equivalent circuit of the isolated system with a number of different electrical devices connected to it.

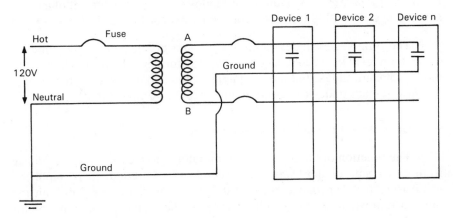

Figure 15-17 Parallel capacitance added to ground by each device added to the isolated power transformer.

As each new device is added to the isolated power system, it produces additional capacitance to ground. Since these capacitances are in parallel, the capacitance is additive. Since capacitive impedance is proportional to the reciprocal of capacitance $(1/j\omega c)$, as each device is added to the isolated power system, the impedance to ground is reduced. A point can be reached when enough capacitance on the power system to ground is reached that if a low-impedance fault were to take place, excessive currents would flow to ground. The current will be limited by the total impedance to ground from the power sides of the isolation transformer. The smaller this impedance is, the greater the current that will flow if there is a short to ground.

Now we can see why the LIM is an impedance-measuring device. An isolated power system can be compromised by the added capacitance to ground of the devices that are plugged into it. Thus LIMs measure impedance and are set to alarm when the impedance to ground is such that 2 mA will flow in the ground circuit if a "short" comes on the system. The impedance to ground under this condition is 60 kΩ.

At this point it is advantageous to define some terms that are used in conjunction with line isolation monitors.

Total hazard index: The total hazard index is defined as the total current, in milliamperes, that would flow to ground through a low-impedance fault if it were connected between either isolated conductor and ground. The total hazard index is made up of two components: the monitor hazard index and the fault hazard index. A LIM is usually set to alarm for a total hazard index of 2 mA.

Monitor hazard index: The monitor hazard index is the hazard current due to the LIM with no other devices plugged into the isolated power transformer.

Fault hazard index: The fault hazard index is the hazard current of a given isolated system with all devices connected except the LIM.

The total hazard index limits the amount of equipment and length of cable than can be used per isolation transformer. Thus the lower the monitor hazard index, the more equipment that can be used on the system and still remain under the 2-mA limit for total hazard current.

Operation of the Line Isolation Monitor

The function of the line isolation monitor is to measure the impedance from either side of the transformer secondary to ground. It is important to measure this impedance because it is this impedance that will limit the fault current when there is a fault to ground. The ideal line isolation monitor should represent an infinite impedance to ground and therefore have a zero monitor hazard current. It should also have an output current proportional to the total hazard index of the system. The zero monitor hazard current ideal can be approached with today's new types of LIMs, but the second requirement, of current output proportional to the ground current which is predicted but does not exist at the time the measurement is being made, is a difficult engineering problem to solve.

The analysis of different types of LIMs will be limited to detectors suitable for two-wire single-phase ac systems, which is typical of the power used in most areas of the hospital. We will consider resistive and capacitive ground-fault impedances in eight different configurations. Inductive ground-fault impedances are rare and therefore will not be considered here.

We will consider the following eight types of R, C, and RC combinations that form ground-fault impedances:

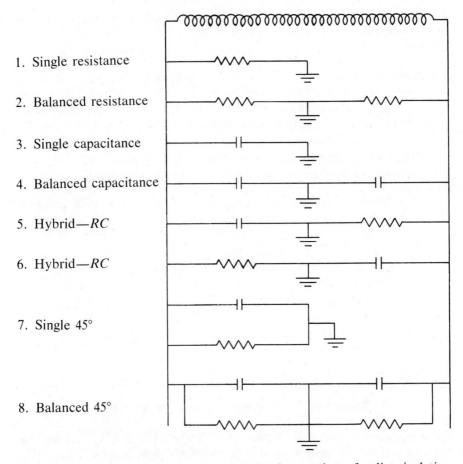

1. Single resistance

2. Balanced resistance

3. Single capacitance

4. Balanced capacitance

5. Hybrid—RC

6. Hybrid—RC

7. Single 45°

8. Balanced 45°

To provide some insight into the theory of operation of a line isolation monitor, we will consider the first type of LIM used, the static LIM.

Figure 15-18 is a diagram of a simple detector system. We assume

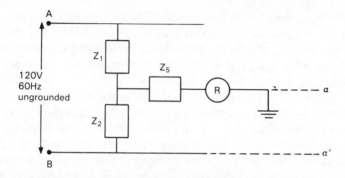

Figure 15-18 Simple static line isolation monitor. R, relay.

that the detector current is the same whether a fault occurs from one side of the isolation transformer secondary or the other. This consideration determines the relationship between Z_1 and Z_2 in the following way.

1. Detector impedance is the same from one power line to the other to ensure equal detection sensitivity to the same fault from either line. Z_1 and Z_2 can be resistive, capacitive, or both. For the latter case, the magnitude of the impedances should be the same so that the phase angle of the detector will be $-45°$.
2. The overall detector impedance should have a phase angle of $-45°$ to ensure equal sensitivity to purely resistive or capacitive faults.

The Thévenin equivalent circuit for the LIM looking in at points a and a' is shown in Figure 15-19. If there is a fault to ground, where Z_f is

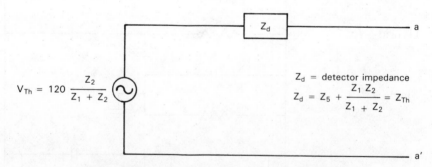

$$V_{Th} = 120 \frac{Z_2}{Z_1 + Z_2}$$

Z_d = detector impedance

$$Z_d = Z_5 + \frac{Z_1 Z_2}{Z_1 + Z_2} = Z_{Th}$$

Figure 15-19 Thévenin equivalent circuit for static LIM.

the impedance of the fault, Figure 15-20 is the equivalent circuit for the fault condition. In most static LIMs, the detector is a relay adjusted to trip when fault conditions could result in a current flow of 2 mA. The relay is used to activate a signal lamp and an audible alarm to attract the attention of those working in the area. This condition arises when a fault impedance

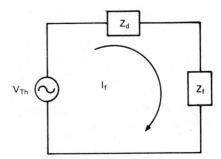

Figure 15-20 Thévenin equivalent circuit
for a fault, Z_f, to ground on an isolation
transformer using an LIM.

of 60,000 Ω shunts either isolated line to ground. The maximum current
that will flow if a second short to ground comes on the system is 2 mA.
This is the basic configuration that conformed to standard NFPA 56 before
1970.

If there is a fault or a change in the impedance from either side of
the isolation transformer to ground, the current through the detector can
easily be computed using the equivalent circuit of Figure 15-20. The detector
can be set to respond to a given current level corresponding to the impedance
of the single fault or change in impedance to ground.

Balanced Fault Condition

When we have a fault or change in impedance from both sides of the
isolation transformer to ground, situations might occur that can go undetected
by the static LIM. Consider the fault conditions in Figure 15-21. When
there are faults or impedances from both sides of the isolation transformer
to ground, as shown in Figure 15-21, the ground connection acts as a

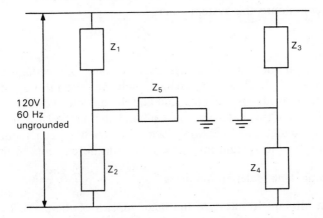

Figure 15-21 Balanced fault condition showing how a static LIM might not
detect this class of fault condition on an islation transformer circuit.

common node between the impedances Z_3 and Z_4. This means that each of these impedances are connected together at a common ground point. This is illustrated in Figure 15-22. In this configuration, the faults to ground, together with the impedances of the LIM, make up a Wheatstone bridge. The current through the detector will depend on the values of the impedances of the LIM portion of the detector system. The LIM acts as part of the Wheatstone bridge and the faults to ground become the other arms of the bridge circuit.

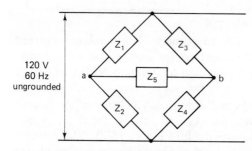

Figure 15-22 Balanced-bridge representation of the balanced fault condition of an isolation transformer with a LIM.

The voltage across the detector portion of the bridge will be the voltage difference between points a and b in the bridge circuit. If the impedances Z_3 and Z_4 are not proportional, the bridge is "unbalanced" and there is a detectable voltage difference between nodes a and b. This voltage is a function of the impedances making up the arms of the bridge and is given by

$$\Delta V = \frac{Z_2}{Z_1 + Z_2} - \frac{Z_4}{Z_3 + Z_4} \tag{15-2}$$

Equation (15-2) is for the unbalanced condition. It is possible that a balanced fault can come onto the system where

$$\frac{Z_1}{Z_2} = \frac{Z_3}{Z_4} \tag{15-3}$$

Under this condition the bridge is balanced and the voltage across the detector is zero. If Z_3 and Z_4 change such that the ratio Z_3/Z_4 remains unchanged, the bridge is still balanced and the detector voltage is still zero. Balanced faults of this nature produce a blind spot in the LIM and represent a potentially dangerous situation since isolation has been reduced and a second fault on the system may now allow unsafe currents to flow.

Dynamic Line Isolation Monitors

In an attempt to alleviate the inadequacies of the static LIM, a line isolation monitor that could detect balanced faults was developed in Canada and called a *dynamic line isolation monitor*.

Initially, dynamic LIMs were characterized by the use of switching elements using fairly slow mechanical switches, switching from 2 to 37 times per second. Figure 15-23 illustrates a dynamic LIM schematic. With

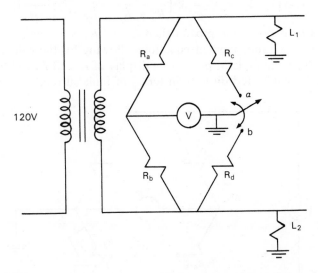

Figure 15-23 Dynamic LIM using mechanical switching to detect balanced faults.

the switch at point a, the equivalent circuit for the LIM can be drawn as in Figure 15-24. The voltage difference $(V_1 - V_2)$ is given by

$$\Delta V = V_1 - V_2 = v\left(\frac{R_b}{R_a + R_b} - \frac{L_2}{L_2 + R_c//L_1}\right) \qquad (15\text{-}4)$$

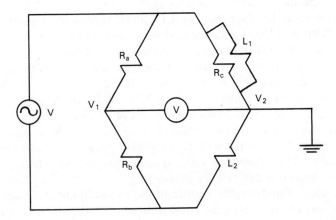

Figure 15-24 Equivalent circuit for the dynamic LIM of Figure 15-23 when the switch is in position a.

With the switch at point b the equivalent circuit becomes that given in Figure 15-25. The voltage difference for this case is given by

$$\Delta V = V_1 - V_2 = V\left(\frac{R_a}{R_a + R_b} - \frac{L_1}{L_1 + L_2//R_d}\right) \qquad (15\text{-}5)$$

As long as R_c is not much greater than L_1 and R_d is not much greater than L_2, there are two independent equations allowing for the calculation of L_1 and L_2. The dynamic LIM provides a display of L_1 and L_2 and the detector meter can be calibrated directly in terms of fault current.

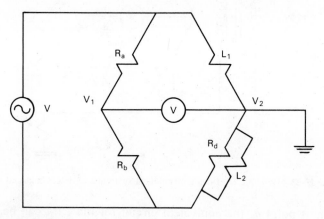

Figure 15-25 Equivalent circuit for the dynamic LIM of Figure 15-23 with the switch in position b.

Even though this second-generation line isolation monitor solves the problem of detecting the presence of balanced faults to ground, this type of monitor produces other problems that we must consider. The first of these problems is the fact that the LIM itself represents a partial fault to ground. With nothing else connected to the isolated power system except for the LIM, the LIM impedance to ground is such that for a second low-impedance fault from either side of the isolated power line to ground, 1 mA of current will flow in the system. If the system has a capacity of 2 mA, LIM alarms at 2 mA; then half the capacity of the isolated system is used by the LIM. This limits the amount of medical equipment that can be placed into service on one isolated system and might require that the operating room be supplied with two separate isolated systems rather than with just one. This is obviously an expensive proposition; thus a LIM with a very low hazard index is very desirable.

The second undesirable feature of the dynamic line isolation monitor is that the switching creates electrical transients on the isolated system that can interfere with physiological monitoring equipment. Sometimes the noise generated by the dynamic LIM makes it difficult to use monitoring equipment, and in extreme cases of LIM-generated noise, the LIM must be switched off for a reading to be taken. Before this is done, it has been found that interference from the LIM can be substantially reduced or eliminated by applying either fresh electrodes to the patient or fresh paste if nondisposable electrodes are being used. If changing the electrodes does not reduce the interference problem and the LIM must be turned off for appropriate monitoring of the patient, it is important that the clinical engineer ensure that the clinical personnel are aware that the LIM is disabled and that they understand that line faults will no longer produce an alarm.

The extent of difficulty encountered by the clinical engineer relative to LIM-generated interference will vary widely with the installation and design of the power system and the physiological monitoring equipment.

Third-Generation Line Isolation Monitors

To overcome the difficulties found in static and dynamic LIMs, a third-generation line isolation monitor, such as the Iso-Gard* LIM, was developed. This form of LIM continuously monitors both sides of the transformer line with no switching. The device contributes only about 25 μA to the hazard index. Let us look briefly at the theory of operation of a third generation LIM.

Consider the system illustrated in Figure 15-26. V_1 and V_2 are the voltages between the corresponding lines and ground and Y_1 and Y_2 are

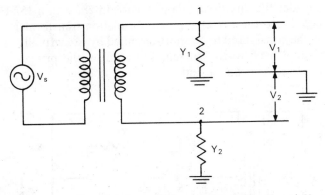

Figure 15-26 Circuit model for a third-generation LIM. V_1 and V_2 are the voltages to ground of each line; Y_1 and Y_2 are the leakage admittance.

leakage admittances. A reduced equivalent circuit for the system is shown in Figure 15-27. Computing the hazard indices from each side of the line

* Iso-Gard is a trademark of the Square D Company.

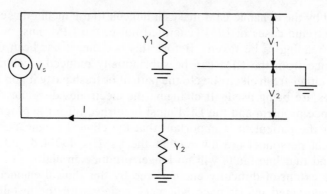

Figure 15-27 Reduced equivalent circuit for the circuit in Figure 15-26.

to ground, we have

$$\text{If line 1 were grounded, } HI_1 = V_s Y_2.$$

$$\text{If line 2 were grounded, } HI_2 = V_s Y_1.$$

The total hazard index is given by the expression

$$HI = HI_1 + HI_2$$

Define

$$Y_0 = Y_1 + Y_2$$

$$HI = HI_1 + HI_2 = V_s Y_0 \tag{15-6}$$

We now consider the circuit shown in Figure 15-28, with a 152-Hz current source injected into the center tap of the isolation transformer. Since the primary may be considered to be short-circuited by V_s, virtually zero impedance is reflected back to the current source from the primary. Thus the ΔV appearing between line 1 and the ground is equal to the ΔV between

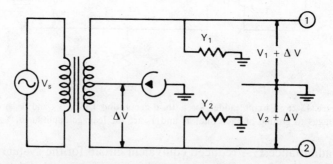

Figure 15-28 Third-generation LIM with 152-Hz signal injected into the secondary center tap of the isolation transformer.

line 2 and the ground. Holding ΔV as a multiple of V_s, we have

$$\Delta V = KV_s$$

$$K = \text{constant}$$

Then

$$\Delta I = \Delta V Y_0 = KV_s Y_0 = K(\text{HI})$$

Calculating the hazard index as in the preceding example, we have

$$\text{HI}_1 = V_s Y_2 = (V_1 + V_2)Y_2$$

$$\text{HI}_1 = V_1 Y_2 + V_2 Y_2$$

$$V_1 Y_1 = V_2 Y_2$$

$$\text{HI}_1 = V_1 Y_2 + V_1 Y_1 = V_1(Y_1 + Y_2) = V_1 Y_0$$

HI_2 can be determined in a manner similar to HI_1. If ΔV is held at KV_1, $\Delta I = \Delta V Y_0 = KV_1 Y_0 = \text{KHI}_1$. Thus HI_1, HI_2, or the total hazard index may be displayed. The current source is usually at 152 Hz to make ΔV easier to measure. A 60-Hz notch filter is used to block V_s from the measurement. The capacitive component of the measured ΔI must then be multiplied by 60/152 before it can be added to the resistive component of ΔI and displayed.

The foregoing process can be summarized as an application of Thévenin's theorem as follows. A ground-fault current can be predicted for any point in a network made up of linear impedance elements by measuring the voltage existing at that point due to the normal sources in the network and dividing that value by the value of the impedance (reciprocal sum of all leakage admittances) to ground. This impedance is measured at that point by use of a separate measurement current source and with all normal sources in the network reduced to zero (considered to be shorted out). The latter condition is accomplished by an impedance measurement made at a different frequency and consequently is unaffected by the line frequency voltages present.

PERFORMANCE TESTING OF ISOLATED POWER SYSTEMS

The testing of an isolated power system can be broken down into testing of its two components, the isolation transformer and the line isolation monitor. The testing of the isolation transformer consists of determining the line-to-ground impedance of each of the power conductors. This can be done by first disconnecting all equipment from the isolated system, including the LIM. At one of the receptacles, connect a 0 to 1 milliammeter between either line to ground and measure the current. Using Ohm's law, we obtain the impedance by dividing the system voltage by the measured current. The impedance should be greater than 500 kΩ from either side of the power system to ground. For a 120-V system, the measured current should be no greater than 240 μA.

In testing the line isolation monitor, it is important to determine the impedance at which the LIM trips. To do this, an adjustable resistance is inserted from one line to ground with the resistance set at its maximum value. This value of resistance should be greater than 60,000 Ω. Figure 15-29 is a diagram of the test procedure. The resistance is then gradually

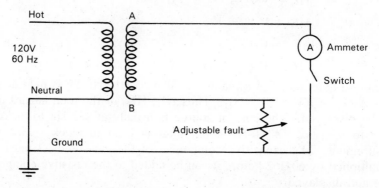

Figure 15-29   Circuit diagram illustrating a test setup for a line isolation monitor.

reduced, decreasing the line-to-ground resistance. At some value of resistance the LIM will alarm. At this point a milliammeter is connected from the other line to ground. This acts as a low-impedance fault and the current through the meter is the current at which the LIM alarmed. This should be within 10% of the current limit for the LIM. In some cases the resistor itself can be calibrated to read fault trip current directly. Figure 15-30 is typical of the commercial devices that can be used for LIM testing that operate on the same ideas presented here. Figure 15-31 presents models of an inspection form that can be used by the clinical engineer or his or her designate for recording the test results on the LIM.

Periodic testing of isolated power systems should be part of the safety and performance program for medical equipment. The "push to test" button should be depressed and the associated alarms and silence functions verified about once a week. When testing the LIM internally, the point on the scale reached when the alarm is triggered, and whether the alarm goes off, should be noted. The LIM may have failed or be incorrectly calibrated if there is a visible or audible alarm and the value on the hazard current meter does not indicate being at the alarm threshold during the test. LIM testing, using your own test device or a commercial device, should be performed about once a month with all breakers closed and all breakers in

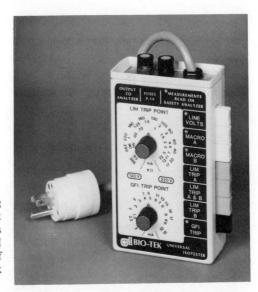

Figure 15-30 Test device used for testing isolated power systems and ground-fault circuit interrupters. Test device provides an adjustable fault for line *A* or *B*, LIM trip point line *A* or *B* for line voltage of 120 and 220 V. (Courtesy of Bio-Tek Instruments, Inc., Burlington, Vt.)

the open position. This will provide a history for the permanently installed isolated system. Should there be any gross changes in the performance of the system, corrective action should be taken.

When dealing with isolated power systems and the line isolation monitors, the clinical engineer should convey to the clinical personnel that an alarm situation is not one for *panic*. Although the urgency of the alarms seems to indicate that a life-threatening situation exists, it usually does not. Its significance is simply that some form of electrical abnormality probably exists which may increase the electrical hazard risk if a second fault occurs before the first fault is cleared.

When an alarm condition exists, do not forget the patient. Determining the cause of the alarm can wait until the patient no longer requires attention. The relative severity of the fault can be determined from the meter on the LIM. If the needle is just into the threshold of alarm area, it is possible that the "fault" is nothing more than the accumulated effect of the inherent capacitance of each of the devices reducing the impedance to ground. If the needle is at the upper extreme, this would indicate that there is a substantial fault on the line.

Without jeopardizing the patient, disconnect each device, one at a time, and note if any one device substantially changes the total hazard index on the LIM meter. Nonelectrical devices that are attached to the isolated power ground can be ignored, as they are not responsible for the fault condition.

Isolated Power System Inspection Form

ACTION
Not Needed ☐ Needed ☐ Taken ☐
Removed From Service ☐
Control No.
Date of Inspection:
Next Inspection Due:

PANEL: LOCATION _____ MANUFACTURER _____

      OUTPUT: _____ VOLTS: _____ AMPERES: _____kVA

      POWER RECEPTACLE TYPE: _____

LIM: LOCATION _____ MANUFACTURER _____

      MODEL _____ SERIAL NO. _____

      TYPE: Dynamic ☐   Static ☐   ALARM LEVEL _____ mA

LOCATION OF PRIMARY OVERCURRENT PROTECTION: _____

LOCATION OF OTHER REMOTE INDICATORS: _____

PERMANENTLY CONNECTED LOAD: _____

ROOM LAYOUT

PANEL LAYOUT

	OK	ACTION NEEDED	ACTION TAKEN (Date & Initials)
1. LAMPS: _____			
2. METERS: _____			
3. ALARMS & SILENCING FEATURE: _____			
4. FUSES: _____			
5. CIRCUIT BREAKERS: _____			
6. GROUNDING JACKS: _____			
7. LINE VOLTAGE: _____			
8. ALARM LEVELS: _____			

RESISTANCE CONNECTED TO	RESISTANCE AT ALARM THRESHOLD	LIM METER READING	METER CONNECTED TO	HAZARD CURRENT
Line 1			Line 2	
Line 2			Line 1	

Figure 15-31 Typical clinical engineering test and inspection form for an isolated power system. (From *Health Devices,* Emergency Care Research Institute, Plymouth Meeting, Pa.)

Isolated Power System Inspection Form, page 2

	OK	ACTION NEEDED	ACTION TAKEN (Date & Initials)
9. SYSTEM LEAKAGE:			
LIM Reading, no connected load:			

* LEAKAGE CURRENT METER READINGS:

LIM Operating	Meter in Line 1	
	Meter in Line 2	
LIM Disabled	Meter in Line 1	
	Meter in Line 2	

	OK	ACTION NEEDED	ACTION TAKEN
*10. GROUNDING JACKS (Operational):			
*11. GROUNDING OF EXPOSED METAL:			
12. RECEPTACLES (Circle unacceptable data):			

Other inspection form used for parallel-blade receptacles

BREAKER NO.											
RECEPTACLE NO.											
PHYSICAL CONDITION											
* POWER CONTACTS											
* GROUND CONTACT											

*Need not be tested at each inspection. Should be tested before: _____ (Date)

	OK	ACTION NEEDED	ACTION TAKEN (Date & Initials)
OTHER TESTS (Describe):			

COMMENTS & DESCRIPTION OF DEFICIENCIES (Refer to item numbers): _____

☐ INSPECTION TAG AFFIXED

☐ EQUIPMENT SERVICE REQUEST FORM COMPLETED

INSPECTED BY: _____

Figure 15-31 *(cont.)*

ISOLATED POWER IN PATIENT CARE AREAS

Isolated power systems were originally used in the operating suite to reduce the risk of ignition of flammable anesthetic agents that were currently being used. We have seen that isolated power systems can limit the flow of current in a fault to 2 mA. An arc of this current level generated by a fault to ground would not contain sufficient energy to ignite flammable vapors. This was one of the motivations behind the development of the standard 56A, "Standard for the Use of Inhalation Anesthetics," by the National Fire Protection Association.

Some of the other benefits of an isolated power system are the reduced macroshock hazard provided by the system on a first-fault basis and the fact that heavy grounding conductors are not necessary to prevent first-fault currents from developing high-voltage differences between grounded but exposed chassis and conductive surfaces. This is due to the fact that ground currents are small in this situation.

Because of these positive benefits it was thought that using isolated power in patient care areas would provide a "safer" electrical environment for both the patient and clinical personnel. In one sense this is true, but only if the economics of the situation can be totally ignored. There are two considerations here, cost/benefit and risk/benefit. The cost of installation of an isolated power system at each patient bedside is many times the cost of a grounded power system. There is also the cost of maintaining these systems once they are installed as well as the education of clinical personnel on their use.

There is also the consideration of the microshock hazard for certain patient areas. Even though the isolated power system can reduce the macroshock hazard, it has little effect on microampere currents that flow in the system. If there is a fault on the line, currents from 400 to 2000 μA can flow in the ground line of the system and thus offer no protection against microshock.

Presently, the standard NFPA 56A specifies that isolated power systems, sometimes referred to as *ungrounded power systems*, be used in hospitals under the following conditions:

a. Where the possibility of wet conditions is present.
b. Where three or more line-operated devices are normally connected to the patient.
c. Where invasive procedures that cannot be terminated quickly are performed.
d. In flammable anesthetizing locations.

The primary advantage of using an isolated power system is its ability to reduce the probability of bringing a flammable atmosphere to the ignition point with an electrical spark. Secondarily, there is the assurance of power continuity in the event of a first line-to-ground fault. The ability of the system to reduce potential differences between grounded surfaces due to ground line current flow in a fault condition is of small practical value when good grounding is assured via electrical safety practices followed in the hospital.

PROBLEMS

1. Define the following terms:
   (a) Essential electrical system
   (b) Emergency electrical system
   (c) General electrical system
   (d) Life safety branch of the electrical system
   (e) Critical branch of the electrical system
   (f) Life support branch of the electrical system
   (g) Critical care area, controlled
   (h) Critical care area, uncontrolled
   (i) General care area
   (j) Anesthetizing location, hazardous
   (k) Anesthetizing location, other than hazardous

2. (a) Describe the gross physiological effects of electric current on humans as a function of increasing current level, for patients with uncompromised skin protection and patients with indwelling catheters.
   (b) Compare the relative level of hazard for each class of patients.

3. Hospitals use three-wire 120-V electrical power for general use. What are the three wires, and what is the function of each wire? Why is a three-wire 120-V power system "safer" than a two-wire power system? Sketch a three-wire system and all the component parts.

4. Can a three-wire power system protect a patient from macroshock? Microshock? Describe under what conditions a patient may not be protected by such a three-wire system.

5. Draw a diagram of an isolated three-wire power system. Explain what an isolation transformer is and what function it serves in an isolated power system. Where in the hospital environment is an isolated power system used? What is the rationale for using it? (Include in your answer such information as what happens in this system under fault conditions. Include the theory of operation as to how an isolated power system provides certain measures of protection under these conditions.)

6. What is a line isolation monitor (LIM)? Why is it used in conjunction with an isolated power system?
   (a) Explain the difference between a "static" LIM and a "dynamic" LIM.
   (b) Why is a dynamic LIM used in preference to a static LIM?
   (c) Outline the basic theory of operation of the LIM.

7. What is leakage current? Why is it of concern in patient care areas? Does an isolated power system reduce the leakage current hazard? Draw an equivalent circuit to illustrate your answer.

8. An isolated power-line pair has a capacitance to ground of 0.025 μF. How much resistive leakage to ground can be tolerated before the total hazard index of 2 mA is reached? Assume a LIM hazard index of 0.25 mA.

9. A "low-leakage-design" electrical cable consists of three 1-mm-diameter conductors situated at the corners of an equilateral triangle 5 mm on a side. The insulation is polytetrafluoroethylene with a dielectric constant of 2.1. Calculate

the capacitive leakage current from the hot to the ground wire per foot of cable for a 120-V ac power source.

10. Given the isolated power system in Figure 15-32, which includes a line isolation monitor, where there is a 50-kΩ capacitive fault on the system due to equipment on the line. At the same time there is a short in a piece of surgical equipment where surgical personnel also represent a 15-kΩ fault to ground. Determine the current through the detector under these conditions.

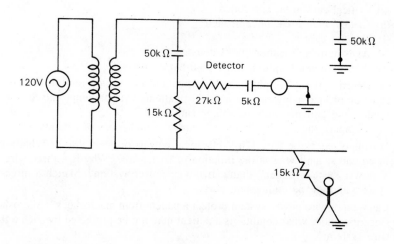

Figure 15-32

11. For the two static LIM circuits given in Figure 15-33, calculate the fault hazard index, relay current, and the total hazard index for the two different fault conditions shown.

12. Two 120-V ac receptacles, R_1 and R_2, are 10 ft apart in a patient's room. R_1 is 30 ft from the distribution panel. The distribution panel is 100 ft from the power transformer. All the wiring is done with AWG No. 10 wire. Two instruments are connected to the line by 10 ft of three-wire No. 16 power cord with the chassis grounded. A short between the hot wire and the chassis occurs in the device plugged into R_2. What will be the maximum potential difference between the two chassis before the breaker is tripped out, clearing the fault?

   (a) Describe how this could be a hazard to a patient who is connected to both of these devices.

   (b) Draw an equivalent circuit for this situation and show how the hazard is presented to the patient.

   (c) Show how the use of an isolation transformer might have reduced the hazard to the patient.

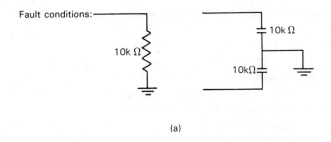

Fault conditions:

10k Ω

10k Ω

10kΩ

(a)

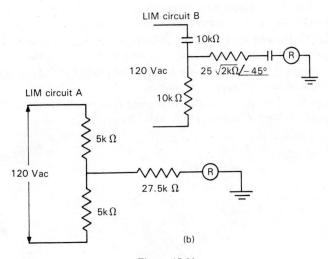

LIM circuit B

10kΩ

120 Vac   25 $\sqrt{2k\Omega/-45°}$

10k Ω   R

LIM circuit A

5k Ω

120 Vac   27.5k Ω   R

5k Ω

(b)

Figure 15-33

REFERENCES

1. Dalziel, C. F., and W. R. Lee, "Lethal electric currents," *IEEE Spectrum,* Feb. 1969, pp. 44–50.

2. Dalziel, C. F., "Electric shock hazard," *IEEE Spectrum,* Feb. 1972, pp. 41–50.

3. Dalziel, C. F., "Electric shock," in *Advances in Biomedical Engineering* (New York: Academic Press, 1973), pp. 223–248.

4. "Electrical safety," *Health Devices,* Jan. 1974.

5. Friedlander, G. D., "Electricity in hospitals: elimination of hazards," *IEEE Spectrum,* Sept. 1971, pp. 40–51.

6. Hopps, J. A., "Shock hazards in operating rooms and patient-care areas," *Anesthesiology,* Vol. 31, No. 2, 1969.

7. Kuster, N. L., "The ground detector problem in hospital operating rooms," *Trans. EIC,* Jan. 1958.

8. McKinley, D. W. R., "An electronic ground detector," *Trans. EIC,* Jan. 1958.

9. NFPA 56A—1973, "Standard for the Use of Inhalation Anesthetics" (Boston: National Fire Protection Association, 1973).

10. NFPA 76B—1977, "Standard for the Safe Use of Electricity in Patient Care Areas of Hospitals" (Boston: National Fire Protection Association, 1977).

11. Shepherd, Marvin D., "Insulation and ground testing in safety programs," *Med. Electron. Data,* May–June 1976, pp. 43–48.

12. Square D, "Hospital Isolating Systems," The Square D Company, Bulletin D-41/30M, June 1976.

13. Stanley, Paul E., "Hospital Electrical Systems," in *CRC Handbook of Clinical Engineering,* Barry N. Feinberg (ed.) (Boca Raton, Fla.: CRC Press, 1980).

14. Starmer, C. F., and R. E. Whalen, "Current density and electrically induced ventricular fibrillation," *Med. Instrum.,* July 1973, pp. 158–161.

15. Turner, James B., "Isolated power and the LIM: maligned and misunderstood," *Actual Specifying Engineer,* May 1974.

16. Von der Mosel, H. A., et al., "Transformer isolated electrical supply systems in hospitals," *Med. Electron. Data,* May–June 1970.

16

ELECTRICAL PERFORMANCE AND SAFETY TESTING OF ELECTROMEDICAL EQUIPMENT

We have touched on the concept of electrical safety in Chapter 15 when dealing with the topics of macroshock and microshock. We have seen that one of the most important aspects of electrical environmental safety is that a good ground be maintained at all times in the patient areas. Actually, this holds true for the entire hospital. The ground wire will carry away fault current so that the chassis voltage cannot rise to a dangerous level above ground potential. It will also carry away leakage current in the same way, so that chassis voltage once again does not rise to a level that would be dangerous to electrically susceptible patients.

Since grounding, both equipment ground and power ground, is important to the patient's electrical environment, they both must be tested as part of an electrical safety program. The rest of the electrical environment includes all permanently built-in equipment, wiring, receptacles, metal surfaces which are part of the electrical equipment or exposed metal parts, and the plugs and line cords of the devices. Besides the power side of a device, there must be testing of the level of device leakage current. In terms of electrical safety, variations on the testing for proper grounding and leakage current levels constitute the majority of the electrical inspections performed on medical devices.

MECHANICAL TESTING OF ELECTRICAL OUTLETS

Let us now consider the power delivery point in the patient area, which usually consists of the outlets in the vicinity of the patient. To start with, these should be a good-quality three-prong duplex wall receptacle that meets the National Electrical Manufacturers' Association (NEMA) ground retention force requirements. These force requirements are important, as they assure that the plugs on medical devices do not fall out of the receptacle, possibly placing the patient in danger.

As with medical devices, the first step in a safety inspection is to inspect the receptacle visually, looking for possible damage, such as cracks or burns on the ground or power contacts. The receptacle should also be checked for a loose mounting box and loose screws. Since there usually are a number of electrical receptacles in a patient room, it is wise to draw a diagram on the inspection sheet of the location of the receptacle under test. If a receptacle is found to be defective at this point, it should be tagged for replacement before further tests are performed on it.

Further testing of the outlet requires tension testing. Good grounding contacts and holding force on the wall outlet are essential for safety of the patient and other medical personnel. The holding tension provided by a set of contacts in the outlet can be measured with a spring-loaded tester that measures the ounces of force required to extract the device after it is inserted into the outlet. A minimal mechanical retaining force for each of the three contacts in a receptacle is about 115 g or 4 oz (see NFPA 76B). This permits a total mechanical holding force of ¾ lb between the plug and the receptacle. Tension-testing devices can be purchased from a number of companies that sell hospital testing equipment.

One might be tempted to think that if the tension in the outlet were higher than the minimum stated above, this is a good situation. This is not the case, however, as people in their own ways will destroy a plug, electrical cable, or the receptacle when the tension is too high. It is possible that someone will yank on a cord to pull it out of a receptacle. If the plug is held in place, the cord may be pulled out of the plug or a conductor partially opened. If there is a strain relief on the plug, a tug of war will ensue.

A second scenario is when portable electronic devices are on a cart and plugged into the outlet. Then someone rolls the device away from the patient without unplugging it. If the receptacle tension is too high the plug will remain in the outlet and the device will be yanked off the cart, with disastrous results. There is also the possibility that this now-infamous someone will trip on an electrical cord. In this case, if the plug does not come out of the outlet, the person may be injured and/or the receptacle will be damaged along with the plug.

It is suggested in the standard NFPA 76B that the receptacles be tested in general care areas and in wet locations once a year and in critical care areas twice a year.

ELECTRICAL TESTING OF ELECTRICAL OUTLETS

Electrical testing of a wall receptacle consists of determining whether the power is at the receptacle and if its polarity is correct. Consider Figure 16-1. The proper polarization of the common electrical outlet means that the hot, neutral, and ground wires are connected to their respective positions as shown in Figure 16-1. Miswiring of an outlet can happen during the original construction of the area or when a broken outlet is replaced. The outlet can easily be tested using an outlet tester consisting of three LEDs or lights. This device can test for 8 of 64 possible states of the outlet. Since the light is a binary device, it is either on or off, and each of the three contacts in the outlet can have four states (open, ground, neutral, and hot). Testing of the outlet for possible reverse wiring or loss of grounding is important, as these are potentially hazardous conditions and should be corrected immediately.

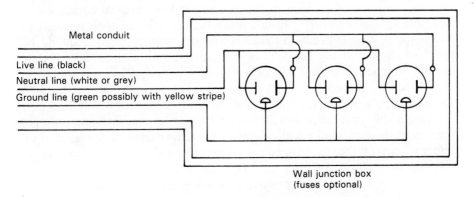

Metal conduit

Live line (black)

Neutral line (white or grey)

Ground line (green possibly with yellow stripe)

Wall junction box
(fuses optional)

Figure 16-1 Wiring for a common single-phase electrical outlet.

TESTING OF ELECTROMEDICAL EQUIPMENT

A testing program for medical electronic equipment is composed of five parts:

1. Documentation
2. Visual inspection
3. Safety
4. Performance testing
5. Tagging

Documentation and tagging of an electromedical device are part of the equipment control record system. Such a system can be the most simple 3 × 5 card file system up to the elegance of a computer data base system.

The minimum information that should be contained in a document created for a new device coming into the hospital includes:

1. Type of equipment
2. Name of the manufacturer
3. Serial number
4. Model number
5. Location of the device in the hospital

Other possible items to be included are where service can be obtained, the salesperson's name, and warranty information.

There should be two copies of the instruction manual and a service manual with the device. One instruction manual should stay with the device and the second should be put on file in the clinical engineering department of the hospital. A service manual should be obtained for each device so that the information needed to provide service for medical equipment is available to the engineer. To force manufacturers to sell their manuals to the hospital, two instruction manuals and one service manual should be required as part of the purchase specifications, with a clause that if they are not supplied, payment will be withheld.

Since no piece of medical equipment should come into the hospital for use in patient areas without being first inspected for safety and performance, this procedure should always be part of the incoming protocol for electromedical equipment. Safety and performance testing is also an integral part of the electromedical equipment safety program that is an ongoing function in the hospital. Thus the procedures outlined here are valid for ongoing inspection protocols for equipment in place in the hospital as well as for incoming equipment.

The physical condition of the device is usually a good barometer of how well the medical equipment has been treated. Thus the first part of a safety inspection is to inspect the equipment visually. One should look for dents, cracked or bent controls, and cracked case or meter faces. The power cord should be inspected for bent blades in the plug and cracked or frayed insulation. Strain relief must be present at the plug and the line cord entry into the chassis of the device. Any leads that are part of the equipment must be inspected for cracked or frayed insulation. The results of the visual inspection should be recorded on a device inspection form.

ELECTRICAL SAFETY TEST PROCEDURES

There are three test procedures that are common to all electrical devices in the hospital:

1. Ground wire resistance
2. Leakage current measurement
3. Isolation tests

The tests that will be described here can be performed using commercially available test equipment. Figure 16-2 is typical of the type of electrical test devices on the market specifically designed for medical device electrical safety testing. The test equipment for measuring the ground pin to chassis

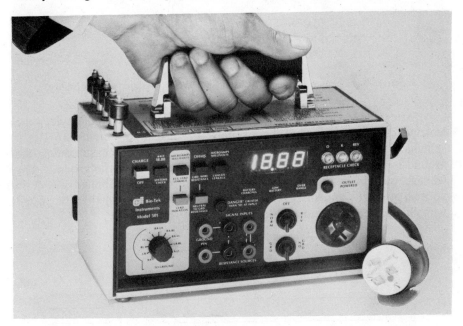

Figure 16-2 Electrical safety analyzer, typical of the test instruments specifically designed for medical device electrical safety testing. (Courtesy of Bio-Tek Instruments, Inc., Burlington, Vt.)

resistance should be set up as shown in Figure 16-3. With electrical safety test equipment or an ohmmeter that can read values less than 1 Ω, measure the resistance from the ground pin on the plug to several locations on the chassis. The lowest and highest resistances should be recorded on an inspection form. The resistance from any point on the chassis to the ground pin on the plug should not exceed 0.15 Ω as specified in the standard NFPA 76B. This test will determine if there is a break in the ground wire or if the internal device ground connection is bad or corroded. The implication of too high a ground wire resistance is that when current flows in the ground wire, the IR drop can produce a voltage on the chassis of the device that may cause problems in areas with electrically sensitive patients.

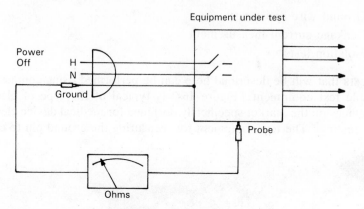

Figure 16-3 Schematic diagram illustrating the testing of ground wire resistance.

After the measurement of grounding resistance of the device, the leakage currents from the chassis must be measured. Figure 16-4 is a schematic of the measurement configuration. For this measurement, the ground is opened and the microammeter is inserted into the ground. The leakage current is then read with the device off, with applied power in normal position, and in reverse polarity. The device will then be tested with the power on again with normal and reverse polarity.

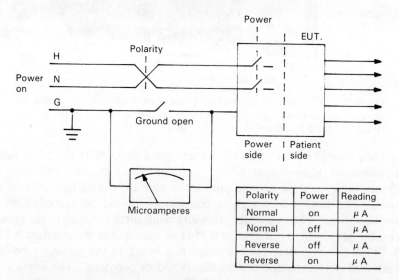

Polarity	Power	Reading
Normal	on	μ A
Normal	off	μ A
Reverse	off	μ A
Reverse	on	μ A

Figure 16-4 Schematic of the test configuration for the measurement of leakage current. The leakage current is measured in the ground line with line polarity normal and reversed.

The rationale for testing chassis leakage current with reverse polarity is not obvious and takes some explanation. First, when chassis leakage current is measured in reverse polarity where the hot wire of the device is connected to the neutral of the power receptacle, and the neutral of the device is connected to the hot wire of the receptacle, the leakage current may be higher than in the normal condition. The test for leakage current in reverse polarity is performed because it is possible that the power receptacle into which the particular piece of medical equipment is plugged may be wired incorrectly, with the polarity reversed. To guard against this potential hazard, it is important that the device under test have low leakage current under both conditions.

The level of acceptable leakage current for a particular device depends on the use of the device. For equipment that is not to be used on patients, the leakage current should not exceed 500 μA. This includes such things as maintenance and housekeeping appliances used in the vicinity of a patient. The current limit for devices intended for use on a patient should not exceed 100 μA. These limits are the same regardless of whether the leakage current is measured in normal or reverse polarity.

The leakage current must also be measured from the patient leads on those devices which have patient leads. An ECG system is typical of this kind of device. Figure 16-5 indicates the test configuration. In this case the leakage current between all patient leads connected together and ground is measured with power applied, normal polarity. The leakage current should not exceed 100 μA as specified in NFPA 76B. The leakage current between any two lead pairs should be tested and must not exceed 50 μA for the ground wire open and closed.

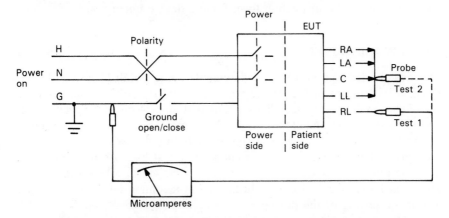

Figure 16-5 Schematic of the test configuration for the measurement of leakage current from ECG patient leads. Test 1, reference lead to ground; test 2, signal leads to ground.

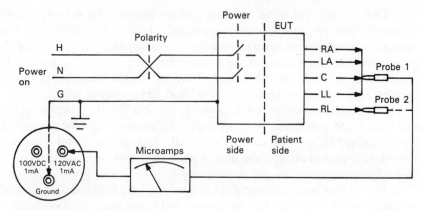

Figure 16-6 Schematic of the test configuration for the measurement of patient isolation. Ac isolation impedance to ground, patient side.

New equipment with patient leads is usually an "isolated input" system. This means that this device will not act as a low-impedance path to ground via the patient leads. This is important in case line voltage in some way comes in contact with the patient's body. If the patient lead input to the device acts as a very high impedance path, the patient will be protected against possible fatal electrical currents.

Figure 16-6 illustrates how patient lead isolation is tested. In this case, the full power line voltage is applied to each lead and ground; the measured leakage current should be less than 20 μA in each case. The test should be made with the device's normal patient cables and with the device turned on and operating. All of this leakage current data should be recorded on the inspection form for the device under test.

PERFORMANCE TESTING

Besides the leakage current measurements made on all electrical devices used in patient areas, the device must also conform to the manufacturer's specification for the device. This means that all performance specifications as set forth by the manufacturer should be met or exceeded by the device being tested. Even though a defibrillator is brand new, it may not work or deliver the appropriate level of energy to a test load, or an ECG system may be out of specification in terms of its bandwidth. Therefore, records on a particular medical device begin at the incoming inspection and performance testing of the device.

Many of the devices used in the hospital are detection and display systems such as ECG, blood pressure, and respiration equipment. As part of the performance testing of devices such as these, simulated inputs to the systems can be used to determine the performance of the device while saving time for the clinical engineer by providing calibrated voltages and frequencies that can be used to test the device.

Typical of the medical waveform simulators on the market is the ECG/ blood pressure simulator pictured in Figure 16-7. This type of simulator provides the capability of testing the performance of ECG and blood pressure monitors by generating physiologically representative waveforms at a wide variety of calibrated rates and amplitudes. The different modes of operation and waveform generation provide for testing of conventional monitors as well as those with arrhythmia detection and analysis.

Blood pressure waveforms are simulated in nine static steps from -50 to $+240$ mmHg. There are also time-varying waveforms that simulate a typical arterial pressure waveform of a magnitude of 120/80 mmHg. Another section of the simulator produces a simulated thoracic impedance and is used to test the performance of respiratory monitors that sense changes in impedance caused by volume change in the thorax during respiration.

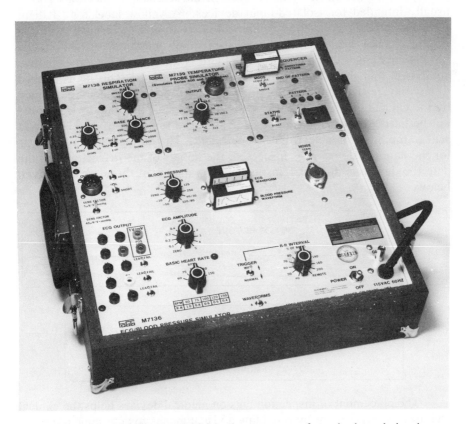

Figure 16-7 Multifunction ECG/blood pressure waveform simulator designed to test dynamically the performance of specific medical devices. (Courtesy of Fogg Systems Co., Aurora, Colo.)

Some of the performance testing of various devices was touched upon in previous chapters. Because of the large number of medical devices used in the hospital, it is not possible to go into the performance testing of all of them in this text.

TAGGING OF MEDICAL EQUIPMENT

The tagging required for medical equipment should be part of the equipment control program implemented by the clinical engineering department in the hospital. Emergency Care Research Institute (ECRI) recommends a series of tags and color codes for electromedical equipment which can serve as a convenient guide to be used directly or as a framework in developing your own system.

A green label "Patient Care SUITABLE for Intracardiac Connection" indicates that the device can be used in intracardiac connections and is usually intended for isolated-input devices. A yellow label for "Patient Care Equipment" is used on equipment likely to come in contact with the patient but not used in an intracardiac mode of operation. A white tag is for "Nonpatient Equipment." Last, there is a red tag used for defective equipment that should not be used on or near patients. This tag is applied to any device that has been found to be unsafe for the reason determined by the clinical engineer and warns the medical staff that the device should not be used. If possible, the device should be removed from service and sent to the hospital's electronics department. The tags should have spaces for the date the device was inspected, the initials of the inspector, and the date the device is due for reinspection. Examples of these tags are given below.

```
┌─────────────────────────────────────────────────────────────┐
│                    PATIENT CARE EQUIPMENT                      │
│                      DO NOT REMOVE TAG                         │
│                                                               │
│  Date inspected _____ Date next inspection _____  │
│                                                               │
│  Inspector _____                          │
└─────────────────────────────────────────────────────────────┘
```

```
┌─────────────────────────────────────────────────────────────┐
│                    DEFECTIVE – DO NOT USE!                     │
│                                                               │
│     Date _____ Inspector _____        │
│                                                               │
│                  DO NOT REMOVE THIS LABEL                     │
└─────────────────────────────────────────────────────────────┘
```

The placement of inspection tags on medical devices helps the clinical engineering department by providing a visual means to determine if a piece of equipment has been inspected. It is not uncommon for an equipment salesperson, nurse, or physician to bypass the clinical engineer and bring a piece of personal equipment into the hospital for use without it being safety inspected first. Unfortunately, while the equipment is in the hospital, the hospital is responsible if it is used on a patient and an injury results from its use.

Another benefit of the tagging system is that medical personnel using the equipment feel more confident in using it if it has been safety inspected. Also, the medical personnel can provide feedback to the clinical engineer if they notice that a device due date for inspection has passed and the device has not yet been inspected. It is important to impress on the users of patient-oriented medical equipment that they should only use equipment that is tagged and carries a current inspection date.

The frequency of testing of medical devices within the sphere of the clinical engineer can be broken down into four classifications:

1. Clinical devices should be inspected at least twice a year or as needed, depending on the frequency of use and the importance of the device.
2. Nonclinical devices, which include electric beds, should be inspected annually or as required by intensity of use.
3. The electrical system, which includes receptacles, should be tested at least annually or more frequently as needed.
4. Any device that has been involved in any kind of incident or has been the object of abuse (such as being dropped) should be inspected for safety even if it is not due yet for inspection. Any device that is questionable in its performance or function should be tagged as not to be used.

A good electrical safety, inspection, and documentation program is the keystone to reduction of electrical hazards and accidents in the hospital. The link with the medical personnel is the rest of the defense against accidents. To make maximum use of the information link that the user of medical equipment provides, it is essential that users be educated with respect to problems in electrical and electronic equipment that can cause electric shock. They must not be allowed to develop a false sense of security that the device they are using is safe just because there is a recent date on an inspection tag for the device. The point must be made that the inspection can guarantee only that the equipment was in proper working condition at the time of the inspection. It should be pointed out that a defect can develop in a device after it has been inspected. Also, if the device has been abused by a user or dropped, it is possible that the device may have become hazardous to the patient even though there is an inspection tag testifying that the device is safe to use.

Since the daily user of medical equipment in the patient environment is the first line of defense in preventing equipment from being misused or abused and in detecting electrical faults in the device, it is important that the user is educated to recognize problems or changes in the condition of the device. The educational background of most medical personnel is oriented toward administering to the patient, not to electrical engineering. Thus their mode of thinking about electrical devices is different from that of the clinical engineer or biomedical engineer.

Medical personnel should be sensitized to observing the physical condition and performance of the devices they use. They should report immediately to the clinical engineer such things as frayed or broken cords on any electrical device, any deviation from expected or normal performance of the equipment, and any shocks produced by the device, no matter how minor. Finally, they should report any change in the physical appearance of the device, such as dents that were not there at the last use of the device; controls, knobs, or switches that are missing or do not function properly; and cracks in meter faces. It is the combination of good clinical engineering practice, safety inspections, and documentation combined with an educated medical staff that produces a program that will be the most effective in bringing the number of electrical-device-related accidents asymptotically closer to the ideal of zero accidents.

PROBLEMS

1. Design a safety inspection form for electromedical equipment.
2. Set up an incoming inspection protocol for electromedical equipment.
3. Set up a protocol for a hospital policy on personal patient electrical equipment that pertains to what can and cannot be brought into the hospital. Cite any codes and standards that are applicable.
4. Set up a protocol for a hospital policy on personal physician equipment used in the hospital.

REFERENCES

1. Billin, A. Gilbert, "Electrical Safety in the Hospital" (Rochester, N.Y.: The Ritter Company).
2. Link, Leo J., and Barry Feinberg, "Electrical safety in hospitals," *J. Hosp. Res.,* American Sterilizer Company, 1976.
3. "Patient safety," Application Note AN718 (Waltham, Mass.: Hewlett-Packard, 1971).
4. Simmons, David (ed.), *Medical and Hospital Control Systems, The Critical Difference* (Boston: Little, Brown, 1972).

17

CODES AND STANDARDS
THAT PERTAIN TO THE
CLINICAL ENGINEER

The health care industries in general and hospitals in particular are the most regulated industries in the United States. Some interesting statistics on this are that there are more than 28 organizations and governmental agencies through over 130 committees that have produced about 1200 standards dealing with health care devices and practices. Thus regulations governing the operation of the hospital are part of the territory, and in particular, there are a number that directly affect the clinical engineer and his or her function in the hospital.

It is not the intent of this chapter to provide a comprehensive exposition and analysis of all the codes, standards, and regulations that govern the hospital in general and concern the clinical engineer in particular—on the contrary, as this would produce a book in itself. In a sense, codes, standards, and regulations are "living" documents, as they are always in a state of change and subject to interpretation. Thus, only some of the more relevant codes and standards will be discussed and then only generally since they are probably in a state of change while you are reading these words. It is the object of this chapter to present a brief summary of a few codes, standards, and regulations so that the reader with specific areas of interest will be able to find information on that subject in the document cited.

Reference has been made to several codes throughout this book, but in order to evaluate their interpretation, we will first define what is meant by codes, standards, regulations, and specifications. The majority of the codes that pertain to the function of the clinical engineer have to do with electrical safety.

Codes: a system of principles or regulations or a systematized body of law. An accumulation of a system of regulations and standards. Probably the most familiar code to the clinical engineer is the *National Electrical Code®*, put out by the National Fire Protection Association (NFPA). Although the NFPA is the most widely known of the bodies that generate codes that relate to the hospital, it is not a law-generating body, nor does it have a governmental mandate to generate these codes. We will see just how these codes become law later in the chapter.

Regulations: rules normally promulgated by the U.S. government and codified in the various U.S. codes.

Standards: documents that are perceived to establish a minimum level of performance or to standardize such items as test methods or materials. They are usually accepted as an appropriate level of performance by a duly constituted authority. Typical of this type of standard is NFPA 76B, "Standard for the Safe Use of Electricity in Patient Areas of Hospitals."

Specifications: documents used to control the procurement of equipment. These types of documents usually cover design criteria, system performance, materials, and technical data.

There are three general types of standards, particularly for medical devices:

1. *Voluntary standards:* developed under the consensus process, where manufacturers, users, consumers, and government come together voluntarily in open session. These may become de facto mandatory standards due to industry usage.
2. *Mandatory standards* (regulations): required by law, such as the 1976 Medical Device Amendment to the Food, Drug, and Cosmetic Act.
3. *Proprietary standards:* developed either by a company for its own internal use or by a trade association for use by its members. Proprietary standards can serve as a basis for voluntary or mandatory standards if the appropriate exposure is given to these standards and if a consensus is reached among all parties.

Activity in the medical device standards area grew from the late 1960s with the device classification of the Food and Drug Administration (FDA). In the voluntary standards area, work was done by the following groups:

1. Institute of Electrical and Electronic Engineers (IEEE)
2. American Society for Testing and Materials (ASTM)
3. American Dental Association (ADA)
4. National Fire Protection Association (NFPA)

5. Underwriters' Laboratory (UL)
6. Association for the Advancement of Medical Instrumentation (AAMI)
7. Canadian Standards Association (CSA)

Underwriters' Laboratory is an independent laboratory initially started by a group of insurance underwriters in 1894. Currently, UL is sponsored by membership in other areas, not solely insurance companies. At this time, UL is dedicated to investigating devices, products, equipment, and materials as to their hazards to life and property. UL develops standards of safety for devices sold to the public and then conducts tests and evaluations to determine whether the device meets its standards. If it does, the device is listed as approved and the UL mark can be attached to the product.

The UL standard 544 pertains to electromedical and dental equipment. The document contains sections on the construction of equipment, performance (tests), ratings, markings, caution and warning notices, other required notices, and a glossary. It contains many detailed standards relative to components which constitute the construction of devices; thus one who, for example, is concerned with a heating element of a device may refer to UL544 and see what design and construction factors must be considered.

Performance aspects of leakage current and grounding are included in the standard. Cable insulation is discussed and major emphasis is placed on how to test devices which are double insulated, or how to test for x-radiation. The manual seems to be oriented toward the manufacturers of equipment, but the clinical or biomedical engineer should be familiar with these standards as they pertain to the hospital. UL listing of a device that passes UL tests implies a level of clinical safety, not a level of clinical performance.

The Joint Commission on the Accreditation of Hospitals (JCAH) establishes standards for the operation of hospitals that hospitals must comply with if they are to become accredited by the JCAH. The JCAH publishes a document with the title "Accreditation Manual for Hospitals." [1] This manual outlines performance or duties in a health care facility with respect to all areas and departments by setting standards of operation and performance for the various areas of the hospital. What follows is only a summary of the JCAH material, so for more information about the standards and the interpretation the JCAH has for each of the standards, one should go directly to the "Accreditation Manual for Hospitals." The important provisions of the JCAH hospital accreditation standards of which a clinical engineer should be aware are excerpted below.

"1. *Anesthesia services:* Precautions shall be taken to ensure the safe administration of anesthetic agents.
2. *Building and grounds safety:* The hospital buildings and grounds shall be designed, constructed, equipped, and furnished in a manner that

protects the lives and ensures the physical safety of its patients, personnel, and visitors.

3. *Functional safety and sanitation:* There shall be a multidisciplinary hospital safety committee to adopt, implement, and monitor a comprehensive hospital-wide safety program. The program shall contain requirements related to the staffing, equipping, operation, and maintenance of the hospital, which should be designed to produce safe characteristics and practices and to eliminate, or reduce to the extent possible, hazards to patients, hospital staff, and visitors.

Comprehensive safety systems shall be installed, and practices, policies, and procedures instituted, to minimize hazards to patients, hospital staff, and visitors."

This section of the accreditation manual is, by far, the most important section to the clinical engineer, as it pertains to the following items:

a. Hospital safety committee
b. Electrical safety
c. Fire warning and safety systems
d. Compressed-gas cylinders
e. Handling and storage of nonflammable gases
f. Handling and storage of flammable gases and liquids
g. Engineering and maintenance
h. Patient and personnel safety devices and measures
i. Smoking regulations
j. Hospital security

"4. *Nuclear medicine services:* There shall be quality control policies and procedures governing nuclear medicine activities that ensure diagnostic and therapeutic reliability and safety of patients and personnel. Records required by federal, state, and local authorities, as well as records consistent with competent practice of nuclear medicine, should be maintained.

5. *Radiology services:* There shall be written policies and procedures, including safety rules, for the radiology services. Special care units should be designed and equipped to facilitate the safe and effective care of patients."*

The National Fire Protection Association (NFPA) has put forth a number of standards and codes that affect the hospital directly. The best known of these is article 517 of the *National Electrical Code®*, "Health Care Facilities." This section of the NEC contains definitions useful in working with the electrical environment of the hospital. It covers the topics

---

* Excerpted from the Joint Commission of the Accreditation of Hospitals, *Accreditation Manual for Hospitals* (Chicago: JCAH, 1983).

of wiring specification for various areas of the hospital, overcurrent protection, grounding, and the electrical capacity that should be supplied to different areas of the hospital. It presents material on emergency power systems and the electrical requirements for general and critical patient care areas, defining voltage differentials allowed between conductive surfaces in these areas.

Many special requirements of these patient care areas are covered, such as number of outlets at each patient location, grounding requirements, and wire sizes for the current load. Conventional and isolated power systems are taken up as well as the maximum ground impedance allowed in these systems. The power distribution system and branches dealing with the essential electrical systems, emergency system, life safety branch, critical branch, and equipment systems are also defined and specified.

SAFE USE OF ELECTRICITY IN PATIENT CARE AREAS OF HOSPITALS

NFPA 76B is a standard that takes up the topics of electrical power systems in patient care areas, including the availability of electrical power, construction and installation requirements, operational safety requirements, testing intervals, and documentation. Within this standard, the hospital requirements for electrical appliances are discussed broaching the topics of policies and procedures, safe appliance policy and criteria for hospitals, and tests conducted for hospitals. This section of NFPA 76B discusses leakage current limits as well as leakage current tests, which include diagrams of test configurations.

The last chapter of this standard sets forth the requirements for manufacturers of patient-care-related electric appliances. Here the general requirements for manufacturers of patient-care-related electric appliances are set forth together with manufacturers' tests for safety on patient-care-related electric appliances.

Standard for Use of Inhalation Anesthetics: NFPA 56A specifies the safe practices that should be used when using anesthetic gases. This standard provides for the use of conductive flooring, an isolation transformer, and a grounded electrical system when inflammable anesthetics such as cyclopropane are used in the operating room. Also, it specifies the environment for the use of nonflammable anesthetic gases such as halothane.

High-Frequency Equipment in Hospitals: NFPA 76C sets forth the safe practices associated with the use of such devices as electrosurgery and diathermy equipment.

There are many other codes and standards that are too numerous to consider in this text but should be familiar to the clinical engineer. Only a few of the major ones have been cited here. The codes and standards that apply to a particular hospital depend on what the local jurisdiction governing the hospital has adopted as law as well as compliance with the JCAH. There are so many standards that apply to the hospital that there are times when they are in conflict with one another.

THE OCCUPATIONAL SAFETY AND HEALTH ACT

OSHA has been in effect since April 28, 1971. Its stated purpose is "to assure so far as possible every working man and woman in the Nation safe and healthful working conditions and to preserve our human resources." The provisions of this act apply to every private employer that is engaged in business affecting commerce and has employees. Private hospitals are subject to the provisions of OSHA, and public hospitals may or may not be depending on legal interpretation of the law in the state. The OSHA standards are contained in five volumes that are available from the U.S. government. Of the five volumes that constitute the act, the following are of most interest to hospitals:

> *Volume I:* General Industry Standards
> *Volume IV:* Other Regulations and Procedures
> *Volume V:* Compliance Operations Manual

Volume I contains general industry standards that are applicable to the hospital as a workplace. Volume IV discusses the administrative procedures that apply to all covered employers, and Volume V contains the regulations followed by the Occupational Safety and Health Administration in carrying out the law.

Hospitals and other health care institutions covered under OSHA should be aware of their responsibilities as employers under this law. They need to familiarize themselves with whatever safety and health standards set forth by OSHA apply to them.

FDA MEDICAL DEVICE REGULATIONS

Of importance to the clinical engineer relative to government approval of medical devices for sale to the medical market is the Medical Device Amendments of 1976 to the Food, Drug, and Cosmetic Act. This law amends the powers of the Food and Drug Administration (FDA) so that it now regulates nearly every facet of the manufacture and sale of medical and diagnostic devices that are involved in interstate commerce. The government accomplishes this via premarket control, which includes the following items:

1. Premarket approval requirements
2. Mandatory performance standards
3. Good manufacturing practices

Continued control after initial approval is maintained by the FDA through postmarket controls. These controls are manifested as:

1. Defect notification requirements
2. Requirements for records and reports
3. Authority to ban risky or deceptive devices

The FDA, via the Medical Device Amendments of 1976, classifies medical devices intended for human use into three classes:

(A) Class I, General Controls
  (i) A device for which the controls authorized by law are sufficient to provide reasonable assurance of the safety and effectiveness of the device.
  (ii) A device for which insufficient information exists to determine that the controls referred to in clause (i) are sufficient to provide reasonable assurance of the safety and effectiveness of the device or to establish a performance standard to provide such assurance, but because it
    (I) is not purported or represented to be for a use in supporting or sustaining human life or for a use which is of substantial importance in preventing impairment of human health, and
    (II) does not present a potential unreasonable risk of illness or injury,
    it is to be regulated by the controls referred to in clause (i).

(B) Class II, Performance Standards
  A device which cannot be classified as a class I device because the controls authorized by law by themselves are insufficient to provide reasonable assurance of the safety and effectiveness of the device, for which there is sufficient information to establish a performance standard to provide such assurance, and for which it is therefore necessary to establish for the device a performance standard . . . to provide reasonable assurance of its safety and effectiveness.

(C) Class III, Premarket Approval
  A device which because
  (i) it (I) cannot be classified as a class I device because insufficient information exists to determine that the controls authorized by law are sufficient to provide reasonable assurance of the safety and effectiveness of the device and (II) cannot be classified as a class II device because insufficient information exists for the establishment of a performance standard to provide reasonable assurance of its safety and effectiveness, and

(ii) (I) is purported or represented to be for a use in supporting or sustaining human life or for a use which is of substantial importance in preventing impairment of human health, or

(II) presents a potential unreasonable risk of illness or injury,

is to be subject, in accordance with section 515 [premarket approval section], to premarket approval to provide reasonable assurance of its safety and effectiveness. If there is not sufficient information to establish a performance standard for a device to provide reasonable assurance of its safety and effectiveness, the Secretary may conduct such activities as may be necessary to develop or obtain such information.

From the brief excerpt from the Medical Device Amendments of 1976, we see that it is primarily oriented toward the manufacturers of medical devices. This does not let hospitals, in general, off the hook as they assume new responsibilities under the act. In particular, the internal notification, recall, reporting, and record maintenance systems of the hospital will need to be sufficient to respond adequately to notifications from the FDA or from a manufacturer concerning hazardous devices. This highlights the importance of the record-keeping function for medical devices performed by the engineering staff of a hospital.

CONCLUSION

This text has brought together in a unified manner much of the science and engineering that is part of the profession of clinical engineering. The clinical engineer is one of the newest members of the health care delivery team. The importance of this hospital staff position is growing with increasing utilization of the high-technology base of the engineering profession as it applies to medicine and the delivery of health care.

As health care, in terms of technology utilization, depends more and more on decisions made by the clinical engineer, it is important for the clinical engineer to seek the professional and legal status of becoming registered in his or her state as a Professional Engineer (PE). In addition to registration, there is certification in clinical engineering, which is a peer-group evaluation of one's competency in the field of clinical engineering performed by the International Certification Commission. Because of the complexity of health care delivery, which is based on engineering and science, it is possible that hospitals in the future will require that their clinical engineer be a Registered Professional Engineer and be certified in clinical engineering.

PROBLEMS

1. Define the following terms:
   (a) Code
   (b) Standard

(c) Regulation

(d) Specification

2. What are the major federal agencies that are involved in medical standards? Explain their purpose and function.

3. What are the primary organizations that influence and establish standards regarding medical devices?

4. Which bureaus of the FDA play a role in medical device regulation? What are the different classifications of medical devices as set forth by the FDA, and why do these different classifications exist?

5. What are the two JCAH standards that relate most directly to clinical engineering practice?

6. Summarize the JCAH standard relating to electrical safety.

7. An operating theater in which explosive anesthetics are used contains permanent x-ray and lighting fixtures. Also used are a heart bypass pump, a defibrillator, an ECG monitor, and an electrosurgery unit. Describe the type of power to be provided to each of these electrical devices, the type of grounding required, the nature of the receptacles, and the specifications of the line isolation monitor. Use as references the most recent copy of the *National Electrical Code®* and other appropriate NFPA standards on the safe use of electricity in health care facilities and other standards that are appropriate to this situation.

8. In a hospital, a faulty floor polisher fused the line conductors in a receptacle in the corridor outside an intensive care unit. As a result, the circuit breakers tripped out for that area. Because of this, all the respirators in the ICU stopped! Fortunately, emergency cables could be connected from another part of the ICU to supply power to the respirators. As the hospital's clinical engineer, you have been asked to determine whether the electrical contractor who installed the wiring has violated the *National Electrical Code®*, and if so, to demonstrate that this is the case so as to advise legal counsel on what has happened.

REFERENCES

1. Joint Commission of the Accreditation of Hospitals, *Accreditation Manual for Hospitals* (Chicago: JCAH, 1983).

2. *National Electrical Code®*—1984, Article 517, "Health Care Facilities" (Boston: National Fire Protection Association, 1984).

3. NFPA 56A, "Standard for the Use of Inhalation Anesthetics" (Boston: National Fire Protection Association, 1973).

4. NFPA 76C, "High Frequency Electricity in Health Care Facilities" (Boston: National Fire Protection Association, 1975).

5. "Occupational Safety and Health Act of 1970, Law and Explanation" (Chicago: Commerce Clearing House, Inc., 1971).

6. NFPA 76B, "Standard for the Safe Use of Electricity in Patient Areas of Hospitals" (Boston: National Fire Protection Association)

INDEX